Advanced Treatment Technologies for Wastewater Reuse

Advanced Treatment Technologies for Wastewater Reuse

Editor: Gabriel Craig

MURPHY & MOORE
www.murphy-moorepublishing.com

www.murphy-moorepublishing.com

MURPHY & MOORE

Cataloging-in-Publication Data

Advanced treatment technologies for wastewater reuse / edited by Gabriel Craig.
p. cm.
Includes bibliographical references and index.
ISBN 978-1-63987-757-7
1. Water--Purification. 2. Water--Purification--Technological innovations.
3. Sewage--Purification--Environmental aspects. 4. Water reuse. I. Craig, Gabriel.
TD430 .A38 2023
628.162--dc23

Murphy & Moore Publishing
1 Rockefeller Plaza,
New York City,
NY 10020, USA

ISBN 978-1-63987-757-7

Contents

Preface

The purpose of the book is to provide a glimpse into the dynamics and to present opinions and studies of some of the scientists engaged in the development of new ideas in the field from very different standpoints. This book will prove useful to students and researchers owing to its high content quality.

Wastewater treatment is a process used to remove contaminants from the wastewater in order to convert it into an effluent that can be returned to the water cycle. Some common wastewater treatment technologies include advanced oxidation processes (AOPs), nanotechnology, membrane filtration technology, and automatic variable filtration (AVF) technology. There are various researches being performed in the field of nanotechnology for developing advanced treatments that can decompose toxic substances. The upcoming technologies for wastewater treatment and reuse include fabrication of nanomaterial membranes, using catalysts for decomposing noxious compounds, and using mixed oxides, bimetallic nanoparticles, carbon compounds, metals and zeolites. This book includes some of the vital pieces of works being conducted across the world, on various topics related to wastewater treatment and reuse technologies. The various studies that are constantly contributing towards advancements in this field have been examined in detail. The book will serve as a valuable source of reference to graduate and postgraduate students.

At the end, I would like to appreciate all the efforts made by the authors in completing their chapters professionally. I express my deepest gratitude to all of them for contributing to this book by sharing their valuable works. A special thanks to my family and friends for their constant support in this journey.

Editor

1

Carbon Nanotubes-Sponge Modified Electro Membrane Bioreactor (EMBR) and their Prospects for Wastewater Treatment Applications

Ali M. Almusawy [1], Riyad H. Al-Anbari [1], Qusay F. Alsalhy [2,*] and Arshed Imad Al-Najar [1]

1 Civil Engineering Department, University of Technology, Alsinaa Street 52, Baghdad 10066, Iraq; 40393@student.uotechnology.edu.iq (A.M.A.); riyadhassan2003@yahoo.com (R.H.A.-A.); aarshad_emad@yahoo.com (A.I.A.-N.)

2 Membrane Technology Research Unit, Chemical Engineering Department, University of Technology, Alsinaa Street 52, Baghdad 10066, Iraq

* Correspondence: qusay_alsalhy@yahoo.com or 80006@uotechnology.edu.iq

Abstract: A novel membrane bioreactor system utilizes Multi-Walled Carbon Nanotubes (MWCNTs) coated polyurethane sponge (PUs), an electrical field, and a nanocomposite membrane has been successfully designed to diminish membrane with fouling caused by activated sludge. The classical phase inversion was harnessed to prepare Zinc Oxide/Polyphenylsulfone (ZnO/PPSU) nanocomposite membranes using 1.5 g of ZnO nanoparticles (NPs). The prepared nanocomposite membrane surface was fully characterized by a series of experimental tools, e.g., Scanning electron microscope (SEM), Atomic force microscopy (AFM), contact angle (CA), pore size, and pore size distribution. The testing procedure was performed through an Activated Sludge-Membrane Bioreactor (ASMBR) as a reference and results were compared with those obtained with nanotubes coated sponge–MBR (NSMBR) and nanotubes coated sponge-MBR in the presence of an electrical field (ENSMBR) system. Observed fouling reduction of the membrane has improved significantly and, thus, the overall long-term was increased by 190% compared with the control ASMBR configuration. The experimental results showcased that sponge-carbon nanotubes (CNTs) were capable of adsorbing activated sludge and other contaminants to minimize the membrane fouling. At a dosage of 0.3 mg/mL CNT and 2 mg/mL of SDBS, the sponge-CNT was capable of eliminating nitrogen and phosphorus by 81% and >90%, respectively.

Keywords: membrane bioreactor; electrochemistry; carbon nanotube; membrane fouling; wastewater

1. Introduction

The rapid growth, witnessed in recent decades in industrialization and population, has given rise to release massive organic contamination into the aquatic system [1,2]. Versatile treatment technologies have been harnessed to cope with the endless catastrophic consequences occasioned by the presence of these organic components. Advanced oxidation (e.g., ozonation and UV-photocatalysis) and membrane filtration processes are ubiquitous examples of these techniques. Despite the relatively advanced state of these technologies, there is a huge debate about which one can outweigh among the other. In this context, membrane bioreactor (MBR) systems have emerged as a potent alternative to the traditional biological processes for the reclamation and removal of nutrients and carbon sources [3–6]. The leading characteristics of MBR are represented by the low sludge generation, typical retention, and exemplary separation efficiency with minimal space requirements. However, like any other membrane separation process, external and internal membrane fouling is an inevitable phenomenon associated with MBR systems. Fouling topics have been covered by an enormous research activity since the first membrane's inception. Among the available methods, process modification through nanomaterials has offered an unlimited research area to diminish fouling with the aid of electrochemistry.

Among the wide diversity of available nanostructures, the carbon-based nanomaterials, e.g., one-dimensional carbon nanotubes (CNTs), have attracted the scientific community owing to their optical and mechanical characteristics, tunable electrical properties, light weight, and high specific area [7]. In the water treatment industry, electrochemistry has been involved in oxidizing the pollutants in an aqueous form. These outstanding characteristics have been employed for the manufacturing of highly porous networks for pollutants' adsorption [8]. The way has been paved toward their application in electrodes manufacturing for adsorbing organic contaminants and degrading them through electrochemical oxidation [9]. The combination of electrochemistry with CNT filters has been reported as an efficient way for adsorptive trapping and electro-oxidant degradation of organic pollutants. [10] Claimed that CNT-based electric filter technology can effectively eliminate certain contaminants such as pharmaceuticals, colourants, viruses, bacteria, and organic aromatic compounds from aquatic environments.

Nevertheless, the high specific surface area/volume ratios of zero-dimensional nanomaterials have been harnessed for infinite applications of organic wastewater treatment [11,12]. The zinc oxide (ZnO) nanoparticles are one of these extraordinary nanomaterials employed for revising the surface characteristics of the hydrophobic membrane's material [13]. Many pieces of literature claimed that ZnO NMs could impart highly hydrophilic and antifouling features on the target membranes [14–17]. Based on the nano additives ratio, the performance of the ZnO modified nanocomposite membranes may vary against organic foulants [18].

This work shed the light on an efficient wastewater treatment system that combined an electro-membrane bioreactor (EMBR) and CNTs modified commercial polyurethane sponges as a filter. Herein, both CNTs and polyurethane sponges play significant roles in designing the water purification process. The CNTs were functioning as high-performance electro-catalysts for pollutants degradation (through electron transfer). In this context above, ZnO-modified mixed matrix membranes were fabricated at a dissimilar loading ratio to determine the optimum performance. This CNTs-sponge-MBR (NSMBR) system was designed for removal of NO_3, PO_4-P, and Chemical Oxygen Demand (COD), as real wastewater, in order to mitigate membrane fouling. The tests were conducted by playing with operational conditions.

2. Experimental

2.1. Materials

Polyphenyl sulfone (PPSU-Radel R-5000), with a Mw of 50 KDa, was purchased from Solvay Advanced Polymers (Brussels, Belgium). N-methyl-2-pyrrolidone (NMP) and sodium dodecylbenzene sulfonate (SDBS) were purchased from Sigma–Aldrich (St. Louis, MO, USA). ZnO nanoparticles (10–30 nm diameter) were purchased from SSNano, (Houston, TX, USA). Multi-walled carbon nanotubes have been obtained from Suzhou Jiedi Nano Technology Co., Ltd., (Suzhou, China). The polyurethane sponge was purchased from Shanghai YueXia Industrial Co., Ltd., (Shanghai, China). All chemicals were used as it is without further purification.

2.2. Nanocomposite Membrane Preparation

A ZnO-PPSU nanocomposite membrane was prepared via the classical phase separation technique. Briefly, 15 wt.% of PPSU was dissolved in 85 wt.% NMP and stirred overnight until a homogeneous dope solution was achieved. Only one loading weight (1.5 g) of ZnO NPs was added to the casting solution here in the current work. The membrane was cast via an automatic thin-film applicator at a 200-μm clearance gap. Following that, the prepared films were immersed in a deionized water (DI) coagulation bath until complete phase separation is achieved. The membrane was rinsed thoroughly and placed in a glycerol solution to save the membrane's structure. The amount of ZnO adopted was based on our previous work results where the nanocomposite membrane prepared at 1.5 g has showcased the optimum performance. A more detailed description could be found elsewhere [19].

2.3. Fabrication of the CNT Sponge Filter

The conductive sponge has been produced according to a procedure given in a preceding literature [12]. The CNTs ink preparation process was initiated by dispersing 0.3 mg/mL CNTs and 2 mg/mL SDBS in Di water for 15-min. SDBS was employed to encourage homogeneous CNTs dispersion in the aqueous solution and to avoid agglomeration. Later, a piece of polyurethane sponge was immersed in the CNTs suspension and then dried in an oven at 50 °C for 30 min. The high porosity of the sponge encourages the solution's rapid penetration into the sponge structure [20]. The color change of the sponge sample endowed a convincing evidence for the successful coating process onto the sponge surface. The surface of the sponge became relatively rougher when compared to the uncoated sample. Herein, CNTs could function as a high-performance electro-catalyst for adsorption and oxidation of pollutants due to their excellent electrical conductivity. It could also facilitate the functionalization of polyurethane sponges as a conductive anode. The coated sponge was cut to fit the membrane's module cell with dimensions of 10 cm × 15 cm × 4 cm. This coating process was repeatedly conducted three times at least to enhance the CNT loading, and, ultimately, the sponge's electrical conductivity. Thereafter, the prepared CNTs-sponges were treated for 6 h with 4 mol/L HNO_3 solution at room temperature, aiming to eliminate the remaining surfactant. Finally, the CNT sponges were moved to the NSMBR configuration ready for the tests. The schematic diagram of the fabrication process was depicted in Figure 1 below. It was reported that CNTs–sponge could be sufficiently conductive to function as an electrode in small scale applications. It is also possible to host the conductive CNTs with other support, such as cotton.

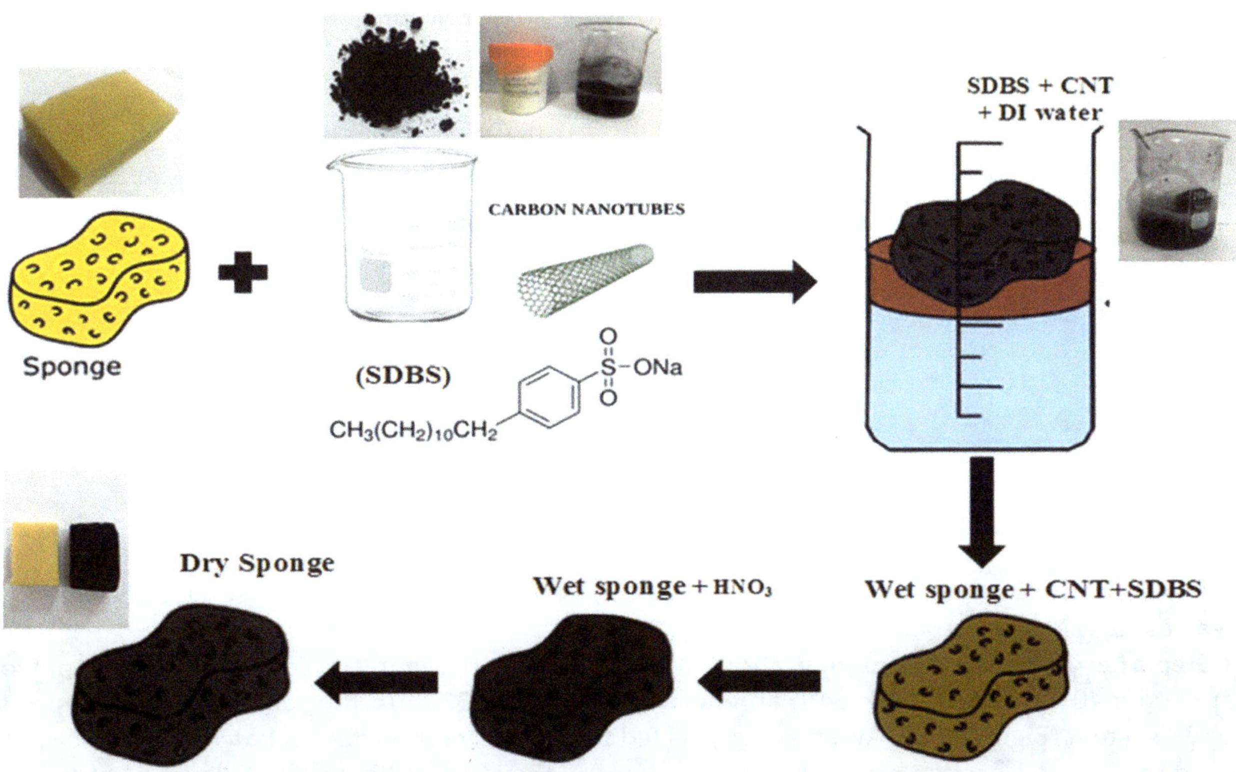

Figure 1. Schematic diagram of the fabrication process.

2.4. Membrane Characterization

Scanning electron microscopy (SEM) has visualized the surface and cross-sectional morphological changes. A TESCAN VEGA3 SB instrument, SEM image (EO-Service, Kohoutovice, Czech Republic). Prior to imaging, using freeze smashing in liquid nitrogen, the membrane cross-sections were prepared while all membranes were sputtered with a thin chromium coat.

An atomic force microscope (AFM) (model AA3000, Angstrom Advanced Inc., Stoughton, MA, USA) was used for inspections of surface topography of the nano-composite membrane. The membrane pore size and its distribution can be tested by AFM. The most common implementation of this technique uses a cantilever with a sharp tip to scan over the sample surface to get an image at an atomic level. By utilizing the AFM test, van der wall forces and charges of the surface between the tip and the surface of the membrane could be determined. AFM characterization can also give information about the fouling and antifouling behaviour of the membrane since these properties depend on the surface smoothness. Obtained images were processed by using an IMAGER 4.31 statistical program.

A contact angle instrument (CAM 110-O4W, Taiwan) was employed for measurements of the wettability of the PPSU-ZnO membranes. Approximately 4 μL of distilled water was injected on the surface of the membrane and the instrument program captured the contact angle (CA) between the drop and the surface. There were five triplicates taken and average values of CA were presented with a standard deviation.

2.5. MBR Experimental Setup

Initially, a real wastewater source obtained from a local domestic wastewater treatment plant (WWTP) has been utilized for the activated sludge-membrane bioreactor (ASMBR) system. All characteristics of the wastewater were summarized in Table 1. The activated sludge was grown in a local WWTP while the biomass was acclimatized to the wastewater for two months to ensure stable conditions before the membrane filtration experiments. The mixed liquor suspended solids (MLSS) concentration, the sludge retention time, and organic loading rate were maintained at 10,000 mg/L, 24-days, and 0.2 g COD/g MLSS-day, respectively. The effective membrane area was 0.000887 m^2 with a module cell dimension having 10 cm × 15 cm. All experiments were performed under the conditions of −0.4 bar and 25 ± 2 °C.

Table 1. Characteristics of domestic wastewater.

Pollutant	Influent Concentration (mg/L)	Maximum Allowable Limit (mg/L)
1	450–800	100
NH_3	150–180	0.02–0.06
NO_3	25–70	50
P	16–24	3
EC	2393–5102	200–800 μS cm^{-1}
TDS	500–860	100
TSS	89–113	20
Pb	0.005	0.05
Cd	0.0025	0.1
pH	7.2–7.4	6.5–8.5

The second step was divided into two stages, NSMBR and Nano-sponge electro-membrane bioreactor (ENSMBR), respectively.

Step one: The proposed configuration of the ASMBR pilot plant, employed in the current work, comprised a 40-L aeration tank with an effective volume of 32 L. In the schematic diagram of the AS-MBR system (Figure 2), the wastewater was fed into the reactor through a batch feeding to control the feed rate while a suction pump was used to control the effluent flow rate. Air was pumped from the base of the aerobic tank to provide the microorganisms with the necessary oxygen. Moreover, to establish a cross-flow velocity throughout the flat-sheet membrane surfaces, it aimed to diminish

MBR fouling. Herein, an intermittent filtration mode was adopted, i.e., 10-min suction and 1-min repose (non-suction).

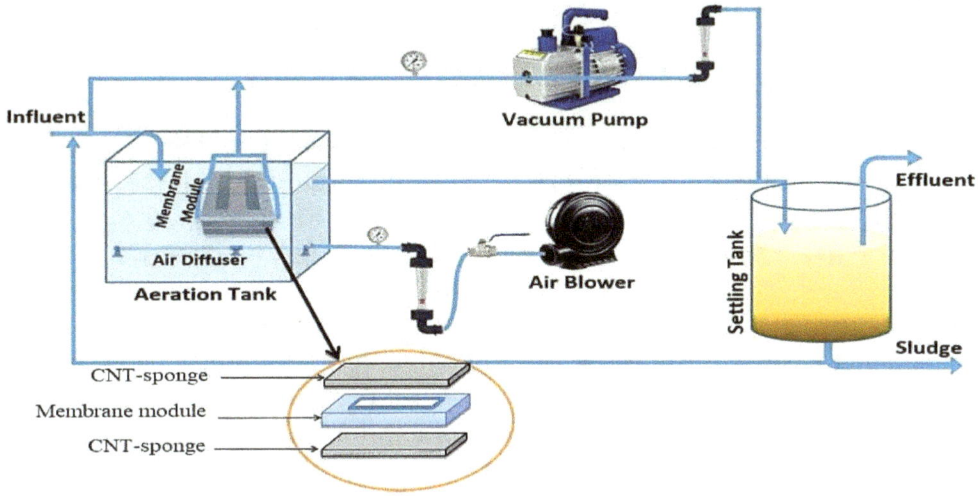

Figure 2. Schematic diagram of the lab-scale AS-MBR system.

In step two, a similar two tanks with a working capacity of 40 L and an effective volume of 32 L have been employed for the NSMBR. During the first stage (NSMBR), the test was performed without applying an electric field. The NSMBR configuration consisted of a tank with a membrane module and an aeration mechanism. It should be noted here that the CNTs-sponge was set up to surround the membrane module. Hence, it played as a barrier to contaminants, which reduced the fouling of the membrane.

In the second stage, the ENSMBR configuration consisted of the same components as the NSMBR configuration except for the electrical field application. Herein, two electrodes were employed (anode and cathode). The anode was made of an aluminum rod whereas the cathode was made of lightweight iron mesh around the CNTs-sponge (Figure 3-Left). The electrodes have been connected to an electronic external DC power supply (Yaxun, PS 1502 AD, Guangzhou, China). A timer (Yueqing Landir Electric co., Ltd., model DH48S, Wenzhou, China) was connected to a DC power source to monitor the intermittent DC supply mode (Figure 3-Right).

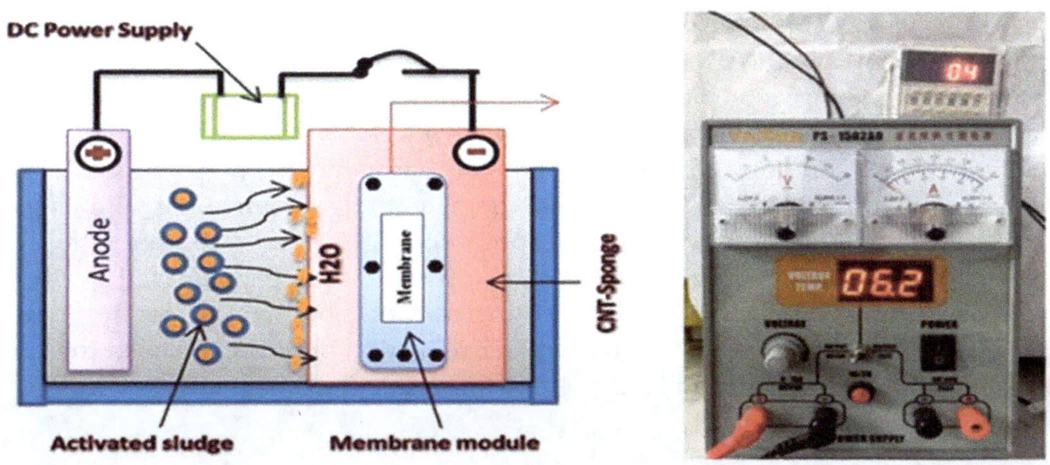

Figure 3. (**Left**) Schematic diagram of carbon nanotubes (CNT)-sponge configuration, (**Right**) DC power supply and timer model.

In the ENSMBR, the aeration system consisted of a porous air diffuser located in the vicinity of the lower center of the electro-bioreactor below the membrane module. It has continuously supplied air to keep the necessary dissolved oxygen (DO) level above 5 mg O_2/L with the support of an air pump. The supplied air also disturbed the sludge and was intended to reduce membrane fouling [21].

2.6. Long-Term Experimental Procedure

The reactor has been operated continuously in two stages. The first stage was carried out for 60 days without wasting the biosolids (SRT = 20 days) while continuing to be supplied with wastewater. Herein, the aeration in the aerobic tank remained continuous. Meanwhile, the ZnO-PPSU membrane was operated at a constant transmembrane pressure. At this stage, no backwashing of the membrane would be performed before the operation of the NSMBR. Meanwhile, during the 90 days of the second stage, the ENSMBR was connected to a power supply in a continuous mode. A constant voltage gradient of 2 V/cm was adopted to be applied at this stage. In order to use the ENSMBR configuration, a preliminary experimental work (stage 2) was carried out to define optimal electro-kinetic conditions. This is essential to define the effective DC field and the intermittence of the exposure period to the DC field. In addition, this is needed to prevent impeding the biological treatment. For this purpose, a series of voltage gradients (1, 2, and 4 V/cm) were applied on a 500-mL activated sludge samples and the optimum value was determined. As a consequence, this value (2 V/cm) was harnessed in the experimental work.

3. Results and Discussion

3.1. Carbon Nanotube-Sponge Characterization

MWCNTs was harnessed in this work due to their electro-catalyst characteristics. This promising property could be beneficial for adsorption and oxidation of various pollutants. The coating of MWCNTs into the polyurethane (PU) sponge was visualized by scanning electron microscopy (SEM), as depicted in Figure 4. Figure 4A displayed the high porosity of the PU-Sponge microstructure. As can be seen, there were no CNTs on the walls of the virgin sponge sample that exhibited open-cells with high interconnectivity and are highly desired for the impregnation of CNTs, as reported by [22]. These characteristics are the major reasons standing behind application of the sponge as a template to produce a porous structure. The successful impregnation of CNTs on the sponge surface was illustrated in Figure 4B. This conductive CNTs layer is expected to endow featured characteristics on the ENSMBR system to mitigate fouling. After the coating, the CNT-coated sponge has kept the mechanical characteristics of the uncoated sponge. It has preserved its flexibility and could be bent arbitrarily to any degree with stretchable and compressible structure. These excellent characteristics, e.g., large surface area, high porosity, and high flexibility, make CNT-sponges a favorable candidate as an adsorbent for the pre-concentration of environmental contaminants [23]. The CNTs were found to have inner diameters of 10–20 nm and an external diameter of 30–40 nm along with lengths of tens to hundreds of microns. A clear uniform distribution of the CNTs was observed at the surface of the polyurethane surface, indicating a successful coating process. As depicted in Figure 4B, a thick layer of CNTs was observed at the sponge surface, giving rise to a rougher surface comparing with the virgin sponge surface. Additionally, there was a clear reduction in the pore size of the sponge's inner structure, and morphology was recognized.

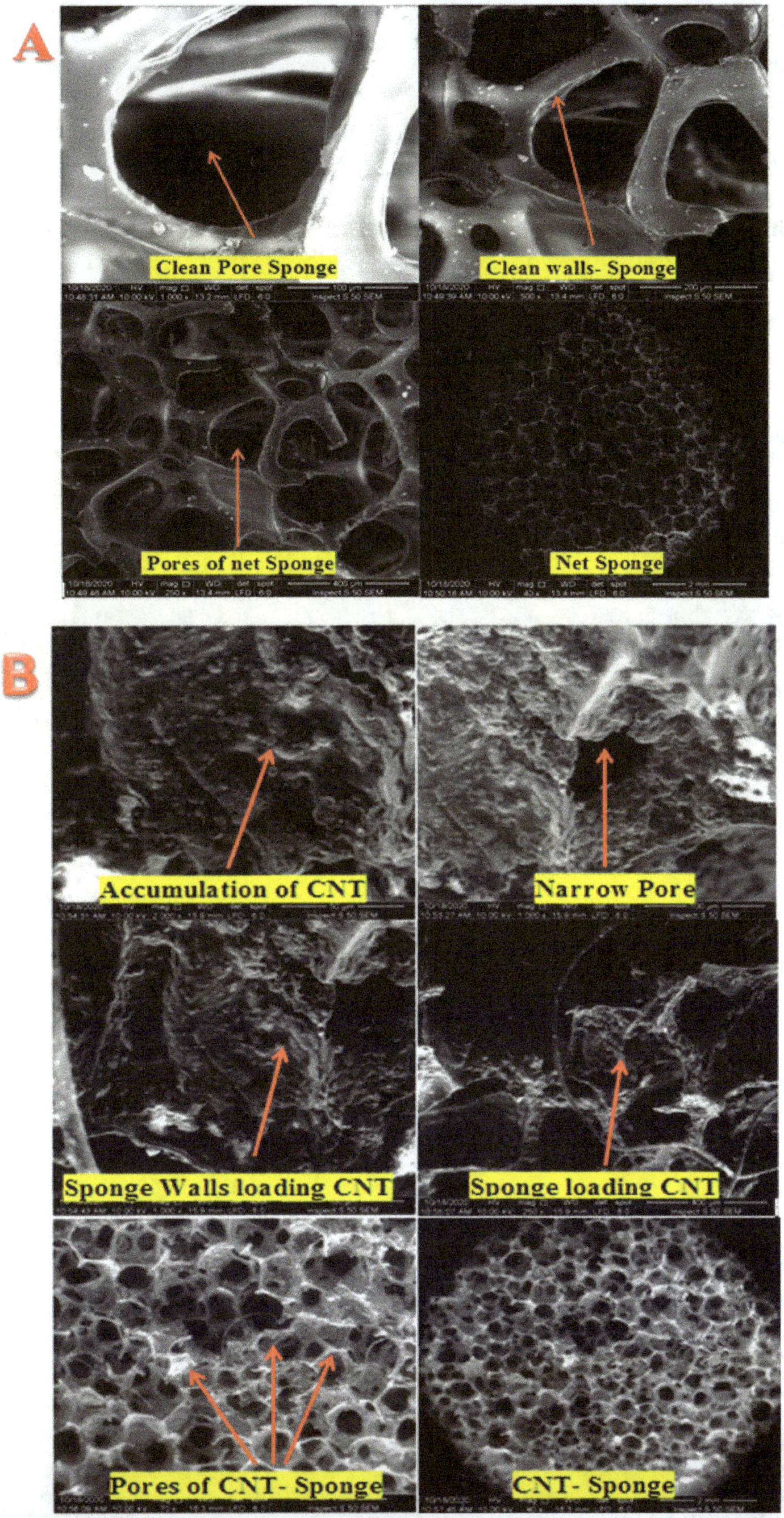

Figure 4. (**A**) SEM images of virgin sponge, (**B**) SEM images of CNTs coated sponge.

3.2. Membrane Surface Characteristics

As previously mentioned, the amount of ZnO adopted was based on our previous work results where the nanocomposite membrane prepared at 1.5 g has showcased the optimal performance. The full characteristics of the nanocomposite membranes were depicted and listed in Figure 5 and Table 2, respectively. A typical cross-section structure consisted of two layers: a sponge structure with a big macrovoids layer near the bottom surface and a finger-like structure near the top surface (Figure 5A). This impact of nanoparticles addition was more explicit on the top surface of the membranes. A semi-porous surface with low pore density and clear homogenous spreading of ZnONPs has been observed at the surface (Figure 5B). Surface roughness characteristics were scanned by AFM (Figure 5C). The roughness parameters, Ra and Rms, were found to be 32.5 and 37.6 nm, respectively (Table 2). The mean pore size and cumulative distribution of pores have been illustrated in Figure 5D.

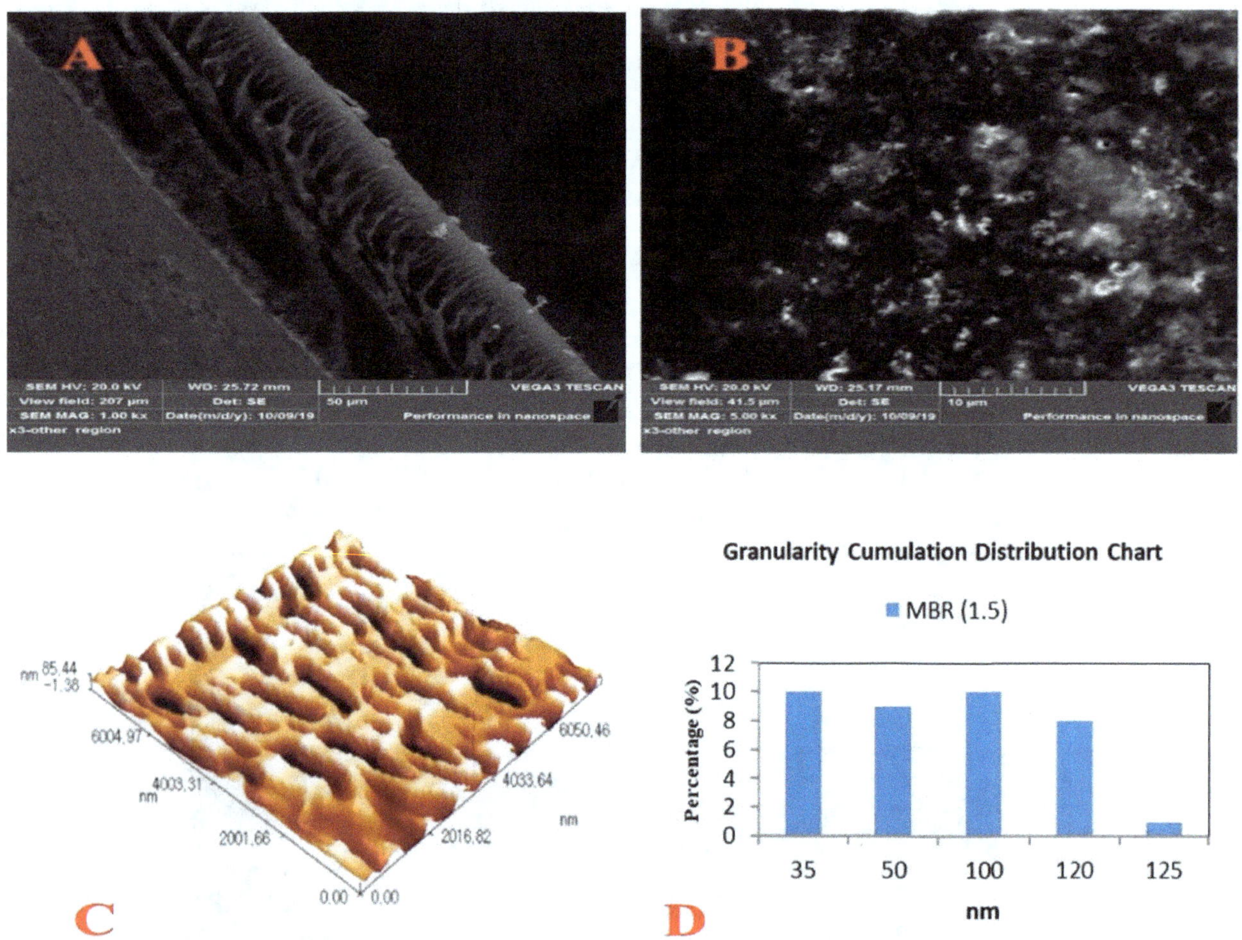

Figure 5. Surface characteristics of the nanocomposite membrane (**A**) SEM cross-section, (**B**) SEM surface, (**C**) AFM, and (**D**) granularity cumulation distribution of pores.

Table 2. 1.5 PPSU/ZnO membrane surface characteristics.

Characteristic	Value
Mean Pore Size (nm)	72.57 nm
Granularity Accumulation Distribution of pores (nm)	35–120 nm
Mean Roughness, Ra (nm)	32.5 nm
Rms	37.6 nm
Contact angle	48.67° ± 2.5

3.3. Membranes Performance and Long-Term Operation

The first step of operation was conducted without utilizing the conductive sponge and DC supply (reference stage). As given in our previous work (19), results demonstrated that the prepared nanocomposite membrane using 1.5 g of ZnO NPs has enclosed the longest operation and was chosen for the ASMBR stage. Hence, the entire improvement in the permeate flux due to the addition of 1.5 g ZnO NPs was 450% higher compared with that of control PPSU membrane (not presented results).

Figure 6 manifested the long-term performance of the three configurations tested here in this work. During step two, the conductive sponge without applied electrical field was harnessed in NSMBR configuration (Stage I). At this stage, the influence of CNTs-sponge was distinguishable for the biological treatment and antifouling properties of the nanocomposite membrane. The permeate flux before day 30 was almost the same for all systems, see Table 3. The prolonged operation of the membrane was enhanced from 31 days for ASMBR to 49 days for the NSMBR configuration. This influence was imparted by the conductive CNTs-sponge, which enhanced the quality of feed water passed later through the membrane. With this configuration, the membrane was operated until 90 days before the need for cleaning due to the fouling reduction obtained (Figure 6). The CNT-Sponge adsorbed the organic matter and microorganisms' products. Therefore, the membrane fouling decreased. After that, (2 V) voltage was applied to the CNT sponge anode and the aluminium road cathode to induce the electro-oxidation of the pollutants. The impact of electricity and CNT-sponges on the prolonged operation of nanocomposite membranes prepared from 1.5 g ZnO NPs in ENSMBR (stage II) systems presented the next improvement step for stage I. Application of the CNT-Sponge and the DC with the MLSS solution has initiated electrocoagulation to mitigate fouling of the membrane. It is worth mentioning here that the decline that occurred in the permeate flux of NSMBR was sharp, which is opposite to that of the ENSMBR that deteriorated gradually up to 96 days. In the ENSMBR configuration, the effect of the Nano-sponge with the electric field was prominent on the long-term operation and reached 57 days.

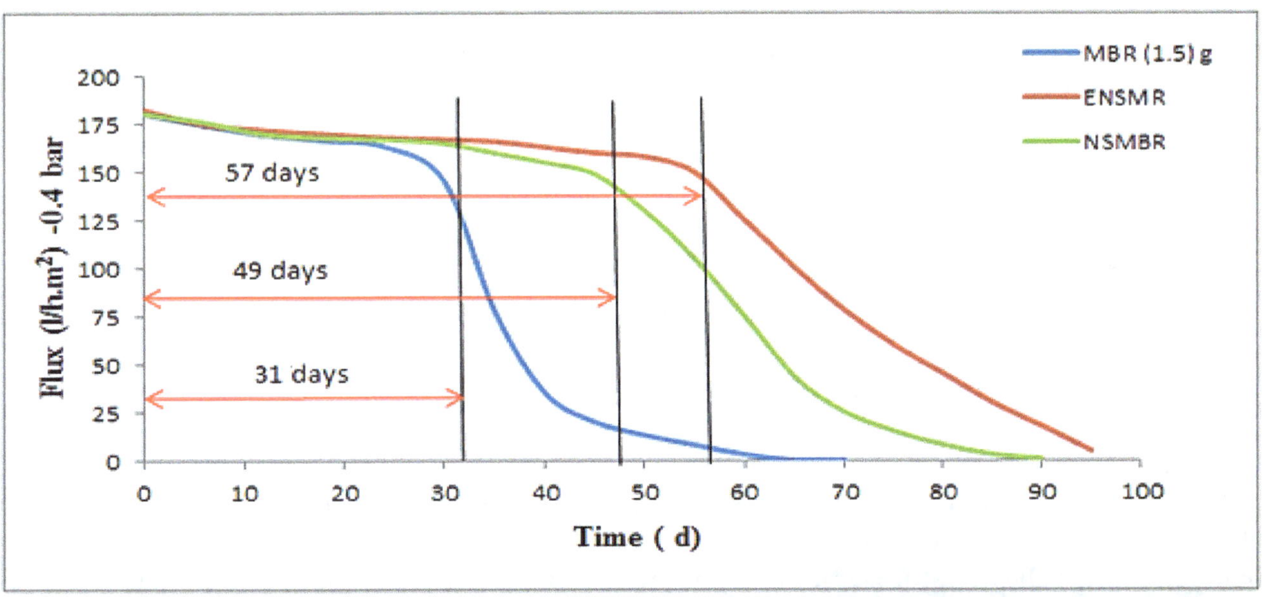

Figure 6. Step II, The effect of the electro field on the lifetime of the PPUS/ZnO NPs membranes in the NSMBR and ENSMBR system.

Table 3. Periods of the membrane life span and the deadline for step II.

Configurations	(Flux) i (L/m^2 d) = 0 d	(Flux) n (L/m^2 d) = 30 d	(Flux) n (L/m^2 d) = 60 d	(Flux) f (L/m^2 d) = 90 d	Long Term (d)
MBR (1.5 gZnO)	180	150	0.5	0	31
NSMBR	180	167	50	1	49
ENSMBR	180	170	140	50	57

Based on the results mentioned above, it was clear that the electric field had a direct impact on the activated sludge, which is thought to control the repulsion/attraction with the electrodes due to the charge difference. The majority of the sludge in the solution was trapped by the Nano-sponge and diminished the solution viscosity. Both causes reduced the accumulation of the sludge on the membrane and mitigated membrane fouling.

3.4. Biological Treatment

3.4.1. COD Removal

Figure 7 shows the capability of the CNTs-sponge to degrade the COD concentration (600–800 mg/L). As can be seen, the rate of electro-oxidation of organic matter has been raised by increasing the cell potential from 1.0 V to 2 V. A further increased change in cell potential over 2.0 V induced a reduction in sponge efficiency due to the presence of many side reactions (e.g., water oxidation and/or by-product oxidation). In addition, the blockage of CNTs' active sites by adsorbed by-products resulted from organic oxidation. For instance, the optimum cell potential of 2 V was sufficiently acceptable following other recorded literature values for the electro-oxidation of organic matters. Here, the CNTs would easily oxidize the organic matter molecule's dimethylamine group at a fairly low applied potential (e.g., +0.5 V vs. Ag/AgCl) [24]. The load amount of CNTs is considered a significant parameter for removing COD. As shown in Figure 7A, COD removal has increased as the nanotubes loading amount increased. Only 28% of COD could be removed at an imposed cell potential of 2 V if the CNTs loading was 0.1 mg/mL. However, when the CNTs load has increased to 0.3 mg/mL, the removal of COD by electro-oxidation could achieve almost double its value (52%) when compared to 0.1 mg/mL CNTs' load. The enhanced accessibility of the organic matter molecules into the CNTs' active centers, at a higher CNTs load, may support the explanation of these results.

Meanwhile, the concentration of the surfactant (SDBS) may also influence the oxidative efficiency of the conductive sponge (Figure 7B). Raising the SDBS concentration could positively participate in the removal of organic matters by electro-oxidation. At 2 V applied cell potential, the oxidation kinetics for the 1.0, 1.5, 2, and 2.5 mg/mL SDBS concentration have showcased 37%, 52%, 80%, and 60%, respectively. Despite its chemical structure, SDBS is an effective surfactant for dispersing MWCNTs in DI water.

Finally, the removal efficiency of COD in the electro-bioreactor zones has witnessed an enhancement from 75%–90% in stage I to 85%–95% in stage II. This occurred due to the role of the electro-coagulation and CNT-sponge adsorption during stage II. These results appeared to be reasonable and agreed with others reported in the literature [25]. The application of the DC field to the mixed liquor would also diminish the load of the dissolved organic matters onto the membrane fouling due to the repulsion forces. This had been reflected by the good performance of the NSMBR and ENSMBR configurations in stage II where the fouling rate has manifested a significant decrease.

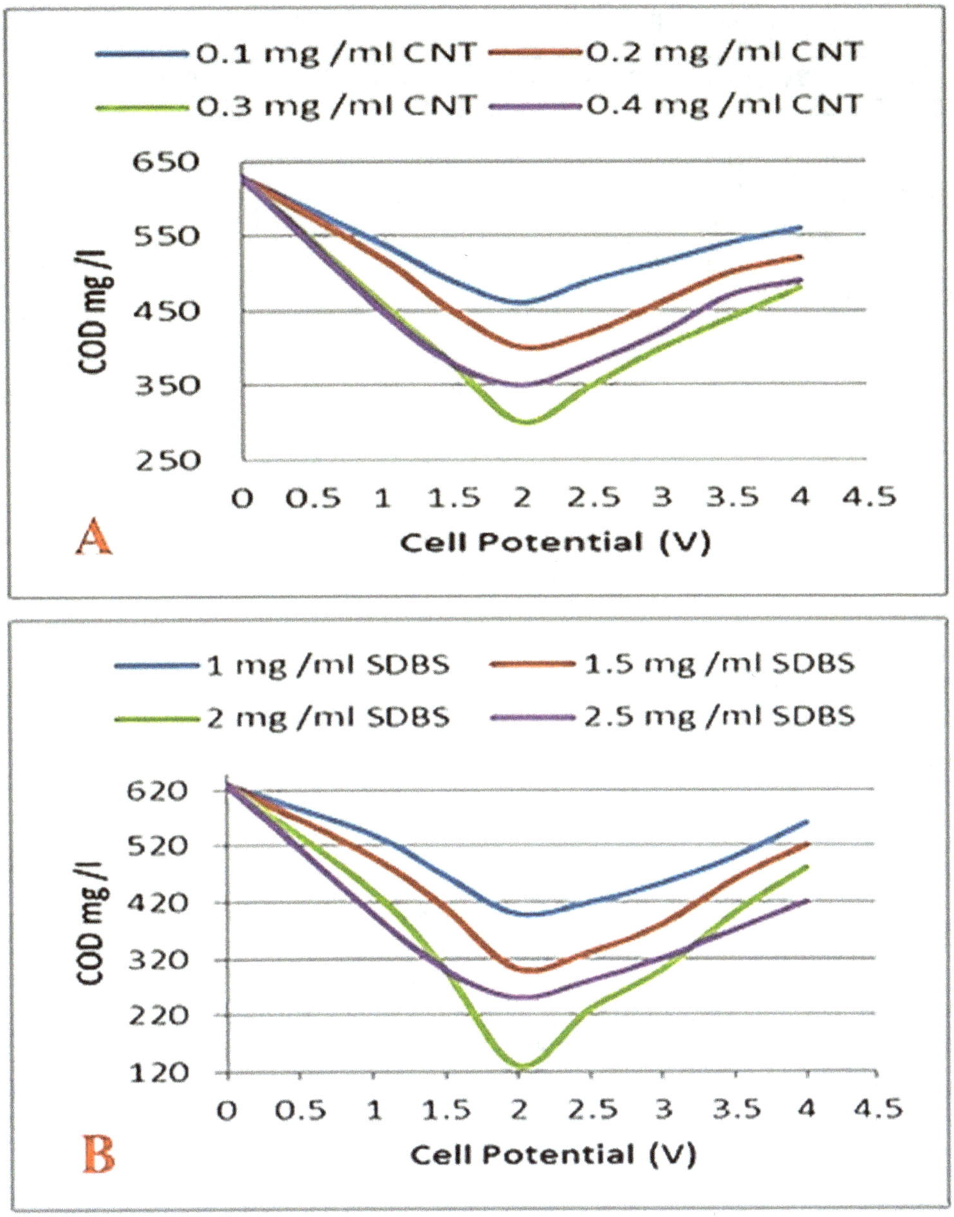

Figure 7. Electro-oxidation COD as a function of (**A**) carbon nanotube (CNT) loading, (**B**) surfactant concentration, and total cell potential.

3.4.2. Impacts of MWNTs on Contaminants Removal

Removal of NH_4-N, NO_3-N, and PO_4-P was examined through the ASMBR, NSMBR, and ENSMBR systems, and results were listed in Table 4 below. In the control reactor (step I), the concentration of NH_4^+-N and PO_4-P was not significantly eliminated in the ASMBR because the activated sludge was unable to remove nitrogen and phosphorus compounds as a standalone configuration. At the second step, the NH_4^+–N and phosphorus were efficiently eliminated in the NSMBR. This was ascribed to the superior performance of the CNT-sponge in the adopted configuration. The removal efficiencies of total nitrogen (TN) and soluble orthophosphorus (SOP) at the end of the reaction were 70% and >90%, respectively.

Table 4. Nutrient removals in membrane bioreactor (MBR), nanotubes coated sponge–MBR (NSMBR), and electro-membrane bioreactor (ENSMBR).

Experimental Runs and Applied Voltage Gradient (2 V)				
		MBR	**NSMBR**	**NSEMBR**
NH_4-N (mg/L)	Influent	31.4 ± 3.1	26.4 ± 2.8	28.5 ± 3
	Effluent	18.7 ± 1.5	11 ± 1.45	6.8 ± 0.6
	Removal (%)	**40.4**	**58.3**	**76.14**
NO_3-N (mg/L)	Influent	40.6 ± 3.6	19.3 ± 2.3	21.8 ± 2.6
	Effluent	14.2 ± 0.86	5.8 ± 0.6	4.1 ± 0.54
	Removal (%)	**65**	**70**	**81.2**
PO_4-P (mg/L)	Influent	11 ± 1.67	9.4 ± 1.3	11.6 ± 1.2
	Effluent	7.3 ± 0.5	3.1 ± 0.4	1.1 ± 0.2
	Removal (%)	**33.6**	**67**	**90.5**

Applying the DC field to the mixed liquor had also contributed to the degradation of the dissolved organic matters. Moreover, many organic compounds could be adsorbed by CNTs and, ultimately, led to a reduction in the concentrations of the contaminants.

According to the experimental results, when the dose of MWNTs increased from 0 mg/L to 0.1 mg/mL, there was no substantial difference in the concentration variations of NH_4^+–N, NO_3–N, and NO_2–N at any time. Besides, there were no adverse effects on the phase of transformation or elimination. However, incorporating 0.3 mg/mL MWNTs has given rise to the highest efficiency for TN (70%) and TP (90.6%) removal.

4. Capital Cost

The economic feasibility of the development of the MBR process has been studied in this work to confirm the feasibility of adding PUs, PUs/MWCNTs, and electrical power on the performance of the developed MBR. Table 5 depicts the capital cost of the three different configurations (e.g., MBR, NSMBR, and ENSMBR) according to the several parameters such as energy oxygen required, energy consumption, and cost of the electrodes. The conventional MBR was operated for 31 days while the two developed MBR have a long-term operation of 49 and 57 days for NSMBRs and ENSMBRs, respectively, as shown in Figure 6. The capital cost of the three MBR configurations was evaluated initially for one month. From Table 5, it can be noticed that the capital cost of both NSMBRs and ENMBRs operations recorded a cost reduction of more than 100%.

Table 5. Capital cost differences among the MBR, NSMBR, and ENSMBR.

Parameter		**MBR**	**NSMBR**	**ENSMBR**
Energy oxygen required (1 Amp/month)		10$ (4 O_2 mg/L)	10$ (4 O_2 mg/L)	10$ (4 O_2 mg/L)
Energy consumption (124 Amp/month) (15 min ON/45 min OFF)		0$	0$	14$/month
Electrodes cost	anode (Sponge)	0$	3$	3$
	cathode	0$	2$	2$
Total cost/month		10$	15$	29$
operation time (d)		31	49	57
Factor operation		31/31 = 1	49/31 = 1.58	57/31 = 1.83
Membrane replace/clogging		yes	No	No
Price membrane replaced		40$	0$	0$
Total cost		50$	−7.7$	−23.1$

0.33$Amp/d. Total cost SMBR = [1 − (Factor operation − 1)] × total cost = (1 − (1.58 − 1)) × 15 = 6.3$.

5. Conclusions

Affordable, efficient, and low-energy water purification technologies are highly desirable to handle present-day environmental concerns. Throughout this work, a low-cost methodology was evolved to impart an effective degradation for organic contaminants. This was conducted by harnessing carbon nanotubes as conductive nanomaterials within a polyurethane sponge structure. The nanotubes have been used as a high-potential electro-catalyst for adsorption and oxidation of the pollutant besides functionalization of the sponge to be employed as a conductive anode. In the meantime, the phase inversion was employed to prepare Zinc Oxide PPSU nanocomposite membranes at 1.5 g of loading weight. This additive ratio was selected based on the optimum results obtained in our previous work. A prepared nanocomposite membrane surface was fully characterized by serious experimental tools, e.g., SEM, AFM, contact angle, pore size, and pore size distribution. Results disclosed that the mean pore size was 72.5 nm with a wide pore size distribution that ranged from 35 to 120 nm. Results of hydrophilicity, Ra and Rms were 48 nm, 32.5 nm, and 37.6 nm, respectively. The testing procedure was performed through an ASMBR as a reference and results were compared with those obtained with nanotubes coated sponge–MBR (NSMBR) and nanotubes coated sponge-MBR in the presence of an electrical field (ENSMBR) system. The conductive sponge was attached to the ASMBR system as a pre-treatment to induce the removal of the organic matter and alleviate membrane fouling. Application of the electrical field (2 V) has imparted further enhancement for the treatment process. Results manifested that the long-term operation of ENSMBR was 57 days compared to 31 and 39 days for MBR and NSMBR, respectively. The fouling reduction was 78% compared to the MBR, which showed 60%. Meanwhile, at 0.3 mg/mL, the removal of COD by electro-oxidation could achieve almost double its value (52%) when compared to 0.1 mg/mL CNTs load. Removal of PO_4 by ENSMBR was almost triple (90.6%) than that obtained by MBR (33.6%). In addition, NO_3 was eliminated by 81%. These results indicated that application of ENSMBR could have a promising potential when applied for wastewater treatment.

Author Contributions: Data curation, A.M.A.; Formal analysis, A.M.A.; Investigation, Q.F.A., A.M.A. and A.I.A.-N.; Methodology, Q.F.A., R.H.A.-A. and A.I.A.-N.; Supervision, Q.F.A. and R.H.A.-A.; Visualization, A.M.A.; Writing—original draft, Q.F.A. and A.M.A.; Writing—review & editing, Q.F.A. and R.H.A.-A. All authors have read and agreed to the published version of the manuscript.

References

1. Liu, Y.; Li, J.; Zhouabc, B.; Li, X.; Chen, H.; Chen, Q.; Wang, Z.; Li, L.; Wang, J.; Cai, W. Efficient electricity production and simultaneously wastewater treatment via a high-performance photocatalytic fuel cell. *Water Res.* **2011**, *45*, 3991–3998. [CrossRef] [PubMed]
2. Li, R.; Zhang, L.; Wang, P. Rational design of nanomaterials for water treatment. *Nanoscale* **2015**, *7*, 17167–17194. [CrossRef] [PubMed]
3. Ngo, H.H.; Guo, W.; Xing, W. Evaluation of a novel sponge-submerged membrane bioreactor (SSMBR) for sustainable water reclamation. *Bioresour. Technol.* **2008**, *99*, 2429–2435. [CrossRef]
4. Alsalhy, Q.F.; Almukhtar, R.S.; Alani, H.A. Treatment of oil refinery wastewater by membrane bioreactor (MBR). *Arab. J. Sci. Eng.* **2016**, *41*, 2439–2452. [CrossRef]
5. Alsalhy, Q.F.; Al-Ani, F.H.; Al-Najar, A.E. A new Sponge-GAC-Sponge membrane module for submerged membrane bioreactor use in hospital wastewater treatment. *Biochem. Eng. J.* **2018**, *133*, 130–139. [CrossRef]
6. Al-Ani, F.H.; Majdi, H.S.; Ali, J.M.; Al Rahawi, A.M.; Alsalhy, Q.F. Comparative study on CAS, UCT, and MBR configurations for nutrient removal from hospital wastewater. *Desalin. Water Treat.* **2019**, *164*, 39–47. [CrossRef]
7. De Volder, M.; Tawfick, S.H.; Baughman, R.H.; Hart, A.J. Carbon Nanotubes: Present and Future Commercial Applications. *Science* **2013**, *339*, 535–539. [CrossRef]
8. Jiang, Y.; Wang, W.-N.; Liu, D.; Nie, Y.; Li, W.; Wu, J.; Zhang, F.; Biswas, P.; Fortner, J.D. Engineered Crumpled Graphene Oxide Nanocomposite Membrane Assemblies for Advanced Water Treatment Processes. *Environ. Sci. Technol.* **2015**, *49*, 6846–6854. [CrossRef]
9. Allen, J.B.; Larry, R.F. *Electrochemical Methods Fundamentals and Applications*; John Wiley & Sons: Hoboken, NJ, USA, 2001.

10. Vecitis, C.D.; Schnoor, M.H.; Rahaman, S.; Schiffman, J.D.; Elimelech, M. Electrochemical Multiwalled Carbon Nanotube Filter for Viral and Bacterial Removal and Inactivation. *Environ. Sci. Technol.* **2011**, *45*, 3672–3679. [CrossRef]
11. Luo, M.-L.; Zhao, J.-Q.; Tang, W.; Pu, C.-S. Hydrophilic modification of poly(ether sulfone) ultrafiltration membrane surface by self-assembly of TiO_2 nanoparticles. *Appl. Surf. Sci.* **2005**, *249*, 76–84. [CrossRef]
12. Singh, T.; Srivastava, N.; Lal Singh, N. Application of TiO_2 nanoparticle in photocatalytic degradation of organic pollutants. *Mater. Sci. Forum* **2016**, *855*, 20–32. [CrossRef]
13. Al Aani, S.; Wright, C.J.; Atieh, M.A.; Hilal, N. Engineering nanocomposite membranes: Addressing current challenges and future opportunities. *Desalination* **2017**, *401*, 1–15. [CrossRef]
14. He, X.; Yang, D.-P.; Zhang, X.; Liu, M.; Kang, Z.; Lin, C.; Jia, N.; Luque, R.J.C.E.J. Waste eggshell membrane-templated CuO-ZnO nanocomposites with enhanced adsorption, catalysis and antibacterial properties for water purification. *Chem. Eng. J.* **2019**, *369*, 621–633. [CrossRef]
15. Ong, C.B.; Mohammad, A.W.; Ng, L.Y. Integrated adsorption-solar photocatalytic membrane reactor for degradation of hazardous Congo red using Fe-doped ZnO and Fe-doped ZnO/rGO nanocomposites. *Environ. Sci. Pollut. Res.* **2018**, *26*, 33856–33869. [CrossRef] [PubMed]
16. Darabi, R.R.; Jahanshahi, M.; Peyravi, M. A support assisted by photocatalytic Fe_3O_4/ZnO nanocomposite for thin-film forward osmosis membrane. *Chem. Eng. Res. Des.* **2018**, *133*, 11–25. [CrossRef]
17. Parvizian, F.; Hosseini, S.; Hamidi, A.; Madaeni, S.; Moghadassi, A. Electrochemical characterization of mixed matrix nanocomposite ion exchange membrane modified by ZnO nanoparticles at different electrolyte conditions "pH/concentration". *J. Taiwan Inst. Chem. Eng.* **2014**, *45*, 2878–2887. [CrossRef]
18. Balta, S.; Sotto, A.; Luis, P.; Benea, L.; Van Der Bruggen, B.; Kim, J. A new outlook on membrane enhancement with nanoparticles: The alternative of ZnO. *J. Membr. Sci.* **2012**, *389*, 155–161. [CrossRef]
19. Almusawy, A.M.; Al-Anbari, R.H.; Alsalhy, Q.F.; Al-Najar, A.E.; Drioli, E. Novel membrane bioreactor system using combination of an intermittent electric field and moving biomedia for treatment of genuine wastewater. *Int. J. Environ. Anal. Chem.* **2020**.
20. Liu, Y.; Li, F.; Xia, Q.; Wu, J.; Liu, J.; Huang, M.; Xie, J. Conductive 3D sponges for affordable and highly-efficient water purification. *Nanoscale* **2018**, *10*, 4771–4778. [CrossRef]
21. Alsalhy, Q.F.; Al-Ani Faris, H.; Al-Najar, A.E.; Jabuk, S.I.A. A study of the effect of embedding ZnO-NPs on PVC membrane performance use in actual hospital wastewater treatment by membrane bioreactor. *Chem. Eng. Process. Process Intensif.* **2018**, *130*, 262–274. [CrossRef]
22. Oliveira, I.M.; Silva, S.S.; Malafaya, P.B.; Rodrigues, M.T.; Kotobuki, N.; Hirose, M.; Gomes, M.E.; Mano, J.F.; Ohgushi, H.; Reis, R.L. Macroporous hydroxyapatite scaffolds for bone tissue engineering applications: Physicochemical characterization and assessment of rat bone marrow stromal cell viability. *J. Biomed. Mater. Res. Part A* **2009**, *91*, 175–186. [CrossRef] [PubMed]
23. Camara, M.; Breuil, P.; Briand, D.; Viricelle, J.-P.; Pijolat, C.; De Rooij, N.F. Preconcentration Modeling for the Optimization of a Micro Gas Preconcentrator Applied to Environmental Monitoring. *Anal. Chem.* **2015**, *87*, 4455–4463. [CrossRef] [PubMed]
24. Liu, Y.; Liu, H.; Zhou, Z.; Wang, T.; Ong, C.N.; Vecitis, C.D. Degradation of the Common Aqueous Antibiotic Tetracycline using a Carbon Nanotube Electrochemical Filter. *Environ. Sci. Technol.* **2015**, *49*, 7974–7980. [CrossRef] [PubMed]
25. Yamato, N.; Kimura, K.; Miyoshi, T.; Watanabe, Y. Difference in membrane fouling in membrane bioreactors (MBRs) caused by membrane polymer materials. *J. Membr. Sci.* **2006**, *280*, 911–919. [CrossRef]

2

Current State of Porous Carbon for Wastewater Treatment

Mongi ben Mosbah [1,2], Lassaad Mechi [1,3], Ramzi Khiari [4,5,6] and Younes Moussaoui [2,7,*]

1 Materials, Environment and Energy Laboratory (UR14ES26), Faculty of Sciences of Gafsa, University of Gafsa, Gafsa 2112, Tunisia; mbenmosbah@yahoo.fr (M.b.M.); mechilassaad@yahoo.fr (L.M.)
2 Faculty of Sciences of Gafsa, University of Gafsa, Gafsa 2112, Tunisia
3 Department of Chemistry, Faculty of Sciences, University of Hail, Hail PO Box 2440, Saudi Arabia
4 Higher Institute of Technological Studies, Ksar Hellal 5070, Tunisia; khiari_ramzi2000@yahoo.fr
5 UR13 ES 63—Research Unity of Applied Chemistry & Environment, Faculty of Sciences, University of Monastir, Monastir 5000, Tunisia
6 LGP2, University of Grenoble Alpes, CNRS, Grenoble INP, LGP2, F-38000 Grenoble, France
7 Organic Chemistry Laboratory (LR17ES08), Faculty of Sciences of Sfax, University of Sfax, Sfax 3029, Tunisia
* Correspondence: y.moussaoui2@gmx.fr

Abstract: Porous materials constitute an attractive research field due to their high specific surfaces; high chemical stabilities; abundant pores; special electrical, optical, thermal, and mechanical properties; and their often higher reactivities. These materials are currently generating a great deal of enthusiasm, and they have been used in large and diverse applications, such as those relating to sensors and biosensors, catalysis and biocatalysis, separation and purification techniques, acoustic and electrical insulation, transport gas or charged species, drug delivery, and electrochemistry. Porous carbons are an important class of porous materials that have grown rapidly in recent years. They have the advantages of a tunable pore structure, good physical and chemical stability, a variable specific surface, and the possibility of easy functionalization. This gives them new properties and allows them to improve their performance for a given application. This review paper intends to understand how porous carbons involve the removal of pollutants from water, e.g., heavy metal ions, dyes, and organic or inorganic molecules. First, a general overview description of the different precursors and the manufacturing methods of porous carbons is illustrated. The second part is devoted to reporting some applications such using porous carbon materials as an adsorbent. It appears that the use of porous materials at different scales for these applications is very promising for wastewater treatment industries.

Keywords: porous materials; activated carbon; biomass; activating agent; adsorption; applications; pollutants

1. Introduction

Porous materials are defined as materials with cavities, or channels, called pores. Pores are empty spaces that exist between particles of any shape in materials. These empty spaces form the porosity of the materials, defined as the volume percentage of air in the material that correspond to the total volume occupied by the voids of the material divided by the total volume of the material [1]. The porous space is a continuity of absence of solid matter nested in the continuity of solid matter. It is essentially irregular in its forms and inconsistent also in its qualities, which can be given great complexity. There are two types of porosity: open porosity (on the external environment) and closed porosity. In all cases, it is generally characterized by its volume fraction (or by the density of the porous material) and the pore size distribution. The open porosity notably controls the adhesion properties of the coatings on the surface of the material [2].

Porous materials have attracted the attention of chemists and materials scientists, and the development of new porous materials has accelerated research development in recent decades [3–5]. This attention comes back not only to the commercial interest that porous materials have for their application in various fields such as separation [6,7], catalysis [7–9], adsorption [6,10–15], energy storage and conversion [16–24], and medicine [25,26], but also, because of the scientific interest in the challenges posed by their synthesis, their treatment and their characterization. Indeed, the progress of technology and the demand for porous materials have pushed researchers to develop synthesis methods allowing for controlling the parameters that determine the structural and textural characteristics of these materials. In this context, a lot of works have shown that controlling the pore size is essential for many specific applications. However, this technological and industrial development with the growth of the world population and domestic activities are causing a remarkable increase in water pollution. It is therefore necessary to purify wastewater before it is released into the environment. Thus, several wastewater treatment technologies have been developed, the adsorption of which on porous materials is proving to be one of the most promising techniques for the removal of pollutants from wastewater due to its affordable price and ease of disposal at large-scale use [27–30]. The most widely used adsorbents are activated carbons, and they are considered the most useful for the removal of organic and inorganic pollutants due to their structural variability (macro-, meso-, and micropores), their large specific surface area, and a wide availability of functional groups [29,31,32].

The purpose of this review is to discuss the preparation processes of porous carbons as well as the main carbon precursors used. In addition, the performance and characteristics of the resulting carbons are investigated. Finally, the contribution of activated carbon in the field of wastewater treatment will also be discussed in detail.

2. Classification of Porous Materials

According to the International Union of Pure and Applied Chemistry-USA(IUPAC) [33,34], porous materials are classified according to the diameter of the pores (d_p) that constitute them, and the three classes are seen in Figure 1.

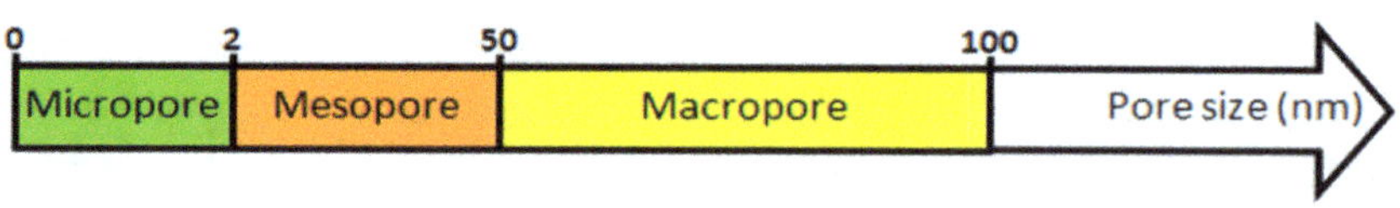

Figure 1. Classification of porosity based on pore diameter according to the International Union of Pure and Applied Chemistry (IUPAC).

2.1. Microporous Materials

Microporous materials are those whose pore diameter is less than 2 nm (d_p < 2 nm). They can be prepared by the sol-gel process and are widely used in catalysis and adsorption.

2.2. Mesoporous Materials

In this category of materials, the pore diameter varies between 2 and 50 nm (2 nm < d_p < 50 nm). A distinction is made between crystalline mesoporous materials and ordered amorphous materials, which are intermediate between that of crystalline microporous solids of the zeolite type and disordered amorphous solids such as silica gel [35–37]. Mesoporous materials have several known applications such as catalysis, filtration, pollution control, optics, and even electronics.

2.3. Macroporous Materials

As these materials have fairly large pores ($d_P > 50$ nm), they find their use in the field of decontamination of polluted water with organic dyes and as a catalyst support for the photodegradation of the pollutant [20,38–41].

3. Graphite and Graphene

Graphite consists of interconnected carbon atoms arranged to hexagonally form flat networks in the form of layers stacked parallel to each other (Figure 2b) [42,43]. These layers are connected by low energy van der Waals type bonds. This explains the laminated morphology of graphite [42,43]. Graphite is characterized by high thermal and electrical conductivity, good chemical stability, and remarkable plasticity [43–50]. A good thermal and electrical conductivity is attributed to the delocalization of the π-electrons of the carbon atoms in graphite, and its plasticity can be explained by the possibility of sliding the sheets layers one onto the other [42,43]. These favorable characteristics make graphite interesting materials for improving the transfer and storage of thermal energy, for preparing anodes of lithium-ion batteries, and for electrochemical applications. In this context, Zhong et al. [51] used pitch-based graphite foams to increase the thermal diffusivity of paraffin wax for a thermal energy storage application. Later, Jana et al. [45] studied the effect of the addition of graphite on the conductivity of a tannin-based resin. They showed, contrary to what was expected a priori, that the smaller graphite particles were much more suitable for obtaining conductive matrices. Indeed, the use of small grains made the viscosity higher because of their higher surface and made it possible to obtain homogeneous foams with a higher density [45]. In 2019, Zhang et al. [52] showed that the use of graphitized mesoporous carbon as anode materials for lithium-ion batteries improved speed performance and had a good cycling stability with a reversible capacity of 248.3 mA h g^{-1} at 1 C after 100 cycles. In the same context, porous graphitic carbon was synthesized from *Eichhornia crassipes* plants collected from the Coimbatore region, Tamilnadu-India, and has been used for the sustainable fabrication of hole-transporting materials electrode in perovskite solar cells [53].

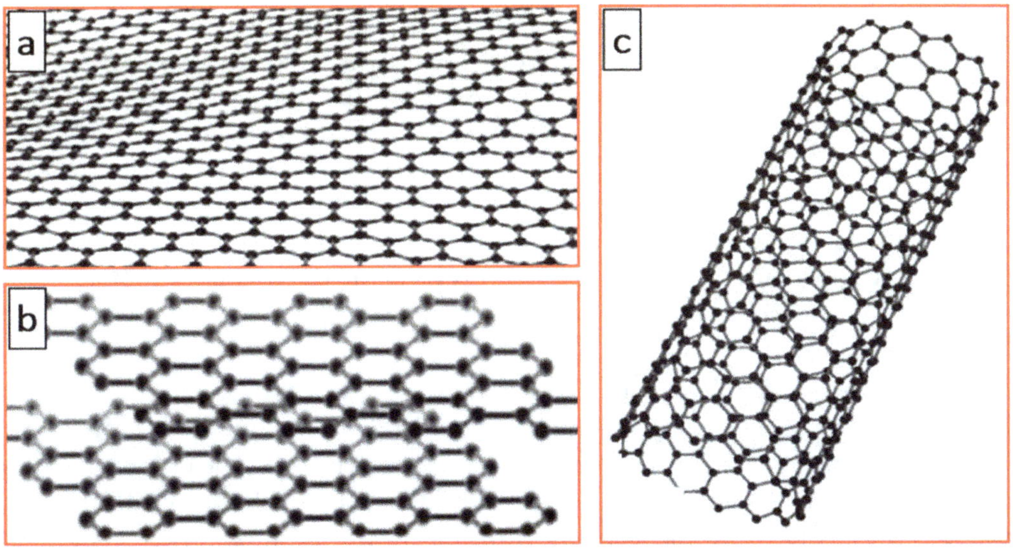

Figure 2. A conceptual model depicting the structure of graphene (**a**), graphite (**b**), and carbon nanotubes (**c**).

Graphene is related to graphite, and it was isolated experimentally for the first time, in 2004, from graphite [54–57]. It is a planar material with a hexagonal structure (Figure 2a), often compared to a honeycomb network, in the form of a two-dimensional (2D) crystal consisting of a simple plane of carbon atoms in sp^2 hybridizations [58]. Graphene represents the basic unit of graphite, and it has a high optical transparency and remarkable rigidity with a tensile strength of more than 130 GPa [59]. It can be wrapped to form many other forms of carbon such as fullerenes, nanotubes, or stacked in graphite. It is important to note that the honeycomb structure of graphene gives it excellent stability with a high mobility of the electrons on its surface [59–61] and can be used in multiple applications in the fields of electronics (flexible screens, high frequency electronics, nanoelectronics, batteries, chargers), the environment (atmospheric pollution sensors, desalination of seawater), and health (targeted therapy). Generally, there are three main techniques for making graphene. The first allows us to isolate a single plane of graphene by the exfoliation of a graphite crystal using adhesive ribbon; this operation is repeated several times so that the film is sufficiently thin before depositing it on a substrate, which more often is silica [55,56,62]. The second method consists of an annealing at high temperature of a silicon carbide substrate, which induces the sublimation of Si, consequently producing a reorganization of the carbon atoms, leading to the formation of graphene planes on the surface [63–65]. The third method is based on chemical deposition on metals from hydrocarbons [44,62,66–69]. The first technique is widely used because it is simple and inexpensive and also because the properties of graphene are easily accessible [55,70,71].

The chemical properties of graphene are similar to those of graphite, and graphene can be used in multiple applications, in particular for field effect transistors due to the high mobility of the surface electrons. Because graphene has chemical stability in contact with electrolytes, mechanical elasticity and good electrical conductivity, it is widely used as a storage medium for lithium [72–75]. It is also used in gas sensors such as NO_2, CO, H_2S, H_2, NO, NH_3, and CO_2 [76–85], or even as a biosensor for the detection of dopamine [86,87], β-nicotinamide adenine dinucleotide [88], paracetamol [87,89], Kojic acid [90], uric acid [86,87], vanillin [91], and caffeic acid [92]. Graphene is also used as a drug carrier [93–97] and as a charge conductor in solar cells as well as a light collector including cells and photodetectors [53,60–62,98–100]. He et al. [101] prepared a three-dimensional graphene-CNTs@ Se aerogel, usingCNTs/selenium that is sandwiched between graphene nanosheets via an environmentally friendly solvothermal method. They showed that due to the 3D sandwich-type protection of the active Se, the cathode offers an initial coulombic efficiency of 92% as well as an excellent cyclic performance. This greatly facilitates the improvement of the specific energy density of Li-Se batteries [101]. Likewise, it has been shown that a graphene-CNTs-Li_2S hybrid aerogel could be effectively used as a high capacity cathode for Li-S batteries with a reversible discharge capacity as large as 1123.6 mA h g^{-1}, and a decrease in capacity of less than 0.02% per cycle [102,103]. The efficiency of these materials is attributed to the 3D structure, which is achieved by using two-dimensional (2D) graphene nanosheets and one-dimensional (1D) carbon nanotubes (CNTs), thus forming highly efficient channels for electron transfer and ionic diffusion [102,103]. This architecture was used for the development of a CoS_2-carbon nanotubes-graphene anode for high performance lithium-ion batteries that can deliver an initial discharge capacity of 993 mA h g^{-1} at 100 mA g^{-1} [104].

4. Carbon Nanotubes

Carbon nanotubes are an allotropic form of carbon (Figure 2c), observed for the first time in 1991 by Japanese researcher Sumio Iijima during the synthesis of fullerenes by an electric arc [105]. They have been characterized as graphene sheets that are wound on themselves in the form of microtubes and will later be renamed multiwall carbon nanotubes (MWCNTs). In 1993, the synthesis of single-walled carbon nanotubes (SWCNTs) was reported by optimizing the method of synthesis by an electric arc [106,107]. The multisheet nanotubes consist of several graphene sheets wound concentrically with a spacing between two sheets of approximately 3.6 Å, slightly greater than the spacing between two sheets in graphite. The diameter of the nanotubes varies, depending on the number of sheets, from 1 to 50 nm, and their lengths can reach 1 μm [43,105,108–110]. SWCNTs consist of a graphene sheet wound on itself, and they form a tube whose diameter is between 0.4 and 3 nm. Carbon nanotubes are therefore composed of one or more sheets of carbon atoms, as in graphite, wound on themselves forming a tube. Carbon nanotubes have good mechanical properties, good adhesion, and excellent electrical conductivity [43,111–113]. Some of these nanotubes have metallic properties [75,114,115], while others are semiconductors [116–121]. SWCNTs and MWCNTs are manufactured by almost the same method; the only distinction can appear on the use of the metal catalyst, generally nickel, iron, or cobalt, which is essential for the synthesis of fullerenes. They can be prepared mainly by arc discharge, laser ablation, and chemical vapor deposition [109,120,122–126], involving one of the following as carbon precursors: xylene, acetylene, toluene, methane, benzene, etc. [43]. On the industrial level, carbon nanotubes, with the particularity that they possess, residing in a small amount of impurities, are used in many forms such as the reinforcing elements of polymers and composites, nanoporous materials for energy storage, passive (nanometric conductors) or active components (diodes and transistors), and systems allowing for the vectorization of drug molecules for the treatment of certain diseases [120–122,127–132].

5. Activated Carbon

Activated carbon is porous, amorphous organic material with a complex structure, characterized by high carbon content [133]. These materials include a wide range of carbonaceous substances with different properties and characteristics (porosity, specific surface, chemical nature of the surface, density, etc.). Activated carbon can be produced from any substance with a high carbon content, whether of a vegetable, fossil, or material of a synthetic nature; examples include date stones [134,135], coffee grounds [133,136,137], almond shell [138], coconut shell [139], corncob wastes [140], Acacia glauca sawdust [141], waste potato residue [142,143], rice husk [144], sunflower piths [145], tomato stem [146], banana peel [147], etc.

5.1. Preparation of Activated Carbon

Generally, activated carbon is prepared in two stages. The first step is the carbonization of raw material to produce a carbon surface. This step is followed by the activation by chemical oxidation or heat treatment in order to further boost the surface of the material obtained (Figure 3).

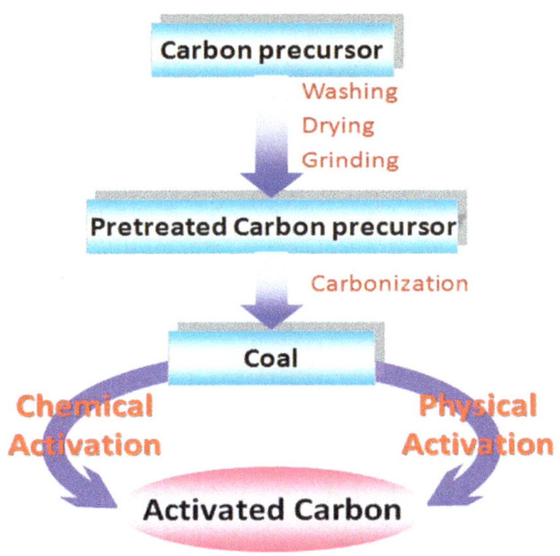

Figure 3. Steps for preparation of activated carbons.

5.1.1. Pyrolysis

Pyrolysis (carbonization) is the thermal decomposition of carbonaceous materials at high temperature between 400 and 800 °C, under an inert atmosphere. The resulting material is an amorphous solid rich in carbon. The remaining carbon atoms are grouped in sheets that arrange themselves irregularly, leaving gaps between them. These interstices give rise to a primary porosity of the charred product. It should be noted that the heat treatment strongly influences the texture of the materials [148]. Indeed, a gentle heating rate slightly modifies the carbon structure compared to the original structure of the material used and regenerates a reduced rate of volatile compounds. In the same context, the pyrolysis temperature and the residence time determine the mass loss and the nature of the surface of the biochar prepared.

The activated carbon can be obtained by the pyrolysis of carbon gels obtained by the sol-gel process. This process involves polycondensation reactions of organic precursors. In 1989, Pekala et al. introduced for the first time the synthesis of carbon aerogel by the polycondensation of resorcinol and formaldehyde in the presence of sodium carbonate as a basic catalyst [149,150]. Subsequently, numerous studies have vested in the preparation of polymers and activated carbon precursors by the condensation of hydroxylated benzene (phenol, catechol, 4-methylcatechol, m-cresol, hydroquinone, phloroglucinol, resorcinol, etc.) with aldehyde (formaldehyde, furfural, etc.) [151–158].

5.1.2. Activation

To improve the textural characteristics of the carbonate, it is necessary to subject it to an activation step. This involves opening the initial porosity and increasing accessibility to the internal structure, as well as creating new pores, eliminating the tar that clogs the pores and creating surface functions. This operation, therefore, leads to a loss of mass, and the estimation of the degree of activation can be determined by the "burn-off":

$$\text{Burn}-\text{off}\,(\%) = ((m_i - m_f)\ 100)/m_i) \tag{1}$$

where m_i is initial mass and m_f is final mass. The burn-off provides information on the rate of activation progress and the nature of porosity. Activation can be carried out by a physical or chemical process.

Physical Activation

Physical activation consists of oxidation of biochar at high temperature (>700 °C) under a stream of an oxidizing gas such as water vapor, oxygen, and carbon dioxide. The reactions are shown in Equations (1)–(4).

$$C + H_2O \rightarrow CO + H_2C + H \tag{1}$$

$$C + O_2 \rightarrow CO_2C + O \tag{2}$$

$$C + 1/2\,O_2 \rightarrow COC + O \tag{3}$$

$$C + CO_2 \rightarrow 2\,COC + CO \tag{4}$$

During physical activation, the development of the porosity of the biochar depends on the nature of the oxidizing gas used, and the pores changes from micro- to mesopores and macropores [27,31,159,160]. The porosity is more developed in the presence of water vapor than in the presence of carbon dioxide. Indeed, the size of the water molecules is smaller than that of carbon dioxide, which gives it a greater diffusion in the pores of carbon, thus causing the enlargement of the micropores in mesopores and macropores. Generally, carbon dioxide promotes the formation of a microporous network, while water vapor generates a mesoporous or macroporous network [29,31,159–161]. Furthermore, the use of oxygen is accompanied by a significant loss in mass [27,159]. It should be noted here that physical activation is preferred because of the low cost of preparation and its simple operation, making it possible to optimize the pyrolysis step and therefore producing better control of the porosity of the activated carbon. In fact, optimization of the yield and surface properties of activated carbon lies in optimizing the temperature, the activation time, and the flow rate of the gas (CO_2, vapor, N_2) [27,160].

Chemical Activation

The chemical activation is carried out with chemical reagents that promote dehydration and a structural reorganization of the pores of the carbonaceous material. The procedure consists of impregnating the carbonaceous material with the chemical agent then undergoing a heat treatment at a temperature between 400 and 900 °C. Various chemical agents can be used, namely salts of alkali metals ($AlCl_3$, $ZnCl_2$), acids (H_3PO_4, H_2SO_4), or bases (KOH, NaOH, K_2CO_3, Na_2CO_3) [134,140,162,163]. Zinc chloride was the main chemical activator used on a large scale. It is now abandoned due to environmental problems linked to the toxicity of residual zinc-derived compounds. Alkali metal hydroxides (KOH, NaOH) are also used successfully to obtain a high surface area reaching 3000 m^2/g [164]. Alkali metal hydroxides are often used to enlarge microporosity, while zinc chloride is used to develop strong microporosity alongside small mesoporosity. It turns out that KOH is the most effective activator (effective in terms of large specific surface area and high microporosity) for preparing activated carbon from biomass [31,165–167]. Furthermore, orthophosphoric acid (H_3PO_4) is widely used in industry to impregnate lignocellulosic precursors and synthetic materials to provide active carbon with smaller pores than those activated with $ZnCl_2$ [168]. Indeed, phosphoric acid plays both the role of an acid catalyst to favor the reactions of the depolymerization of the macromolecules that constitute biomass (cellulose, hemicellulose, and lignin) while promoting the formation of crosslinking through reactions of dehydration, cyclization, and condensation, and a combining agent that subsequently reacts with organic matter to form phosphate esters and polyphosphates, both of which serve as bridges between the fragments of biopolymers that constitute the lignocellulosic biomass [31,169,170]. Furthermore, the addition (or insertion) of phosphate groups leads to a process of expansion of the carbonaceous material which, after the elimination of the acid, leaves the matrix in a developed state with an accessible porous structure [31].

5.2. *Different Forms of Activated Carbon*

Activated carbon is widely used in the industry for a variety of applications [171]. Each application requires a type and form of activated carbon. The most available forms in the markets are seen in Figure 4 and also presented in the following subsections.

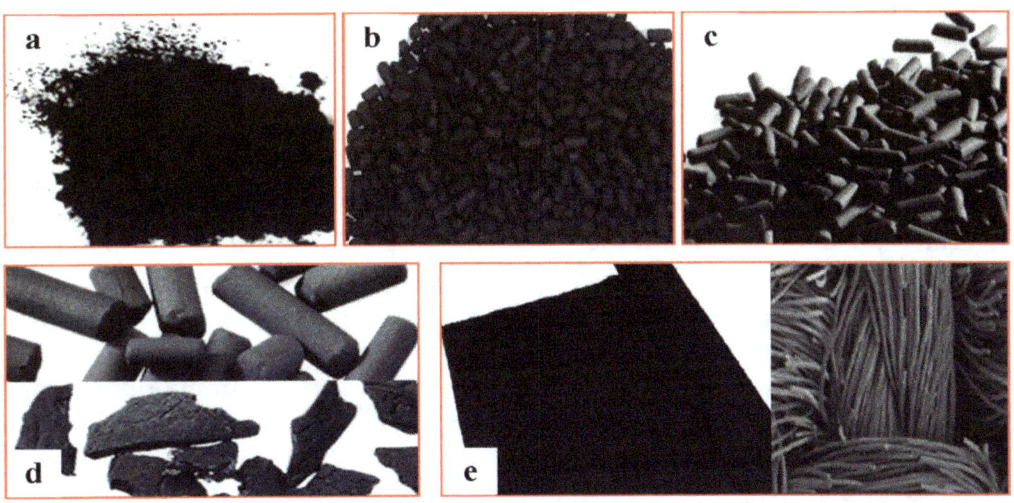

Figure 4. Forms of activated carbon: (**a**) activated carbon powder, (**b**) granular activated carbon, (**c**) extruded activated carbon, (**d**) impregnated activated carbon, (**e**) activated carbon fabric.

5.2.1. Activated Carbon Powder (CAP)

It occurs in the form of fine particles of size less than 100 μm with average pore diameters of between 0.15 and 0.25 μm (Figure 4a). They are characterized by a large external surface and pores of reduced depths. This type of carbon is frequently used for the treatment of liquid effluents in batch mode.

5.2.2. Granular Activated Carbon (CAG)

The granular or granular activated carbon is formed by grains with a size between 0.5 and 10 mm (Figure 4b). These grains are known by a small pore diameter, a large internal surface, and a relatively small external surface. This type of carbon is intended for the treatment of gases and liquid effluents.

5.2.3. Extruded Activated Carbon

To make extruded activated carbon, the carbonaceous raw material is finely ground; a binder is added to this powder in order to obtain a paste which, by an extrusion device, produces extrudates of a determined length and width (Figure 4c). These extrudates are then charred and activated under a gas flow. These materials are used for gas phase applications.

5.2.4. Impregnated Activated Carbon

Activated carbon can also be impregnated with a reagent to optimize its adsorption capacities for certain molecules (Figure 4d). Thus, the removal of certain impurities from gas is facilitated by this type of material, whereas it would have been impossible on a conventional carbon adsorbent.

5.2.5. Activated Carbon Fabric (CAT)

The active carbon tissues are obtained by the carbonization of a synthetic precursor tissue of viscose or polyacrylonitrile type followed by activation Figure 4e). These fabrics are extremely porous materials with a very high specific surface of up to 2000 m^2/g and a distribution of micropores of 90% of the total pore volume. These materials are known for their medicinal applications and in catalysis.

5.3. Application of Activated Carbon for Wastewater Treatment

Activated carbon has long been one of the most widely used materials in the industrial sector due to its essential role in the purification of liquids and the filtration of gases. However, innovations in the manufacture and use of activated carbon are driven by the need to reuse natural resources and prevent environmental pollution. Thus, a lot of progress has been made on the optimization of these carbonaceous materials, both in terms of their manufacture and their regeneration. This has prompted several industries—for example, the metallurgical, petroleum, food, pharmaceutical, chemical, and automotive industries—to use activated carbon for various uses. Currently, the main application of activated carbon lies in the adsorption processes of pollutants in the gas or liquid phases. In fact, activated carbon is in the category of amorphous carbon, which has been chemically improved into a porous structure, mostly microporous with a surface functionality that offers active carbon the properties of excellent materials for the adsorption of various chemical species [43,172–174]. This adsorption phenomenon is due to the existence of van der Waals type physical forces, inducing a higher concentration of the adsorbate at the interface than in the fluid. Adsorption can also be caused by the formation of covalent bonds between the active sites present on the surface of the carbon and the adsorbate, which is similar to a phenomenon of chemisorption. In this regard, a limited overview is presented on the use of active carbon for the adsorption of pollutants in the liquid phase, and in particular, the field of wastewater treatment. The physicochemical surface properties of activated carbon are generally governed by the nature of the starting materials, the activation methods, and the preparation conditions (Table 1).

Table 1. The experimental conditions for activated carbons production, specific surface area (S_{BET}), and adsorption capacities against various toxicants.

Carbon Precursor	Activ. Temp./Activ. Time	Activation	Reagent	S_{BET} (m^2/g)	Dose (g/L)	Pollutant	Q_{max} (mg/g)/C_0 (mg/L)	Ref.
Acacia erioloba	600 °C/1 h	Chemical	H_2SO_4	10.4	1	MB	1.5/25–500	[175]
Acacia glauca	450 °C/1 h	Chemical	H_3PO_4	311.2	2	4-NP	204.7/50–500	[141]
Acrylic fibrous waste	1200 °C	Physical	O_2	280	2	MB	8.7/2–10	[176]
Apricot shell	400 °C/1.5 h	Chemical	H_3PO_4	307	-	Tetracycline	76.9 (FBA)	[177]
Ashitaba waste	800 °C/1.5 h	Chemical	$ZnCl_2$	1228.5	0.01	Congo Red	345.1/0–100	[178]
					0.01	MB	289.2/0–50	
Bamboo	600 °C/4 h	Chemical	H_3PO_4	1400	1	Reactive Black 5	489.9/10–1000	[179]
Bamboo waste + Ms	550 °C/0.5 h	Chemical	$ZnCl_2$	289.5	8	Phenol	25.3/227.5	[180]
Banana peels	MW 700W/10 min	Chemical	KOH/NaOH	1038	2.5	MG	22.5/10	[147]
	800 °C/0.5 h	Chemical	K_2CO_3	1188	0.4	Me-Red	400/40–220	[181]
						MB	454.5/40–220	
Banyan tree	500 °C/1 h	Chemical	KOH	988	2.5	Phenol	26.9/20–100	[182]
Bermuda grass	450 °C/2 h	Chemical	KOH	1833.6	0.5	Cr(VI)	403.2/0–200	[183]
Black cumin seeds	40 °C/48 h	Chemical	H_2SO_4	-	2	MB	16.8/100	[184]
					2	Pb^{2+}	17.9/100	
Black wattle bark waste	700 °C/2 h	Chemical	$ZnCl_2$	414	1	Phenol	85.7/50–500	[185]
Caesalpinia ferrea	600 °C	Chemical	$ZnCl_2$	1480	1.5	Captopril	535/120–2500	[186]
Carya illinoinensis	700 °C/2 h	Chemical + Physical	NaCl, $CaCl_2$, KCl/CO_2	490–808	2	Cu^{2+}, Cd^{2+}, Ni^{2+}, Zn^{2+}	0.011–0.368 mmol/g	[187]
Chamerion angustifolium	450 °C/0.5 h	Physical	-	117.7	0.8	Co^{2+}	9.8/1–40	[188]
						Ni^{2+}	10.1/1–40	
Citrus limetta peels	250 °C/2 h	Chemical	$FeCl_3$	-	1	F^-	4.9/5–30	[28]
	500 °C/2 h	Chemical	$FeCl_3$	-	1	F^-	9.7/5–30	

Table 1. *Cont.*

Carbon Precursor	Activ. Temp./Activ. Time	Activation	Reagent	S_{BET} (m^2/g)	Dose (g/L)	Pollutant	Q_{max} (mg/g)/C_0 (mg/L)	Ref.
Coconut husk	350 °C/3 h	Chemical	KOH	1448	-	F^-	6.7 (FBA)	[139]
Coconut shell	600 °C/1 h	Chemical	NaOH	876	1	MB	200/25–250	[189]
Cocos nucifera shell	275 °C/2 h	Physico chemical	$ZnCl_2/H_2O$	1652	1	Phenol Red	351/500	[190]
						Phenol	264/500	
Coffee grounds	800 °C/1 h	Physico chemical	KOH/CO_2	1865	1	Phenol	211/10–200	[136]
					1	MB	390/50–500	
	950 °C/3 h	Physical	CO_2	2407.8	0.66	MB	680/20–150	[29]
					0.66	MO	612/20–150	
Coffee husk	800 °C/0.5h	Physical	H_2O	383	1	Ni^{2+}	57.1/30–120	[191]
Coffee spent				464	0.8		51.9/30–120	
Corncob	400 °C/1.5 h	Chemical	H_3PO_4	1097	2	NH^{4+}	17/10–40	[140]
Corncob	800 °C/1 h	Chemical	KOH	1054.2	0.02	Hg(II)	2.3/0.02–0.1	[192]
Corncob	800 °C/1.5 h	Chemical	NaOH	2381	0.2	Pb^{2+}	381/10–500	[193]
Corncob	892 °C/40 min	Physical	H_2O	-	0.4	Roxarsone	309.6/40–350	[194]
Date palm petiole	600 °C/2 h	Chemical	NaOH	655	4	Indigo carmine	53.7/10–200	[135]
Durian shell	900 °C/15 min	Physical	CO_2	917	10	AMX	142.7/250	[195]
					10	Tetracycline	126/250	
Eucalyptus residue	400 °C/3 h	Chemical	H_3PO_4	1545	0.2	MB	977/140–260	[196]
Eucalyptus sawdust	700 °C/75 min	Chemical	$FeCl_3$	645.2	1	MB	162.8/50–500	[197]
Fox nutshell	600 °C/1 h	Chemical	$ZnCl_2$	2869	5	Phenol	75.3	[198]
					0.4	MB	968.7	
	700 °C/1 h		H_3PO_4	2636	0.4	Cr(VI)	74.9/35	[199]

Table 1. *Cont.*

Carbon Precursor	Activ. Temp./Activ. Time	Activation	Reagent	S_{BET} (m^2/g)	Dose (g/L)	Pollutant	Q_{max} (mg/g)/C_0 (mg/L)	Ref.
Grape waste	600 °C/1 h	Chemical	$ZnCl_{2.5}$	1455	0.6	MY	386/200–900	[200]
						MB	417/200–900	
Grapefruit peels	800 °C/0.5 h	Chemical	K_2CO_3	1198	0.4	Me-Red	454.5/40–220	[181]
						MB	456.2/40–220	
Guava seeds	750 °C/1.5 h	Chemical	NaOH	2573.6	1	AMX	570.4/400–800	[162]
Jackfruit-PPI-1	600 °C/1 h	Chemical	K_2CO_3	987	1	Disperse Blue 14	178.5/10–120	[201]
Maghara coal	550 °C/1.5 h	Chemical	NaOH	49	0.5	MB	28/10–50	[202]
Mandarin peels	800 °C/0.5 h	Chemical	K_2CO_3	1077	0.4	Me-Red	357.1/40–220	[181]
						MB	400/40–220	
Olive stones	500 °C/3 h	Physical	N_2	9.1	1	MB	769/50–1000	[203]
Opuntia ficus-indica	600 °C/2 h	Chemical	NaOH	331.5	6	4-NP	16.1/10–150	[204]
Palm shell	900 °C/1 h	Chemical	H_2SO_4 + $K_2S_2O_8$	770	-	Bis-A	45.45	[205]
						4-ClP	163.9	
						4-NP	166.7	
						2,4-DNP	250.0	
						2,4,6-TNP	142.0	
PAN	800 °C/0.5 h	Chemical	K_2CO_3	2217	-	MCPA	4.2 mmol/g	[206]
						2,4-D	3.4 mmol/g	
			KOH	2828	-	MCPA	5.1 mmol/g	
						2,4-D	4.4 mmol/g	
Parthenium hysterophorus	500 °C/2h	Chemical	H_3PO_4	256.5	90	Cr	0.7/40–100	[207]

Table 1. *Cont.*

Carbon Precursor	Activ. Temp./Activ. Time	Activation	Reagent	S_{BET} (m^2/g)	Dose (g/L)	Pollutant	Q_{max} (mg/g)/C_0 (mg/L)	Ref.
Peanut shells	700 °C/3 h	Chemical	H_2SO_4	570	20	2,4-DNT-3-SO3 + 2,4-DNT-5-SO3	32	[208]
PET	800 °C/0.5 h	Chemical	K_2CO_3	1206	-	MCPA	2.4 mmol/g	[206]
					-	2,4-D	1.9 mmol/g	
			KOH	1439	-	MCPA	2.6 mmol/g	
					-	2,4-D	2.5 mmol/g	
Phoenix dactylifera L.	500 °C/1 h	Chemical	KOH	817	0.15	Levofloxacin	101.3/50–250	[134]
	300 °C/ 0.5 h	Chemical	H_3PO_4	1225.5	5.25	Rh-B	196	[209]
	500 °C/1 h	Chemical	K_2CO_3	852	-	Ciprofloxacin	2 (FBA)	[210]
						Norfloxacin	1.9 (FBA)	
Pinus eldarica	750 °C/3 h	Chemical	K_2CO_3	1201	0.1	SDS	347/10–125	[211]
Pomelo peels	800 °C/0.5 h	Chemical	K_2CO_3	836	0.4	Me-Red	208.3/40–220	[181]
						MB	222.2/40–220	
Potato peels	400 °C/1 h	Chemical	H_3PO_4	904.5	0.5	Bis-A	445.9/0–300	[212]
Potato peels	500 °C/ 0.5 h	Chemical	H_3PO_4	676	10	Pb^{2+}	8.9/100	[143]
Residue from biomass gasification	600 °C/1 h	Chemical	$ZnCl_2$	259	5	Fe^{2+}	20.5/25–125	[213]
						Cu^{2+}	23.1/25–125	
						Ni^{2+}	18.2/25–125	
Rice husk	950 °C/1 h	Chemical	K_2CO_3	1330	1	NO^{3-}	8.1/15	[144]
Rice husk	800 °C/1.5 h	Chemical	NaOH	2786	0.2	Pb^{2+}	492/10–500	[193]
Sargassum fusiforme	900 °C/0.5 h	Physical	CO_2	1329	0.8	Congo Red	234/50–200	[159]
Sewage sludge	500 °C/1 h	Chemical	$ZnCl_2$	721	0.1	FL, ANT, PYR, BeP, OP, NP, DBP, DEHP	0.07–2.8/ 0.01–0.3	[214]

Table 1. *Cont.*

Carbon Precursor	Activ. Temp./Activ. Time	Activation	Reagent	S_{BET} (m^2/g)	Dose (g/L)	Pollutant	Q_{max} (mg/g)/C_0 (mg/L)	Ref.
Sugarcane bagasse	550 °C/0.5 h	Chemical	$ZnCl_2$	182.9	0.2	Hg^{2+}	11.5/1	[30]
	300 °C/1 h	Chemical	H_3PO_4	7.4	0.2	Hg^{2+}	13.6/1	
Sunflower piths	700 °C/1 h	Chemical	NaOH	2690	1	MB	965/50–1000	[145]
			KOH	2090	1	MB	580/50–1000	
Syzygium cumini seed	900 °C/1 h	Chemical	KOH	747	6	F^-	1.2/20.04	[215]
Tomato stem	700 °C/1 h	Chemical	$FeCl_2$	971	1	Congo red	158.7/100–500	[146]
Tomato waste	600 °C/1 h	Chemical	$ZnCl_2$	1093	0.6	MY	385/200–900	[216]
						MB	400/200–900	
Tyre	750 °C	Chemical	KOH	700	0.5	Bis-A	123/5–80	[217]
Ulva lactuca	800 °C/3 h	Chemical	KOH	345.4	0.06	Cu^{2+}	84.7/5–80	[163]
						Cd^{2+}	84.6/5–80	
						Cr^{3+}	82/5–80	
						Pb^{2+}	83.3/5–80	
Walnut shell	900 °C/1.5 h	Chemical	$ZnCl_2$	1626.9	0.01	Congo Red	281.4/0–100	[178]
					0.01	MB	314.1/0–50	
Waste carpets	700 °C/4 h	Chemical	H_3PO_4	953	8	MB	769.2/50–800	[218]
Waste tires	750 °C	Chemical	KOH	265	0.3	Pb^{2+}	49.7/5–100	[219]
					0.3	Cd^{2+}	10.4/5–100	
					0.3	Cr^{3+}	29.4/5–100	
White sugar	700 °C/2 h	Chemical	NaOH	1144.7	0.8	Rh-B	123.4/25–100	[220]

As a result, the adsorption properties of activated carbon vary from source to source. Biomass is widely used for the production of activated carbon because of its great abundance, its renewability, and its low cost. In fact, various biomass wastes have been used to produce adsorbents by chemical or physical activation. Nowicki et al. [181] studied the influence of the precursor on the structural and sorption properties of methyl red (Me-Red) and methylene blue (MB)of activated carbon obtained by chemical activation with potassium carbonate involving grapefruit, mandarin, pomelo, and banana peels as precursors. They confirmed that this waste can be successfully used as precursors for the preparation of activated carbon that show a well-developed porous structure with specific surfaces of 836, 1198, 1077, and 1188 m^2/g, respectively, optimizing the conditions activation (temperature, time, impregnation ratio). It turns out that the efficiency of removal of Me-Red and MB from aqueous solutions depends on the type of precursor used and that the adsorption capacity increases with the increase in the specific surface area of the adsorbent. Charcoal from grapefruit peels was found to be the most effective adsorbent with a sorption capacity of 456.2and 454.5 mg/g for MB and Me-Red, respectively. In the same context, Baysal et al. [145] found that the adsorption capacity of MB increases from 580 to 965 mg/g when the surface area of activated carbon obtained from sunflower piths increases from 2090 to 2690 m^2/g. The increase in specific surface area was attributed to the nature of the activating agent where the highest value was obtained with sodium hydroxide [138]. Activation by potassium hydroxide mainly caused the development of micropores with a very narrow pore distribution. Thus, the sodium hydroxide appeared to be more reactive, which resulted in the micropores fusing to form mesopores [135,145,193,204]. Indeed, several studies have reported that potassium hydroxide is an activator that promotes the formation of activated carbon with a microporous structure [166,167,183,206,221]. However, some authors obtained a mesoporous structure by activation with potassium hydroxide [134]. Indeed, the structural properties of activated carbon also depend on the temperature and the duration of activation as well as the impregnation ratio. For example, the activated carbon obtained from a near of corn passes from a macroporous structure for a pyrolysis temperature of 800 °C and duration of 1 h [192] to a microporous structure for a pyrolysis temperature of 850 °C and duration of 3 h [221]. Likewise, the specific surface increases from 1054 to 1618 m^2/g [192,221]. In this respect, the variation in the parameters of preparation of the activated carbon offers varied in the textural and structural properties. Consequently, there is a variation of the porosity, both in shape and in dimensions, of the functional groups and of the pH_{ZCN}. These properties with the conditions of application determine the applicability of the activated carbon. Large-pore activated carbon is effectively used for the adsorption of large pollutants and vice versa [135,179,204,206].

Da Paixao Cansado et al. [206] did a comparative study of the adsorption of pesticides on chemically prepared activated carbon involving K_2CO_3 and KOH. The aim was to obtain activated carbon from polymer waste (Polyacrylonitrile (PAN) and Polyethyleneterephthalate (PET)) and to determine their main characteristics, namely the characterization of their porosity, their functional groups on the surface, and their adsorption capacities for 4-chloro-2-methylphenoxyacetic acid (MCPA) and 2,4-dichlorophenoxyacetic acid (2,4-D). The results showed that these pollutants were better adsorbed on the activated carbon obtained with KOH, which has a microporosity and a specific surface area greater than that obtained with K_2CO_3 [206]. In the same context, Laksaci et al. [136] studied the influence of the combination of CO_2 with KOH on the morphological properties and the sorption capacities of adsorbents from coffee grounds for phenol and MB. Activation by KOH under CO_2 flow considerably improves the porous structure of the adsorbents. Indeed, the use of CO_2 makes it possible to increase the diameter of the pores and develop a narrow distribution of the size of the pores [136]. This improves the microporosity of the activated carbon and increases the adsorption for large organic molecules. The results showed that MB adsorbs better than phenol with adsorption capacities of 390 and 211 mg/g, respectively [136].

Activated carbon obtained from date palm petiole via chemical activation using sodium hydroxide showed a surface area of 655 m^2/g, with a mesoporous structure [135]. This mesoporosity is a favorable structural property for use in the elimination of indigo carmine from aqueous solution in batch and continuous column systems. In batch mode, the adsorption was chemical and exothermic, and it underwent pseudo-second-order model kinetics. The adsorption mechanism is related to the physisorption enhanced by chemisorption. The adsorption capacity of the indigo carmine decreases from 53.76 to 32.57 mg/g if the temperature rises to 40°C [135]. However, in continuous mode, this activated carbon provides good efficiency removal of indigo carmine after three repetition cycles [135]. It appears that the adsorption is largely influenced by the *pH* of the solution and the pH_{ZCN}. The effect of *pH* on adsorption can manifest itself through two mechanisms: an electrostatic interaction between the functional groups of the adsorbent (carboxylic, phenolic, lactonic, basic functions, etc.) and the adsorbate, or a chemical reaction between adsorbent and adsorbate. In fact, at $pH<pH_{ZCN}$, the surface of the carbon is positively charged, which promotes the adsorption of the anions, while the adsorption of cations is favored at $pH>pH_{ZCN}$, where the surface of the carbon is negatively charged [28–30,135,159,176,184,196,211,222–224]. Tu et al. [183] studied the sorption properties of Bermuda grass derived activated carbon for Cr(VI) removal. Sorption of Cr(VI) ions depends on *pH* of the solution; the best Cr(VI) uptake was reported at *pH* 2.0 with a maximum adsorption capacity of 403.2 mg/g, and the removal mechanism was predominantly chemisorption. The same observations were reported by Bedada et al. [207] using activated carbon produced from Parthenium hysterophorus weed chemically activated by phosphoric acid. This is to highlight that mostly metal ions (Pb^{2+}, Cd^{2+}, Cr^{3+}, Fe^{2+}, Cu^{2+}, Ni^{2+}, Co^{2+}, Hg^{2+}, etc.) removal is favorable in the *pH* range 2–6 [143,163,188,192,213,219].

In addition, the search for inexpensive alternative resources as well as methods of preparing activated carbon has attracted the attention of scientific researchers. Several researchers are interested in converting waste from the polymer and textile industries into activated carbon as a favorable approach [176,206,217–219,225]. In this context, activated carbon derived from acrylic fiber waste is obtained by physical activation in the presence of air via controlled heat treatment and used as an adsorbent for the removal of MB from aqueous media [176]. The results indicate that the adsorption of MB on activated carbon requires a longer equilibrium time when the initial concentration of dye is increased with an adsorption capacity of 8.7 mg/g [176]. Likewise, Hassan and Elhadidy transformed carpet waste into activated carbon by activation with H_3PO_4 at different impregnation ratios [218]. They found that the specific surface area of activated carbon increases with the impregnation ratio, which reaches 953 m^2/g for an impregnation ratio of 3. This carbon has a mesoporous structure promoting the adsorption of MB where the capacity of adsorption is 769.2 mg/g [218]. A study of the effects of the impregnation rate, the carbonization temperature, and the time on the characteristics of the active carbon produced from tomato waste by chemical activation with $ZnCl_2$, was carried out by Saygili and Güzel [216]. This study showed that the influence of the operating parameters on the properties of the pores of the carbons followed a sequence of carbonization temperature> carbonization time> impregnation rate. The activated carbon produced under optimal conditions has a specific surface of 1093 m^2/g with a mesoporous structure. This carbon has been used effectively for the adsorption of large MB and metanil yellow (MY) molecules, showing maximum monolayer adsorption capacities of 400 mg/g for MB and 385 mg/g for MY [216]. The same observations were made by Wang et al. [167] during the preparation of carbon derived from peanut shell by activation by potassium hydroxide.

Certain inorganic anions constitute a group of pollutants in wastewater that are known by their toxic and carcinogenic effects. Among these ions are the nitrates (NO^{3-}) and fluoride (F^-) ions. It was reported that activated charcoal, derived from Rice husk developed by chemical activation using potassium carbonate as an activating agent followed by modification by urea, could remove the nitrate ions from aqueous solutions with a maximum adsorption capacity of 8.1 mg/g [144]. In addition, activated carbon has been used effectively for defluorination of water. By way of example, the *citrus limetta* peels give, by activation with $FeCl_3$, an activated carbon with a maximum monolayer capacity of 9.7 mg/g under the optimal adsorbent dosage conditions 1 g/L, a contact time of 240 min, and a temperature of 25 °C with fluoride concentration ranging from 5 to 30 mg/L [28].

It should be noted that an understanding of the activation process is a fundamental step for optimizing the preparation of activated carbon. Indeed, various chemical, physical, and physicochemical processes are used for the synthesis of activated carbon from different natural or synthetic resources. Chemical treatment involves the use of activating agents by impregnation followed by pyrolysis, whereas the physical process involves a high temperature heat treatment in the presence of vapor or carbon dioxide. However, the combination of these two methods has also been used for the production of activated carbon. It remains difficult to predict the physicochemical and morphological properties of carbons according to the treatment applied. It seems that the morphological, the physicochemical, the textural, and the surface properties of activated carbon change significantly by changing the activation time and temperature and the nature of the precursor used. The chemical nature of the surface of activated carbon depends on the type of heteroatoms, such as oxygen, nitrogen, sulfur, etc., that are present on the surface. These hetero atoms or groups accumulated on the surface either come from the starting material itself, or are incorporated during the activation process. This is largely related to the nature of the precursor. All these properties are of great importance for the use of carbons as an adsorbent of pollutants in aqueous solutions. Indeed, the adsorption of pollutants onto activated carbon is a complicated process, and various interactions can coexist during the process involving physical actions and/or chemical bonds. Thus, the adsorption mechanism referred to as electrostatic interactions, hydrogen bonds, π–π stacking interactions, pore filling, hydrophobic interactions, ion exchange, and also other interactions (n–π interactions, etc.) and complexation reactions can take place [30,184,187,194,197,203,205,220,226–228]. Figure 5 illustrates the principal interaction methods for adsorption of pollutants on activated carbon.

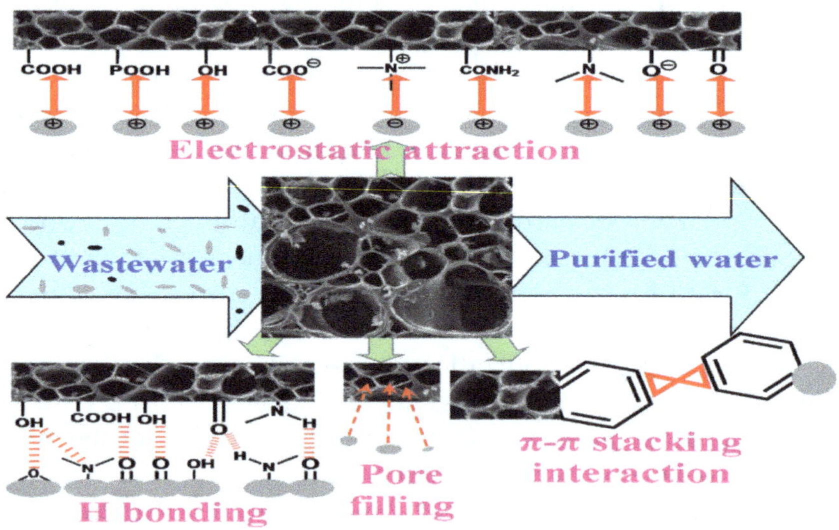

Figure 5. Postulated mechanisms of the interactions of pollutant with activated carbon.

The micropores and mesopores of the porous network of activated carbon contribute to the adsorption of pollutants by pore-filling mechanisms according to their total volume and the size of the pollutant molecules, and this is influenced by the affinity of the adsorbent with respect to the pollutant [229]. The π–π stacking interactions are suggested for the sorption of aromatic compounds on the surface of activated carbon containing aromatic rings, suggesting the presence of graphene in the carbon [205]. Electrostatic interactions are encountered during the adsorption of ionic and ionizable compounds [30,135,230]. Positively charged compounds will tend to sorb onto negatively charged carbon surfaces, while anionic compounds will bind to positively charged sites. These interactions also involve the weak interaction forces and hydrogen bonds between the functional groups (carboxylic, lactonic, phenolic, etc.) present on the surface of the activated carbon and the functional groups of the adsorbate. In fact, the polar groups on the surface of the carbon facilitate the sorption of water and promote the formation of hydrogen bonds between the carbon and the sorbates containing electronegative elements [30,135,230,231]. Furthermore, the ion exchange and complexation processes are induced by the presence of cations or metal ions in the structure of the activated carbon [226,230,232].

6. Conclusions

Carbon materials are one of the most studied areas in the scientific community. Indeed, these materials have remarkable physical and chemical properties, and this explains their uses in various fields such as biology, chemistry of materials, energy storage, electrochemistry, and wastewater treatment. In addition, despite several advances made, researchers face many challenges concerning the synthesis, morphological uniformity, and reproducibility of carbonaceous materials. Graphene and graphite offer excellent electrical mobility that favors their use for the photodegradation of organic pollutants and the manufacture of anodes for lithium-ion batteries and pollution sensors. Thus, materials derived from graphene offer greater flexibility in modifying the properties of the catalyst according to specific requirements. In this respect, the morphological properties of carbon nanotubes combined with their chemical and thermal conduction properties give them an interesting future as a catalyst support. In the same context, activated carbon is widely used in the field of water treatment because of their adsorption properties that result from the diversity of chemical, structural, and textural composition. These properties are related to the precursors and the method of preparation of the activated carbon. In this review, the performance of the retention of pollutants from wastewater was detailed on the basis of recent examples from the literature. It turns out that mastering the methods of preparing carbonaceous materials for larger-scale production and understanding the many properties of these materials are important steps to be taken in order to be able to assist with their applications in the various areas of interest.

Author Contributions: M.b.M. and L.M. designed the manuscript and wrote the main text. R.K. and Y.M. designed and revised the manuscript. All authors have read and agreed to the published version of the manuscript.

Acknowledgments: The authors would like to express their deep gratitude to Mohamed Naceur Belgacem, the director of Grenoble INP-Pagora and Agefpi, for his valuable advice and assistance, and also to the Tunisian Ministry of Higher Education for its financial support.

Abbreviations

2,4-DNP	2,4-Dinitrophenol
2,4-DNT-3-SO3	2,4-dinitrotoluene-3-sulfonate
2,4-DNT-5-SO3	2,4-dinitrotoluene-5-sulfonate
2,4,6-TNP	2,4,6-Trinitrphenol
4-ClP	4-Chlorophenol
4-NP	4-nitrophenol
Activ. Temp.	Activation temperature
Activ. Time	Activation time
AMX	Amoxicillin
Bis-A	Bisphenol A
C_0	Concentration initial
FBA	Fixed bed adsorption
MB	Methylene blue
Me-Red	Methyl red
MG	Malachite green
MO	Methyl orange
Ms	Municipal sludge
MW	Microwave

MWCNTs	Multiwalled carbon nanotubes
MY	Metanil yellow
PAN	Polyacrylonitrile
PET	Polyethyleneterephthalate
pH_{ZCN}	*pH* of the point of zero charge
Q_{max}	Maximum adsorption capacity
RB5	Reactive Black 5
Rh-B	Rhodamine B
S_{BET}	Brunauer–Emmett–Teller area
SDS	Sodium dodecylbenzene sulfonate
SWCNTs	Single-walled carbon nanotubes
TMA	TriMethylAmine

References

1. Rouquerol, F.; Luciani, L.; Llewellyn, P.; Denoyel, R.; Rouquerol, J. *Techniques de L'ingénieur, Analyse et Caractérisation: Texture des Matériaux Pulvérulents ou Poreux*; Inist-CNRS-Paris: Paris, France, 2003; pp. 1050–1057.
2. Besson, J.; Billon, N.; Cantournet, S.; Chastel, Y.; Lorenzon, A.F.G.; Haudin, J.M.; Monasse, B.; Naze, L. *Matériaux pour l'ingénieur*; Edition Mines: Paris, France, 2006; pp. 81–82.
3. Ma, T.Y.; Yuan, Z.Y. *Bioinspired Approach to Synthesizing Hierarchical Porous Materials, Hierarchically Structured Porous Materials: From Nanoscience to Catalysis, Separation, Optics, Energy, and Life Science*, 1st ed.; Su, B.L., Sanchez, C., Yang, X.Y., Eds.; Wiley: Weinheim, Germany, 2012; pp. 173–207.
4. Zhuo, H.; Hu, Y.; Tong, X.; Zhong, L.; Peng, X.; Sun, R. Sustainable hierarchical porous carbon aerogel from cellulose for high performance supercapacitor and CO_2 capture. *Ind. Crop. Prod.* **2016**, *87*, 229–235. [CrossRef]
5. Canal-Rodríguez, M.; Menéndez, J.A.; Arenillas, A. Performance of carbon xerogel-graphene hybrids as electrodes in aqueous supercapacitors. *Electrochim. Acta* **2018**, *276*, 28–36.
6. Sarvi, M.N.; Bee, T.B.; Gooi, C.K.; Woonton, B.W.; Gee, M.L.; O'Connor, A.J. Development of functionalized mesoporous silica for adsorption and separation of dairy proteins. *Chem. Eng. J.* **2014**, *235*, 244–251. [CrossRef]
7. Xin, Y.; Peng, S.; Chen, J.; Yang, Z.; Zhang, J. Continuous flow synthesis of porous materials. *Chin. Chem. Lett.* **2019**, *31*, 1448–1461. [CrossRef]
8. Wei, Y.; Parmentier, T.E.; de Jong, K.P.; Zecevic, J. Tailoring and visualizing the pore architecture of hierarchical zeolites. *Chem. Soc. Rev.* **2015**, *44*, 7234–7261. [CrossRef]
9. Zhong, W.; Liu, H.; Bai, C.; Liao, S.; Li, Y. Base-Free Oxidation of Alcohols to Esters at Room Temperature and Atmospheric Conditions using Nanoscale Co-Based Catalysts. *ACS Catal.* **2015**, *5*, 1850–1856. [CrossRef]
10. Xu, L.; Guo, L.; Hu, G.; Chen, J.; Hu, X.; Wang, S.; Dai, W.; Fan, M. Nitrogen-doped porous carbon spheres derived from d-glucose as highly-efficient CO_2 sorbents. *RSC Adv.* **2015**, *5*, 37964–37969. [CrossRef]
11. Li, D.; Chen, Y.; Zheng, M.; Zhao, H.; Zhao, Y.; Sun, Z. Hierarchically structured porous nitrogen-doped carbon for highly selective CO_2 capture. *ACS Sustain. Chem. Eng.* **2016**, *4*, 298–304. [CrossRef]
12. Xu, A.-R.; Chen, L.; Guo, X.; Xiao, Z.; Liu, R. Biodegradable lignocellulosic porous materials: Fabrication, characterization and its application in water processing. *Int. J. Biol. Macromol.* **2018**, *115*, 846–852. [CrossRef]
13. Zhu, L.; Shen, D.; Luo, K.H. A critical review on VOCs adsorption by different porous materials: Species, mechanisms and modification methods. *J. Hazard. Mater.* **2020**, *389*, 122102. [CrossRef]
14. Yang, F.; Sun, L.; Zhang, W.; Zhang, Y. One-pot synthesis of porous carbon foam derived from corn straw: Atrazine adsorption equilibrium and kinetics. *Environ. Sci. Nano.* **2017**, *4*, 625–635. [CrossRef]
15. Little, I.; Alorkpa, E.; Khan, V.; Povazhnyi, V.; Vasiliev, A. Efficient porous adsorbent for removal of cesium from contaminated water. *J. Porous Mater.* **2019**, *26*, 361–369. [CrossRef]
16. Zhu, Y.; Hu, H.; Li, W.; Zhang, X. Resorcinol-formaldehyde based porous carbon as an electrode material for supercapacitors. *Carbon* **2007**, *45*, 160–165. [CrossRef]
17. Zhang, W.F.; Huang, Z.-H.; Cao, G.P.; Kang, F.; Yang, Y. A novel mesoporous carbon with straight tunnel-like pore structure for high rate electrochemical capacitors. *J. Power Sources* **2012**, *204*, 230–235. [CrossRef]

18. Wu, K.; Liu, Q. Nitrogen-doped mesoporous carbons for high performance supercapacitors. *Appl. Surf. Sci.* **2016**, *379*, 132–139. [CrossRef]
19. Wang, P.; Xu, J.; Xu, F.; Zhao, W.; Sun, P.; Zhang, Z.; Qian, M.; Huang, F. Constructing hierarchical porous carbon via tin punching for efficient electrochemical energy storage. *Carbon* **2018**, *134*, 391–397. [CrossRef]
20. Wang, Y.; Qu, Q.; Gao, S.; Tang, G.; Liu, K.; He, S.; Huang, C. Biomass derived carbon as binder-free electrode materials for supercapacitors. *Carbon* **2019**, *155*, 706–726. [CrossRef]
21. Yu, K.; Wang, J.; Wang, X.; Liang, J.; Liang, C. Sustainable application of biomass by-products: Corn straw-derived porous carbon nanospheres using as anode materials for lithium ion batteries. *Mater. Chem. Phys.* **2020**, *243*, 122644. [CrossRef]
22. Huang, J.; Liang, Y.; Hu, H.; Liu, S.; Cai, Y.; Dong, H.; Zheng, M.; Xiao, Y.; Liu, Y. Ultrahigh-surface-area hierarchical porous carbon from chitosan: Acetic acid mediated efficient synthesis and its application in superior supercapacitors. *J. Mater. Chem.* **2017**, *5*, 24775–24781. [CrossRef]
23. Chen, Z.; Ye, S.; Evans, S.D.; Ge, Y.; Zhu, Z.; Tu, Y.; Yang, X. Confined assembly of hollow carbon spheres in carbonaceous nanotube: A spheres-in-tube carbon nanostructure with hierarchical porosity for high-performance supercapacitor. *Small* **2018**, *14*, 1704015. [CrossRef]
24. Kim, B.; Park, J.; Baik, S.; Lee, J.W. Spent coffee derived hierarchical porous carbon and its application for energy storage. *J. Porous Mater.* **2020**, *27*, 451–463. [CrossRef]
25. Marsh, H.; Rodriguez-Reinoso, F. *Activated Carbon*; Elsevier: Amsterdam, The Netherlands, 2006.
26. Yang, Y.; Wang, H.; Yan, F.-Y.; Qi, Y.; Lai, Y.-K.; Zeng, D.-M.; Chen, G.; Zhang, K.-Q. Bioinspired Porous Octacalcium Phosphate/Silk Fibroin Composite Coating Materials Prepared by Electrochemical Deposition. *ACS Appl. Mater. Interfaces* **2015**, *7*, 5634–5642. [CrossRef] [PubMed]
27. Pallares, J.; Cencerrado, A.G.; Arauzo, I. Production and characterization of activated carbon from barley straw by physical activation with carbon dioxide and steam. *Biomass Bioenergy* **2018**, *115*, 64–73. [CrossRef]
28. Siddique, A.; Nayak, A.K.; Singh, J. Synthesis of FeCl3-activated carbon derived from waste *Citrus limetta* peels for removal of fluoride: An eco-friendly approach for the treatment of groundwater and bio-waste collectively. *Groundw. Sustain. Dev.* **2020**, *10*, 100339. [CrossRef]
29. Chiang, C.-H.; Chen, J.; Lin, J.-H. Preparation of pore-size tunable activated carbon derived from waste coffee grounds for high adsorption capacities of organic dyes. *J. Environ. Chem. Eng.* **2020**, *8*, 103929. [CrossRef]
30. Giraldo, S.; Robles, I.; Ramirez, A.; Florez, E.; Acelas, N. Mercury removal from wastewater using agroindustrial waste adsorbents. *SN Appl. Sci.* **2020**, *2*, 1029. [CrossRef]
31. Danish, M.; Ahmad, T. A review on utilization of wood biomass as a sustainable precursor for activated carbon production and application. *Renew. Sustain. Energy Rev.* **2018**, *87*, 1–21. [CrossRef]
32. Wong, S.; Ngadi, N.; Inuwa, I.M.; Hassan, O. Recent advances in applications of activated carbon from biowaste for wastewater treatment: A short review. *J. Clean. Prod.* **2018**, *175*, 361–375. [CrossRef]
33. Sing, K.S.W. Reporting physisorption data for gas/solid systems with special reference to the determination of surface area and porosity (Recommendations 1984). *Pure Appl. Chem.* **1985**, *57*, 603–619. [CrossRef]
34. Thommes, M.; Kaneko, K.; Neimark, A.V.; Olivier, J.P.; Rodriguez-Reinoso, F.; Rouquerol, J.; Sing, K.S.W. Physisorption of gases, with special reference to the evaluation of surface area and pore size distribution (IUPAC Technical Report). *Pure Appl. Chem.* **2015**, *87*, 1051–1069. [CrossRef]
35. Mnasri, N.; Charnay, C.; Ménorval, L.C.; Moussaoui, Y.; Elaloui, E.; Zajac, J. Silver nanoparticle-containing submicron-in-size mesoporous silica-based systems for iodine entrapment and immobilization from gas phase. *Microporous Mesoporous Mater.* **2014**, *196*, 305–313. [CrossRef]
36. Costa, J.A.S.; Jesus, R.A.; Santos, D.; Manoa, J.F.; Romao, L.P.C.; Paranhos, C.M. Recent progresses in the adsorption of organic, inorganic, and gas compounds by MCM-41-based mesoporous materials. *Microporous Mesoporous Mater.* **2020**, *291*, 109698. [CrossRef]
37. Luo, X.; Guoa, J.; Chang, P.; Qian, H.; Pei, F.; Wang, W.; Miao, K.; Guo, S.; Feng, G. ZSM-5@MCM-41 composite porous materials with a core-shell structure: Adjustment of mesoporous orientation basing on interfacial electrostatic interactions and their application in selective aromatics transport. *Sep. Purif. Technol.* **2020**, *239*, 116516. [CrossRef]
38. Zhang, C.; Zhao, P.; Liu, S.; Yu, K. Three-dimensionally ordered macroporous perovskite materials for environmental applications. *Chin. J. Catal.* **2019**, *40*, 1324–1338. [CrossRef]

39. Zeng, F.; Zhang, X.; Pan, Y.; Xu, M.; Yang, L.; Qu, Y.; Guo, M.; Yuan, C. Graphene-templated growth of vertical MnO nanosheets with open macroporous architectures as anode materials for fast lithium storage. *J. Alloys Compd.* **2018**, *769*, 10–17. [CrossRef]
40. Simaioforidou, A.; Kostas, V.; Karakassides, M.A.; Louloudi, M. Surface chemical modification of macroporous and mesoporous carbon materials: Effect on their textural and catalytic properties. *Microporous Mesoporous Mater.* **2019**, *279*, 334–344. [CrossRef]
41. Lal, S.; Prat, M.; Plamondon, M.; Poulikakos, L.; Partl, M.N.; Derome, D.; Carmeliet, J. A cluster-based pore network model of drying with corner liquid films, with application to a macroporous material. *Int. J. Heat Mass Transf.* **2019**, *140*, 620–633. [CrossRef]
42. Deprez, N.; McLachlan, D.S. The analysis of the electrical conductivity of graphite conductivity of graphite powders during compaction. *J. Phys. D Appl. Phys.* **1988**, *21*, 101–107. [CrossRef]
43. Liu, Y.; Li, X.; Fan, L.; Li, S.; Sari, H.M.K.; Qin, J. A Review of Carbon-Based Materials for Safe Lithium Metal Anodes. *Front. Chem.* **2019**, *7*, 721. [CrossRef]
44. Marinho, B.; Ghislandi, M.; Koning, C.E.; de With, G. Electrical conductivity of compacts of graphene, multi-wall carbon nanotubes, carbon black, and graphite powder. *Powder Technol.* **2012**, *221*, 351–358. [CrossRef]
45. Jana, P.; Fierro, V.; Pizzi, A.; Celzard, A. Thermal conductivity improvement of composite carbon foams based on tannin-based disordered carbon matrix and graphite fillers. *Mater. Des.* **2015**, *83*, 635–643. [CrossRef]
46. Dong, Y.; Meng, M.; Groves, M.M.; Zhang, C.; Lin, J. Thermal conductivities of two-dimensional graphitic carbon nitrides by molecule dynamics simulation. *Int. J. Heat Mass Transf.* **2018**, *123*, 738–746. [CrossRef]
47. Duan, J.; Wu, W.; Nolan, A.M.; Wang, T.; Wen, J.; Hu, C.; Mo, Y.; Luo, W.; Huang, Y. Lithium graphite paste: An interface compatible anode for solid-state batteries. *Adv. Mater.* **2019**, *31*, 1807243. [CrossRef] [PubMed]
48. Luo, H.-M.; Chen, H.; Chen, Y.-Z.; Li, P.; Zhang, J.-Q.; Zhao, X. Simple synthesis of porous carbon materials for high-performance supercapacitors. *J. Appl. Electrochem.* **2016**, *46*, 703–712. [CrossRef]
49. Li, G.; Yin, Z.; Guo, H.; Wang, Z.; Yan, G.; Yang, Z.; Liu, Y.; Ji, X.; Wang, J. Metalorganic Quantum Dots and Their Graphene-Like Derivative Porous Graphitic Carbon for Advanced Lithium-Ion Hybrid Supercapacitor. *Adv. Energy Mater.* **2019**, *9*, 1802878. [CrossRef]
50. Li, G.; Yang, Z.; Yin, Z.; Guo, H.; Wang, Z.; Yan, G.; Liu, Y.; Li, L.; Wang, J. Non-aqueous dual-carbon lithium-ion capacitors: A review. *J. Mater. Chem. A* **2019**, *7*, 15541–15563. [CrossRef]
51. Zhong, Y.; Guo, Q.; Li, S.; Shi, J.; Liu, L. Heat transfer enhancement of paraffin wax using graphite foam for thermal energy storage. *Sol. Energy Mater. Sol. Cells* **2010**, *94*, 1011–1014. [CrossRef]
52. Zhang, J.; Xu, T.; Cong, Y.; Zhang, Y.; Li, X.; Dong, Z.; Li, Y.; Yuan, G.; Zhang, J.; Cui, Z. Improved rate performance and cycling stability of graphitized mesoporous carbon as anode materials for lithium-ion batteries. *J. Mater. Sci.* **2019**, *54*, 648–658. [CrossRef]
53. Pitchaiya, S.; Eswaramoorthy, N.; Natarajan, M.; Santhanam, A.; Asokan, V.; Ramakrishnan, V.M.; Rangasamy, B.; Sundaram, S.; Ravirajan, P.; Velauthapillai, D. Perovskite Solar Cells: A Porous Graphitic Carbon based Hole Transporter/Counter Electrode Material Extracted from an Invasive Plant Species *Eichhornia Crassipes*. *Sci. Rep.* **2020**, *10*, 6835. [CrossRef]
54. Horiuchi, S.; Gotou, T.; Fujiwara, M.; Asaka, T.; Yokosawa, T.; Matsui, Y. Single graphene sheet detected in a carbon nanofilm. *Appl. Phys. Lett.* **2004**, *84*, 2403–2405. [CrossRef]
55. Novoselov, K.S.; Geim, A.K.; Morozov, S.V.; Jiang, D.; Zhang, Y.; Dubonos, S.V.; Grigorieva, I.V.; Firsov, A.A. Electric Field Effect in Atomically Thin Carbon Films. *Science* **2004**, *306*, 666–669. [CrossRef] [PubMed]
56. Berger, C.; Song, Z.; Li, T.; Li, X.; Ogbazghi, A.Y.; Feng, R.; Dai, Z.; Marchenkov, A.N.; Conrad, E.H.; First, P.N.; et al. Ultrathin Epitaxial Graphite: 2D Electron Gas Properties and a Route toward Graphene-based Nanoelectronics. *J. Phys. Chem. B* **2004**, *108*, 19912–19916. [CrossRef]
57. Novoselov, K.S.; Geim, A.K.; Morozov, S.V.; Jiang, D.; Katsnelson, M.I.; Grigorieva, I.V.; Dubonos, S.V.; Firsov, A.A. Two-dimensional gas of massless Dirac fermions in graphene. *Nature* **2005**, *438*, 197–200. [CrossRef] [PubMed]
58. Allen, M.J.; Tung, V.C.; Kaner, R.B. Honeycomb Carbon: A Review of Graphene. *Chem. Rev.* **2010**, *110*, 132–145. [CrossRef]
59. Nair, R.R.; Blake, P.; Grigorenko, A.N.; Novoselov, K.S.; Booth, T.J.; Stauber, T.; Peres, N.M.R.; Geim, A.K. Fine Structure Constant Defines Visual Transparency of Graphene. *Science* **2008**, *320*, 1308. [CrossRef]

60. Varghese, S.S.; Lonkar, S.; Singh, K.K.; Swaminathan, S.; Abdala, A. Recent advances in graphene based gas sensors. *Sens. Actuators B Chem.* **2015**, *218*, 160–183. [CrossRef]
61. You, P.; Liu, Z.; Tai, Q.; Liu, S.; Yan, F. Efficient Semitransparent Perovskite Solar Cells with Graphene Electrodes. *Adv. Mater.* **2015**, *27*, 3632–3638. [CrossRef]
62. De Nicola, F.; Purayil, N.S.P.; Miŝeikis, V.; Spirito, D.; Tomadin, A.; Coletti, C.; Polini, M.; Krahne, R.; Pellegrini, V. Graphene Plasmonic Fractal Metamaterials for Broadband Photodetectors. *Sci. Rep.* **2020**, *10*, 6882. [CrossRef]
63. Emtsev, K.V.; Bostwick, A.; Horn, K.; Jobst, J.; Kellogg, G.L.; Ley, L.; McChesney, J.L.; Ohta, T.; Reshanov, S.A.; Röhrl, J.; et al. Towards wafer-size graphene layers by atmospheric pressure graphitization of silicon carbide. *Nat. Mater.* **2009**, *8*, 203–207. [CrossRef]
64. Guo, Z.; Dong, R.; Chakraborty, P.S.; Lourenco, N.; Palmer, J.; Hu, Y.; Ruan, M.; Hankinson, J.; Kunc, J.; Cressler, J.D.; et al. Record maximum oscillation frequency in C-face epitaxial graphene transistors. *Nano Lett.* **2013**, *13*, 942–947. [CrossRef]
65. Tetlow, H.; Posthuma de Boer, J.; Ford, I.J.; Vvedensky, D.D.; Coraux, J.; et Kanto rovich, L. Growth of epitaxial graphene: Theory and experiment. *Phys. Rep.* **2014**, *542*, 195–295. [CrossRef]
66. Eda, G.; Fanchini, G.; Chhowalla, M. Large-area ultrathin films of reduced graphene oxide as a transparent and flexible electronic material. *Nat. Nanotechnol.* **2008**, *3*, 270–274. [CrossRef] [PubMed]
67. Ago, H.; Ogawa, Y.; Tsuji, M.; Mizuno, S.; Hibino, H. Catalytic growth of graphene: Toward large-area single-crystalline graphene. *J. Phys. Chem. Lett.* **2012**, *3*, 2228–2236. [CrossRef] [PubMed]
68. Mattevi, C.; Kim, H.; Chhowalla, M. A review of chemical vapour deposition of graphene on copper. *J. Mater. Chem.* **2011**, *21*, 3324–3334. [CrossRef]
69. Rabchinskii, M.K.; Ryzhkov, S.A.; Kirilenko, D.A.; Ulin, N.V.; Baidakova, M.V.; Shnitov, V.V.; Pavlov, S.I.; Chumakov, R.G.; Stolyarova, D.Y.; Besedina, N.A.; et al. From graphene oxide towards aminated graphene: Facile synthesis, its structure and electronic properties. *Sci. Rep.* **2020**, *10*, 6902. [CrossRef] [PubMed]
70. Mayorov, A.S.; Gorbachev, R.V.; Morozov, S.V.; Britnell, L.; Jalil, R.; Ponomarenko, L.A.; Blake, P.; Novoselov, K.S.; Watanabe, K.; Taniguchi, T.; et al. Micrometer-scale ballistic transport in encapsulated graphene at room temperature. *Nano Lett.* **2011**, *11*, 2396–2399. [CrossRef] [PubMed]
71. Soldano, C.; Mahmood, A.; Dujardin, E. Production, properties and potential of grapheme. *Carbon* **2010**, *48*, 2127–2150. [CrossRef]
72. Kang, H.-K.; Woo, S.-G.; Kim, J.-H.; Lee, S.-R.; Lee, D.-G.; and Yu, J.-S. Three-dimensional monolithic corrugated graphene/Ni foam for highly stable and efficient Li metal electrode. *J. Power Sources* **2019**, *413*, 467–475. [CrossRef]
73. Liu, W.; Xia, Y.; Wang, W.; Wang, Y.; Jin, J.; Chen, Y.; Paek, E.; Mitlin, D. Pristine or highly defective? Understanding the role of graphene structure for stable lithium metal plating. *Adv. Energy Mater.* **2019**, *9*, 1802918. [CrossRef]
74. Kang, H.K.; Woo, S.G.; Kim, J.H.; Yu, J.S.; Lee, S.R.; Kim, Y.J. Fewlayer graphene island seeding for dendrite-free Li metal electrodes. *ACS Appl. Mater. Interfaces* **2016**, *8*, 26895–26901. [CrossRef]
75. Lin, D.; Liu, Y.; Liang, Z.; Lee, H.W.; Sun, J.; Wang, H.; Yan, K.; Xie, J.; Cui, Y. Layered reduced graphene oxide with nanoscale interlayer gaps as a stable host for lithium metal anodes. *Nat. Nanotechnol.* **2016**, *11*, 626–632. [CrossRef] [PubMed]
76. Suvarnaphaet, P.; Pechprasarn, S. Graphene-Based Materials for Biosensors: A Review. *Sensors* **2017**, *17*, 2161. [CrossRef] [PubMed]
77. Kang, M.-A.; Ji, S.; Kim, S.; Park, C.-Y.; Myung, S.; Song, W.; Lee, S.S.; Lima, J.; An, K.-S. Highly sensitive and wearable gas sensors consisting of chemically functionalized graphene oxide assembled oncotton yarn. *RSC Adv.* **2018**, *8*, 11991–11996. [CrossRef]
78. Hong, H.S.; Phuong, N.H.; Huong, N.T.; Nam, N.H.; Hue, N.T. Highly sensitive and low detection limit of resistive NO_2 gas sensor based on a MoS_2/graphene two-dimensional heterostructures. *Appl. Surf. Sci.* **2019**, *492*, 449–454. [CrossRef]
79. Chou, C.-Y.; Tseng, S.-F.; Chang, T.-L.; Tu, C.-T.; Han, H.-C. Controlled bridge growth of ZnO nanowires on laser-scribed graphene-based devices for NO gas detection. *Appl. Surf. Sci.* **2020**, *508*, 145204. [CrossRef]
80. Hoang, N.V.; Hung, C.M.; Hoa, N.D.; Duy, N.V.; Park, I.; Hieu, N.V. Excellent detection of H_2S gas at ppb concentrations using $ZnFe_2O_4$ nanofibers loaded with reduced graphene oxide. *Sens. Actuators B: Chem.* **2019**, *282*, 876–884. [CrossRef]

81. Pearce, R.; Iakimov, T.; Andersson, M.; Hultman, L.; Spetz, A.L.; Yakimova, R. Epitaxially grown graphene based gas sensors for ultra sensitive NO_2 detection. *Sens. Actuators B Chem.* **2011**, *155*, 451–455. [CrossRef]
82. Shaban, M.; Ali, S.; Rabiaj, M. Design and application of nanoporous graphene oxide film for CO_2, H_2, and C_2H_2 gases sensing. *J. Mater. Res. Technol.* **2019**, *8*, 4510–4520. [CrossRef]
83. Fan, Y.-Y.; Tu, H.-L.; Pang, Y.; Wei, F.; Zhao, H.-B.; Yang, Y.; Ren, T.-L. Au-decorated porous structure graphene with enhanced sensing performance for low-concentration NO_2 detection. *Rare Met.* **2020**, *39*, 651–658. [CrossRef]
84. Alfano, B.; Massera, E.; Miglietta, M.L.; Polichetti, T.; Schiattarella, C.; Di Francia, G. Graphene Decoration for Gas Detection. In *Sensors and Microsystems*; Leone, A., Forleo, A., Francioso, L., Capone, S., Siciliano, P., Di Natale, C., Eds.; AISEM 2017; Lecture Notes in Electrical Engineering: Springer, 2018; Volume 457, pp. 35–40.
85. Chen, Y.; Gao, B.; Zhao, J.-X.; Cai, Q.-H.; Fu, H.-G. Si-doped graphene: An ideal sensor for NO- or NO_2-detectionand metal-free catalyst for N_2O-reduction. *J. Mol. Model.* **2012**, *18*, 2043–2054. [CrossRef]
86. Wang, Y.; Li, Y.; Tang, L.; Lu, J.; Li, J. Application of graphene-modified electrode for selective detection of dopamine. *Electrochem. Commun.* **2009**, *11*, 889–892. [CrossRef]
87. Brownson, D.A.C.; Foster, C.W.; Banks, C.E. The electrochemical performance of graphene modified electrodes: An analytical perspective. *Analyst* **2012**, *137*, 1815–1823. [CrossRef] [PubMed]
88. Lin, W.-J.; Liao, C.-S.; Jhang, J.-H.; Tsai, Y.-C. Graphene modified basal and edge plane pyrolytic graphite electrodes for electrocatalytic oxidation of hydrogen peroxide and β-nicotinamide adenine dinucleotide. *Electrochem. Commun.* **2009**, *11*, 2153–2156. [CrossRef]
89. Kang, X.; Wang, J.; Wu, H.; Liu, J.; Aksay, I.A.; Lin, Y. A graphene-based electrochemical sensor for sensitive detection of paracetamol. *Talanta* **2010**, *81*, 754–759. [CrossRef]
90. Wang, Y.; Zhang, D.; Wu, J. Electrocatalytic oxidation of kojic acid at a reduced graphene sheet modified glassy carbon electrode. *J. Electroanal. Chem.* **2012**, *664*, 111–116. [CrossRef]
91. Mohan, H.K.S.V.; Chee, W.K.; Li, Y.; Nayak, S.; Poh, C.L.; Thean, A.V.Y. A highly sensitive graphene oxide based label-free capacitive aptasensor for vanillin detection. *Mater. Des.* **2020**, *186*, 108208. [CrossRef]
92. Zhang, J.-W.; Wang, K.-P.; Zhang, X. Fabrication of SnO_2 decorated graphene composite material and its application in electrochemical detection of caffeic acid in red wine. *Mater. Res. Bul.* **2020**, *126*, 110820. [CrossRef]
93. Zuchowska, A.; Chudy, M.; Dybko, A.; Brzozka, Z. Graphene as a new material in anticancer therapy-in vitro studies. *Sens. Actuators B Chem.* **2017**, *243*, 152–165. [CrossRef]
94. Cheng, C.; Li, S.; Thomas, A.; Kotov, N.A.; Haag, R. Functional Graphene Nanomaterials Based Architectures: Biointeractions, Fabrications, and Emerging Biological Applications. *Chem. Rev.* **2017**, *117*, 1826–1914. [CrossRef]
95. Rajakumar, G.; Zhang, X.-H.; Gomathi, T.; Wang, S.-F.; Ansari, M.A.; Mydhili, G.; Nirmala, G.; Alzohairy, M.A.; Chung, I.-M. Current Use of Carbon-Based Materials for Biomedical Applications-A Prospective and Review. *Processes* **2020**, *8*, 355. [CrossRef]
96. Russier, J.; León, V.; Orecchioni, M.; Hirata, E.; Virdis, P.; Fozza, C.; Sgarrella, F.; Cuniberti, G.; Prato, M.; Vázquez, E.; et al. Few-Layer Graphene Kills Selectively Tumor Cells from Myelomonocytic Leukemia Patients. *Angew. Chem. Int. Ed.* **2017**, *56*, 3014–3019. [CrossRef] [PubMed]
97. Su, S.; Wang, J.; Vargas, E.; Wei, J.; Martinez-Zaguilan, R.; Sennoune, S.R.; Pantoya, M.L.; Wang, S.; Chaudhuri, J.; Qiu, J. Porphyrin Immobilized Nanographene Oxide for Enhanced and Targeted Photothermal Therapy of Brain Cancer. *ACS Biomater. Sci. Eng.* **2016**, *2*, 1357–1366. [CrossRef]
98. Fadlalla, M.I.; Kumar, P.S.; Selvam, V.; Babu, S.G. Emerging energy and environmental applicationof graphene and their composites: A review. *J. Mater. Sci.* **2020**, *55*, 7156–7183. [CrossRef]
99. Yao, Y.; Shankar, R.; Rauter, P.; Song, Y.; Kong, J.; Loncar, M.; Capasso, F. High-responsivity mid-infrared graphene detectors with antenna-enhanced photocarrier generation and collection. *Nano Lett.* **2014**, *14*, 3749–3754. [CrossRef] [PubMed]
100. Begum, H.; Kim, Y.-B. Improvement of Catalytic Activity of Platinum Nanoparticles Decorated Carbon Graphene Composite on Oxygen Electroreduction for Fuel Cells. *Processes* **2019**, *7*, 586. [CrossRef]
101. He, J.; Chen, Y.; Lv, W.; Wen, K.; Li, P.; Wang, Z.; Zhang, W.; Qin, W.; He, W. Three-Dimensional Hierarchical Graphene-CNT@Se: A Highly Efficient Freestanding Cathode for Li–Se Batteries. *ACS Energy Lett.* **2016**, *1*, 16–20. [CrossRef]

102. He, J.; Chen, Y.; Lv, W.; Wen, K.; Xu, C.; Zhang, W.; Qin, W.; He, W. Three-Dimensional CNT/Graphene–Li_2S Aerogel as Freestanding Cathode for High-Performance Li–S Batteries. *ACS Energy Lett.* **2016**, *1*, 820–826. [CrossRef]
103. He, J.; Chen, Y.; Li, P.; Fu, F.; Wang, Z.; Zhang, W. Three-dimensional CNT/graphene–sulfur hybrid sponges with high sulfur loading as superior capacity cathodes for lithium–sulfur batteries. *J. Mater. Chem. A* **2015**, *3*, 18605–18610. [CrossRef]
104. Xu, C.; Jing, Y.; He, J.; Zhou, K.; Chen, Y.; Li, Q.; Lin, J.; Zhang, W. Self-assembled interwoven CoS_2/CNTs/graphene architecture as anode for high-performance lithium ion batteries. *J. Alloys Compd.* **2017**, *708*, 1178–1183. [CrossRef]
105. Iijima, S. Helical microtubules of graphitic carbon. *Nature* **1991**, *354*, 56–58. [CrossRef]
106. Bethune, D.S.; Klang, C.H.; De Vries, M.S.; Gorman, G.; Savoy, R.; Vasquez, J.; Beyers, R. Cobalt catalysed growth of carbon nanotubes with single-atomic-layer walls. *Nature* **1993**, *363*, 605–607. [CrossRef]
107. Iijima, S.; Ichihashi, T. Single-shell carbon nanotubes of 1-nm diameter. *Nature* **1993**, *363*, 603–605. [CrossRef]
108. Inagaki, M. Advanced Carbon Materials. In *Handbook of Advanced Ceramics*, 2nd ed.; Somiya, S., Ed.; Academic Press: Cambridge, MA, USA, 2013; pp. 25–60.
109. Gupta, S.; Tai, N.H. Carbon materials and their composites for electromagnetic interference shielding effectiveness in X-band. *Carbon* **2019**, *152*, 159–187. [CrossRef]
110. Lee, D.W.; Park, J.; Kim, B.J.; Kim, H.; Choi, C.; Baughman, R.H.; Kim, S.J.; Kim, Y.T. Enhancement of electromagnetic interference shielding effectiveness with alignment of spinnable multiwalled carbon nanotubes. *Carbon* **2019**, *142*, 528–534. [CrossRef]
111. Kumar, S.; Rani, R.; Dilbaghi, N.; Tankeshwar, K.; Kim, K.-H. Carbon nanotubes: A novel material for multifaceted applications in human healthcare. *Chem. Soc. Rev.* **2017**, *46*, 158–196. [CrossRef] [PubMed]
112. Wang, K.; Luo, S.; Wu, Y.; He, X.; Zhao, F.; Wang, J.; Jiang, K.; Fan, S. Super-aligned carbon nanotube films as current collectors for lightweight and flexible lithium ion batteries. *Adv. Funct. Mater.* **2013**, *23*, 846–853. [CrossRef]
113. Qi, F.; He, J.; Chen, Y.; Zheng, B.; Li, Q.; Wang, X.; Yu, B.; Lin, J.; Zhou, J.; Li, P.; et al. Few-layered ReS_2 nanosheets grown on carbon nanotubes: A highly efficient anode for high-performance lithium-ion batteries. *Chem. Eng. J.* **2017**, *315*, 10–17. [CrossRef]
114. Zhang, R.; Cheng, X.B.; Zhao, C.Z.; Peng, H.J.; Shi, J.L.; Huang, J.Q.; Wang, J.; Wei, F.; Zhang, Q. Conductive nanostructured scaffolds render low local current density to inhibit lithium dendrite growth. *Adv. Mater.* **2016**, *28*, 2155–2162. [CrossRef]
115. Deng, Y.; Lu, H.; Cao, Y.; Xu, B.; Hong, Q.; Cai, W.; Yang, W. Multiwalled carbon nanotube interlayers with controllable thicknesses for high capacity and long-life lithium metal anodes. *J. Power Sources* **2019**, *412*, 170–179. [CrossRef]
116. Rutherglen, C.; Burke, P. Nanoelectromagnetics: Circuit and Electromagnetic Properties of Carbon Nanotubes. *Small* **2009**, *5*, 884–906. [CrossRef]
117. Daneshvar, F.; Aziz, A.; Abdelkader, A.M.; Zhang, T.; Sue, H.-J.; Welland, M.E. Porous SnO_2–CuxO nanocomposite thin film on carbon nanotubes as electrodes for high performance supercapacitors. *Nanotechnology* **2018**, *30*, 015401. [CrossRef] [PubMed]
118. Moreno-Fernandez, G.; Ibanez, J.; Rojo, J.M.; Kunowsky, M. Activated carbon fiber monoliths as supercapacitor electrodes. *Adv. Mater. Sci. Eng.* **2017**, 1–8. [CrossRef]
119. Ma, W.L.; Cai, Z.H.; Zhang, Y.; Wang, Z.Y.; Xia, L.; Ma, S.P.; Lia, G.H.; Huang, Y. An Overview of Stretchable Supercapacitors Based on Carbon Nanotube and Graphene. *Chin. J. Polym. Sci.* **2020**, *38*, 491–505. [CrossRef]
120. Yang, Z.; Tian, J.; Yin, Z.; Cui, C.; Qian, W.; Wei, F. Carbon nanotube- and graphene-based nanomaterials and applications in high-voltage supercapacitor. *Carbon* **2019**, *141*, 467–480. [CrossRef]
121. Duong, H.; Cheng, H.; Jewell, D. Advanced Supercapacitors Using Carbon Nanotubes. *Mater. Res. Found.* **2017**, *12*, 207–228.
122. Ying, L.S.; Mohd Salleh, M.A.; Yusoff, H.M.; Rashid, S.B.A.; Razak, J. Continuous production of carbon nanotubes. *J. Ind. Eng. Chem.* **2011**, *17*, 367–376. [CrossRef]
123. Tai, N.H.; Chen, H.M.; Chen, Y.J.; Hsieh, P.Y.; Liang, J.R.; Chou, T.W. Optimization of processing parameters of the chemical vapor deposition process for synthesizing high-quality single-walled carbon nanotube fluff and roving. *Compos. Sci. Technol.* **2012**, *72*, 1855–1862. [CrossRef]

124. Guo, T.; Nikolaev, P.; Rinzler, A.G.; Tomanek, D.; Colbert, D.T.; Smalley, R.E. Selfassembly of tubular fullerenes. *J. Phys. Chem.* **1995**, *99*, 10694–10697. [CrossRef]
125. Ebbesen, T.W.; Ajayan, P.M. Large-scale synthesis of carbon nanotubes. *Nature* **1992**, *358*, 220–222. [CrossRef]
126. Zhou, D.; Anoshkina, E.V.; Chow, L.; Chai, G. Synthesis of carbon nanotubes by electrochemical deposition at room temperature. *Carbon* **2006**, *44*, 1013–1016. [CrossRef]
127. Cao, C.; Zhou, Y.; Ubnoske, S.; Zang, J.; Cao, Y.; Henry, P.; Parker, C.B.; Glass, J.T. Highly stretchable supercapacitors via crumpled vertically aligned carbon nanotube forests. *Adv. Energy Mater.* **2019**, *9*, 1900618. [CrossRef]
128. Zhang, Z.; Wang, L.; Li, Y.; Wang, Y.; Zhang, J.; Guan, G.; Pan, Z.; Zheng, G.; Peng, H. Nitrogen-doped core-sheath carbon nanotube array for highly stretchable supercapacitor. *Adv. Energy Mater.* **2017**, *7*, 1601814. [CrossRef]
129. Hua, C.; Shang, Y.; Wang, Y.; Xu, J.; Zhang, Y.; Li, X.; Cao, A. A flexible gas sensor based on single-walled carbon nanotube-Fe_2O_3 composite film. *Appl. Surf. Sci.* **2017**, *405*, 405–411. [CrossRef]
130. Tyagi, P.; Sharma, A.; Tomar, M.; Gupta, V. A comparative study of RGO-SnO_2 and MWCNT-SnO_2 nanocomposites based SO_2 gas sensors, Sens. *Actuators B Chem.* **2017**, *248*, 980–986. [CrossRef]
131. Gupta, S.; Murthy, C.N.; Ratna Prabha, C. Recent advances in carbon nanotube based electrochemical biosensors. *Int. J. Biol. Macromol.* **2018**, *108*, 687–703. [CrossRef]
132. Farghali, A.A.; Tawab, H.A.A.; Moaty, S.A.A.; Khaled, R. Functionalization of acidified multi-walled carbon nanotubes for removal of heavy metals in aqueous solutions. *J. Nanostruct. Chem.* **2017**, *7*, 101–111. [CrossRef]
133. Getachew, T.; Hussen, A.; Rao, V.M. Defluoridation of water by activated carbon prepared from banana (Musa paradisiaca) peel and coffee (Coffe arabica) husk. *Int. J. Environ. Sci. Technol.* **2015**, *12*, 1857–1866. [CrossRef]
134. Darweesh, T.M.; Ahmed, M.J. Batch and fixed bed adsorption of levofloxacin on granular activated carbon from date (*Phoenix dactylifera* L.) stones by KOH chemical activation. *Environ. Toxicol. Pharmacol.* **2017**, *50*, 159–166. [CrossRef]
135. Khadhri, N.; Saad, M.E.K.; ben Mosbah, M.; Moussaoui, Y. Batch and continuous column adsorption of indigo carmine onto activated carbon derived from date palm petiole. *J. Environ. Chem. Eng.* **2019**, *7*, 102775. [CrossRef]
136. Laksaci, H.; Khelifi, A.; Belhamdi, B.; Trari, M. Valorization of coffee grounds into activated carbon using physic—Chemical activation by KOH/CO_2. *J. Environ. Chem. Eng.* **2017**, *5*, 5061–5066. [CrossRef]
137. Wang, Y.X.; Ngo, H.H.; Guo, W.S. Preparation of a specific bamboo based activated carbon and its application for ciprofloxacin removal. *Sci. Total Environ.* **2015**, *533*, 32–39. [CrossRef] [PubMed]
138. Hashemian, S.; Salari, K.; Yazdi, Z.A. Preparation of activated carbon from agricultural wastes (almond shell and orange peel) for adsorption of 2-pic from aqueous solution. *J. Ind. Eng. Chem.* **2014**, *20*, 1892–1900. [CrossRef]
139. Talat, M.; Mohan, S.; Dixit, V.; Singh, D.K.; Hasan, S.H.; Srivastava, O.N. Effective removal of fluoride from water by coconut husk activated carbon in fixed bed column: Experimental and breakthrough curves analysis. *Groundwater Sustainable Dev.* **2018**, *7*, 48–55. [CrossRef]
140. Vu, M.T.; Chao, H.-P.; Trinh, T.V.; Le, T.T.; Lin, C.-C.; Tran, H.N. Removal of ammonium from groundwater using NaOH-treated activated carbon derived from corncob wastes: Batch and column experiments. *J. Clean. Prod.* **2018**, *180*, 560–570. [CrossRef]
141. Dhorabe, P.T.; Lataye, D.H.; Ingole, R.S. Removal of 4-nitrophenol from aqueous solution by adsorption onto activated carbon prepared from *Acacia glauca* sawdust. *Water Sci. Technol.* **2016**, *73*, 955–966. [CrossRef] [PubMed]
142. Zhang, Z.; Luo, X.; Liu, Y.; Zhou, P.; Ma, G.; Lei, Z.; Lei, L. A low cost and highly efficient adsorbent (activated carbon) prepared from waste potato residue. *J. Taiwan Inst. Chem. Eng.* **2015**, *49*, 206–211. [CrossRef]
143. Osman, A.I.; Blewitt, J.; Abu-Dahrieh, J.K.; Farrell, C.; Al-Muhtaseb, A.H.; Harrison, J.; Rooney, D.W. Production and characterisation of activated carbon and carbon nanotubes from potato peel waste and their application in heavy metal removal. *Environ. Sci. Pollut. Res.* **2019**, *26*, 37228–37241. [CrossRef]
144. Satayeva, A.R.; Howell, C.A.; Korobeinyk, A.V.; Jandosov, J.; Inglezakis, V.J.; Mansurov, Z.A.; Mikhalovsky, S.V. Investigation of rice husk derived activated carbon for removal of nitrate contamination from water. *Sci. Total Environ.* **2018**, *630*, 1237–1245. [CrossRef]

145. Baysal, M.; Bilge, K.; Yilmaz, B.; Papila, M.; Yürüm, Y. Preparation of high surface area activated carbon from waste-biomass of sunflower piths: Kinetics and equilibrium studies on the dye removal. *J. Environ. Chem. Eng.* **2018**, *6*, 1702–1713. [CrossRef]
146. Fu, K.; Yue, Q.; Gao, B.; Wang, Y.; Li, Q. Activated carbon from tomato stem by chemical activation with $FeCl_2$. *Colloids Surf. A* **2017**, *529*, 842–849. [CrossRef]
147. Liew, R.K.; Azwar, E.; Yek, P.N.Y.; Lim, X.Y.; Cheng, C.K.; Ng, J.-H.; Jusoh, A.; Lam, W.H.; Ibrahim, M.D.; Ma, N.L.; et al. Microwave pyrolysis with KOH/NaOH mixture activation: A new approach to produce micro-mesoporous activated carbon for textile dye adsorption. *Bioresour. Technol.* **2018**, *266*, 1–10. [CrossRef] [PubMed]
148. Ioannidou, O.; Zabaniotou, A. Agricultural residues as precursors for activated carbon production-a review. *Renew. Sustain. Energy Rev.* **2007**, *11*, 1966–2005. [CrossRef]
149. Pekala, R.W.; Kong, F.M. Proc. 2nd Int. Symp. of Aerogels (ISA2). *Rev. Phys. Appl.* **1989**, *24 C4*, 33.
150. Pekala, R.W. Organic aerogels from the polycondensation of resorcinol with formaldehyde. *J. Mater. Sci.* **1989**, *24*, 3221–3227. [CrossRef]
151. Yamamoto, Y.; Sugimoto, T.; Suzuki, T.; Mukai, S.R.; Tamon, H. Preparation and characterization of carbon cryogel microspheres. *Carbon* **2002**, *40*, 1345–1351. [CrossRef]
152. Saliger, R.; Bock, V.; Petrcevic, R.; Tillostson, T.; Geis, S.; Fricke, J. Carbon aerogels from dilute catalysis of resorcinol with formaldehyde. *J. Non-Cryst. Solids* **1997**, *221*, 144–150. [CrossRef]
153. Moussaoui, Y.; Elaloui, E.; Ben Salem, R. Etude mécanistique et structurale des résines R/F. *C. R. Chimie* **2012**, *15*, 493–498. [CrossRef]
154. Bock, V.; Emmerling, A.; Fricke, J. Influence of monomer and catalyst concentration on RF and carbon aerogel structure. *J. Non-Cryst. Solids* **2005**, *225*, 69–73. [CrossRef]
155. Wei, W.; Hu, H.; Qin, G.; You, L.; Chen, G. Pore structure control of phenol–formaldehyde based carbon microfiltration membranes. *Carbon* **2004**, *42*, 679–681. [CrossRef]
156. Tonanon, N.; Siyasukh, A.; Tanthapanichakoon, W.; Nishihara, H.; Mukai, S.R.; Tamon, H. Improvement of mesoporosity of carbon cryogels by ultrasonic irradiation. *Carbon* **2005**, *43*, 525–531. [CrossRef]
157. Moussaoui, R.; ben mosbah, M.; Moussaoui, Y.; Elaloui, E. Structural Study of Microporous Xerogels Prepared by Polycondensation of Pyrogallol with Formaldehyde. *J. Adv. Chem.* **2013**, *6*, 878–883. [CrossRef]
158. Moussaoui, R.; ben Mosbah, M.; Elghniji, K.; Elaloui, E.; Moussaoui, Y. Sol-gel synthesis of microporous carbon using Resorcinol and Formaldehyde. *J. Chem. Res.* **2016**, *40*, 209–212. [CrossRef]
159. Ma, M.; Ying, H.; Cao, F.; Wang, Q.; Ai, N. Adsorption of congo red on mesoporous activated carbon prepared by CO_2 physical activation. *Chin. J. Chem. Eng.* **2020**, *28*, 1069–1076. [CrossRef]
160. Rambabu, N.; Rao, B.V.S.K.; Surisetty, V.R.; Das, U.; Dalai, A.K. Production, characterization and evaluation of activated carbons from de-oiled canola meal for environmental applications. *Ind. Crop. Prod.* **2015**, *65*, 572–581. [CrossRef]
161. Guoa, Y.; Tan, C.; Sun, J.; Li, W.; Zhang, J.; Zhao, C. Porous activated carbons derived from waste sugarcane bagasse for CO_2 adsorption. *Chem. Eng. J.* **2020**, *381*, 122736. [CrossRef]
162. Pezoti, O.; Cazetta, A.L.; Bedin, K.C.; Souza, L.S.; Martins, A.C.; Silva, T.L.; Júnior, O.O.S.; Visentainer, J.V.; Almeida, V.C. NaOH-activated carbon of high surface area produced from guava seeds as a high-efficiency adsorbent for amoxicillin removal: Kinetic, isotherm and thermodynamic studies. *Chem. Eng. J.* **2016**, *288*, 778–788. [CrossRef]
163. Ibrahim, W.M.; Hassan, A.F.; Azab, Y.A. Biosorption of toxic heavy metals from aqueous solution by *Ulva lactuca* activated carbon. *Egypt. J. Basic Appl. Sci.* **2016**, *3*, 241–249. [CrossRef]
164. Benaddi, H.; Legras, D.; Rouzaud, J.N.; Beguin, F. Influence of the atmosphere in the chemical activation of wood by phosphoric acid. *Carbon* **1998**, *36*, 306–309. [CrossRef]
165. Heidari, A.; Younesi, H.; Rashidi, A.; Ghoreyshi, A.A. Adsorptive removal of CO_2 on highly microporous activated carbons prepared from Eucalyptus camaldulensis wood: Effect of chemical activation. *J. Taiwan Inst. Chem. Eng.* **2014**, *45*, 579–588. [CrossRef]
166. Nam, H.; Wang, S.; Jeong, H.R. TMA and H_2S gas removals using metal loaded on rice husk activated carbonfor indoor air purification. *Fuel* **2018**, *213*, 186–194. [CrossRef]
167. Wang, S.; Nam, H.; Nam, H. Preparation of activated carbon from peanut shell with KOH activation and its application for H_2S adsorption in confined space. *J. Environ. Chem. Eng.* **2020**, *8*, 103683. [CrossRef]

168. Demiral, H.; Güngor, C. Adsorption of copper(II) from aqueous solutions on activated carbon prepared from grape bagasse. *J. Cleaner Prod.* **2016**, *124*, 103–113. [CrossRef]
169. Gueye, M.; Richardson, Y.; Kafack, F.T.; Blin, J. High efficiency activated carbons from African biomass residues for the removal of chromium(VI) from wastewater. *J. Environ. Chem. Eng.* **2014**, *2*, 273–281. [CrossRef]
170. Danish, M.; Hashim, R.; Ibrahim, M.N.M.; Sulaiman, O. Optimization study for preparation of activated carbon from Acacia mangium wood using phosphoric acid. *Wood Sci. Technol.* **2014**, *48*, 1069–1083. [CrossRef]
171. Girgis, B.S.; El-Hendawy, A.A. Porosity development in activated carbons obtained from date pits under chemical activation with phosphoric acid. *Microporous Mesoporous Mater.* **2002**, *52*, 105–117. [CrossRef]
172. Abdel-Ghani, N.T.; El-Chaghaby, G.A.; ElGammal, M.H.; Rawash, E.-S.A. Optimizing the preparation conditions of activated carbons from olive cake using KOH activation. *New Carbon Mater.* **2016**, *31*, 492–500. [CrossRef]
173. Mahmoudi, K.; Hamdi, N.; Srasra, E. Preparation and characterization of activated carbon from date pits by chemical activation with zinc chloride for methyl orange adsorption. *J. Mater. Environ. Sci.* **2014**, *5*, 1758–1769.
174. Wu, F.C.; Tseng, R.L.; Juang, R.S. Preparation of highly microporous carbons from fir wood by KOH activation for adsorption of dyes and phenols from water. *Sep. Purif. Technol.* **2005**, *47*, 10–19. [CrossRef]
175. Rahman, A.; Hango, H.J.; Daniel, L.S.; Uahengo, V.; Jaime, S.J.; Bhaskaruni, S.V.H.S.; Jonnalagadda, S.B. Chemical preparation of activated carbon from Acacia erioloba seed pods using H_2SO_4 as impregnating agent for water treatment: An environmentally benevolent approach. *J. Cleaner Prod.* **2019**, *237*, 117689. [CrossRef]
176. Naeem, S.; Baheti, V.; Wiener, J.; Marek, J. Removal of methylene blue from aqueous media using activated carbon web. *J. Text. Inst.* **2016**, *108*, 803–811. [CrossRef]
177. Marzbali, M.H.; Esmaieli, M. Fixed bed adsorption of tetracycline on a mesoporous activated carbon: Experimental study and neuro-fuzzy modeling. *J. Appl. Res. Technol.* **2017**, *15*, 454–463. [CrossRef]
178. Li, Z.; Hanafy, H.; Zhang, L.; Sellaoui, L.; Netto, M.S.; Oliveira, M.L.S.; Seliem, M.K.; Dotto, G.L.; Bonilla-Petriciolet, A.; Li, Q. Adsorption of congo red and methylene blue dyes on an ashitaba waste and a walnut shell-based activated carbon from aqueous solutions: Experiments, characterization and physical interpretations. *Chem. Eng. J.* **2020**, *388*, 124263. [CrossRef]
179. Kalra, A.; Hadi, P.; Hui, C.W.; Mackey, H.; Ansari, T.A.; Saleem, J.; McKay, G. Adsorption of Dyes from Water on to Bamboo-Based Activated Carbon—Error Analysis Method for Accurate Isotherm Parameter Determination. *J. Water Sci. Eng.* **2019**, *1*, 1–11.
180. Liu, Y.; Cheng, H.; He, Y. Application and Mechanism of Sludge-Based Activated Carbon for Phenol and Cyanide Removal from Bio-Treated Effluent of Coking Wastewater. *Processes* **2020**, *8*, 82. [CrossRef]
181. Nowicki, P.; Kazmierczak-Razna, J.; Pietrzak, R. Physicochemical and adsorption properties of carbonaceous sorbents prepared by activation of tropical fruit skins with potassium carbonate. *Mater. Des.* **2016**, *90*, 579–585. [CrossRef]
182. Nirmala, G.; Murugesan, T.; Rambabu, K.; Sathiyanarayanan, K.; Show, P.L. Adsorptive removal of phenol using banyan root activated carbon. *Chem. Eng. Commun.* **2019**. [CrossRef]
183. Tu, B.; Wen, R.; Wang, K.; Cheng, Y.; Deng, Y.; Cao, W.; Zhang, K.; Tao, H. Efficient removal of aqueous hexavalent chromium by activated carbon derived from Bermuda grass. *J. Colloid Interface Sci.* **2020**, *560*, 649–658. [CrossRef]
184. Thabede, P.M.; Shooto, N.D.; Naidoo, E.B. Removal of methylene blue dye and lead ions from aqueous solution using activated carbon from black cumin seeds. *S. Afr. J. Chem. Eng.* **2020**, *33*, 39–50. [CrossRef]
185. Lütke, S.F.; Igansi, A.V.; Pegoraro, L.; Dotto, G.L.; Pinto, L.A.A.; Cadaval, T.R.S. Preparation of activated carbon from black wattle bark waste and its application for phenol adsorption. *J. Environ. Chem. Eng.* **2019**, *7*, 103396.
186. Kasperiski, F.M.; Lima, E.C.; Umpierres, C.S.; dos Reis, G.S.; Thue, P.S.; Lima, D.R.; Dias, S.L.P.; Saucier, C.; da Costa, J.B. Production of porous activated carbons from *Caesalpinia ferrea* seed pod wastes: Highly efficient removal of captopril from aqueous solutions. *J. Clean. Prod.* **2018**, *197*, 919–929. [CrossRef]
187. Aguayo-Villarreal, I.A.; Bonilla-Petriciolet, A.; Muñiz-Valencia, R. Preparation of activated carbons from pecan nutshell and their application in the antagonistic adsorption of heavy metal ions. *J. Mol. Liq.* **2017**, *230*, 686–695. [CrossRef]

188. Dwivedi, A.D.; Dubey, S.P.; Sillanpää, M.; Kwon, Y.-N.; Lee, C. Distinct adsorption enhancement of bi-component metals (cobalt and nickel) by Fireweed-derived carbon compared to activated carbon: Incorporation of surface group distributions for increased efficiency. *Chem. Eng. J.* **2015**, *281*, 713–723. [CrossRef]
189. Islam, M.A.; Ahmed, M.J.; Khanday, W.A.; Asif, M.; Hameed, B.H. Mesoporous activated coconut shell-derived hydrochar prepared via hydrothermal carbonization-NaOH activation for methylene blue adsorption. *J. Environ. Manag.* **2017**, *203*, 237–244. [CrossRef] [PubMed]
190. Jain, A.; Balasubramanian, R.; Srinivasan, M.P. Tuning hydrochar properties for enhanced mesopore development in activated carbon by hydrothermal carbonization. *Microporous Mesoporous Mater.* **2015**, *203*, 178–185. [CrossRef]
191. Rodiguez, M.H.; Yperman, J.; Carleer, R.; Maggen, J.; Dadi, D.; Gryglewicz, G.; der Bruggen, B.V.; Hernández, J.F.; Calvis, A.O. Adsorption of Ni(II) on spent coffee and coffee husk based activated carbon. *J. Environ. Chem. Eng.* **2018**, *6*, 1161–1170. [CrossRef]
192. Liu, Z.; Sun, Y.; Xu, X.; Meng, X.; Qu, J.; Wang, Z.; Liu, C.; Qu, B. Preparation, characterization and application of activated carbon from corn cob by KOH activation for removal of Hg(II) from aqueous solution. *Bioresour. Technol.* **2020**, *306*, 123154. [CrossRef]
193. Zhang, Y.; Song, X.; Zhang, P.; Gao, H.; Ou, C.; Kong, X. Production of activated carbons from four wastes via one-step activation and their applications in Pb^{2+} adsorption: Insight of ash content. *Chemosphere* **2020**, *245*, 125587. [CrossRef]
194. Yu, X.; Han, X.; Chang, C.; Hu, Y.; Xu, C.C.; Fang, S. Corncob-derived activated carbon for roxarsone removal from aqueous solution: Isotherms, kinetics, and mechanism. *Environ. Sci. Pollut. Res.* **2020**, *27*, 15785–15797. [CrossRef]
195. Yazidi, A.; Atrous, M.; Soetaredjo, F.E.; Sellaoui, L.; Ismadji, S.; Erto, A.; Bonilla-Petriciolet, A.; Dotto, G.L.; Ben Lamine, A. Adsorption of amoxicillin and tetracycline on activated carbon prepared from durian shell in single and binary systems: Experimental study and modeling analysis. *Chem. Eng. J.* **2020**, *379*, 122320. [CrossRef]
196. Han, Q.; Wang, J.; Goodman, B.A.; Xie, J.; Liu, Z. High adsorption of methylene blue by activated carbon prepared from phosphoric acid treated eucalyptus residue. *Powder Technol.* **2020**, *366*, 239–248. [CrossRef]
197. Chen, C.; Mi, S.; Lao, D.; Shi, P.; Tong, Z.; Li, Z.; Hu, H. Single-step synthesis of eucalyptus sawdust magnetic activated carbon and its adsorption behavior for methylene blue. *RSC Adv.* **2019**, *9*, 22248–22262. [CrossRef]
198. Kumar, A.; Jena, H.M. Removal of methylene blue and phenol onto prepared activated carbon from Fox nutshell by chemical activation in batch and fixed-bed column. *J. Clean. Prod.* **2016**, *137*, 1246–1259. [CrossRef]
199. Kumar, A.; Jena, H.M. Adsorption of Cr(VI) from aqueous solution by prepared high surface area activated carbon from Fox nutshell by chemical activation with H_3PO_4. *J. Environ. Chem. Eng.* **2017**, *5*, 2032–2041. [CrossRef]
200. Sayğili, H.; Güzel, F.; Önal, Y. Conversion of grape industrial processing waste to activated carbon sorbent and its performance in cationic and anionic dyes adsorption. *J. Clean. Prod.* **2015**, *93*, 84–93. [CrossRef]
201. Nagalakshmi, T.V.; Emmanuel, K.A.; Bhavani, P. Adsorption of disperse blue 14 onto activated carbon prepared from Jackfruit-PPI-I waste. *Mater. Today: Proc.* **2019**, *18*, 2036–2051. [CrossRef]
202. Shokry, H.; Elkady, M.; Hamad, H. Nano activated carbon from industrial mine coal as adsorbents for removal of dye from simulated textile wastewater: Operational parameters and mechanism study. *J. Mater. Res. Technol.* **2019**, *8*, 4477–4488. [CrossRef]
203. Al-Ghouti, M.A.; Sweleh, A.O. Optimizing textile dye removal by activated carbon prepared from olive stones. *Environ. Technol. Innovation* **2019**, *16*, 100488. [CrossRef]
204. Elhleli, H.; Mannai, F.; Elaloui, E.; Moussaoui, Y. Nitrophenol Removal from Wastewater onto Prepared Activated Carbon from *Opuntia ficus-indica*. In *Recent Advances in Environmental Science from the Euro-Mediterranean and Surrounding Regions*; Kallel, A., Ksibi, M., Ben Dhia, H., Khélifi, N., Eds.; EMCEI 2017, Advances in Science, Technology & Innovation (IEREK Interdisciplinary Series for Sustainable Development); Springer: Cham, Switzerland, 2018; pp. 1245–1246.
205. Pamidimukkala, P.S.; Soni, H. Efficient removal of organic pollutants with activated carbon derived from palm shell: Spectroscopic characterisation and experimental optimization. *J. Environ. Chem. Eng.* **2018**, *6*, 3135–3149. [CrossRef]

206. da Paixao Cansado, I.P.; Belo, C.R.; Mourao, P.A.M. Pesticides abatement using activated carbon produced from a mixture of synthetic polymers by chemical activation with KOH and K_2CO_3. *Environ. Nanotechnol. Monit. Manag.* **2019**, *12*, 100261. [CrossRef]
207. Bedada, D.; Angassa, K.; Tiruneh, A.; Kloos, H.; Fito, J. Chromium removal from tannery wastewater through activated carbon produced from Parthenium hysterophorus weed. *Energ. Ecol. Environ.* **2020**, *5*, 184–195. [CrossRef]
208. Xu, W.; Zhao, Q.; Wang, R.; Jiang, Z.; Zhang, Z.; Gao, X.; Ye, Z. Optimization of organic pollutants removal from soil eluent by activated carbon derived from peanut shells using response surface methodology. *Vacuum* **2017**, *141*, 307–315. [CrossRef]
209. Danish, M.; Khanday, W.A.; Hashim, R.; Sulaiman, N.S.B.; Akhtar, M.N.; Nizami, M. Application of optimized large surface area date stone (Phoenix dactylifera) activated carbon for rhodamin B removal from aqueous solution: Box-Behnken design approach. *Ecotoxicol. Environ. Saf.* **2017**, *139*, 280–290. [CrossRef] [PubMed]
210. Darweesh, T.M.; Ahmed, M.J. Adsorption of ciprofloxacin and norfloxacin from aqueous solution onto granular activated carbon in fixed bed column. *Ecotoxicol. Environ. Saf.* **2017**, *138*, 139–145. [CrossRef] [PubMed]
211. Valizadeh, S.; Younesi, H.; Bahramifar, N. Highly mesoporous K_2CO_3 and KOH/activated carbon for SDBS removal from water samples: Batch and fixed-bed column adsorption process. *Environ. Nanotechnol. Monit. Manag.* **2016**, *6*, 1–13. [CrossRef]
212. Arampatzidou, A.C.; Deliyanni, E.A. Comparison of activation media and pyrolysis temperature for activated carbons development by pyrolysis of potato peels for effective adsorption of endocrine disruptor bisphenol-A. *J. Colloid Interface Sci.* **2016**, *466*, 101–112. [CrossRef]
213. Runtti, H.; Tuomikoski, S.; Kangas, T.; Lassi, U.; Kuokkanen, T.; Rämö, J. Chemically activated carbon residue from biomass gasification as a sorbent for iron(II), copper(II) and nickel(II) ions. *J. Water Process Eng.* **2014**, *4*, 12–24. [CrossRef]
214. Bjorklund, K.; Li, L.Y. Adsorption of organic stormwater pollutants onto activated carbon from sewage sludge. *J. Environ. Manag.* **2017**, *197*, 490–497. [CrossRef]
215. Araga, R.; Soni, S.; Sharma, C.S. Fluoride adsorption from aqueous solution using activated carbon obtained from KOH-treated jamun (Syzygium cumini) seed. *J. Environ. Chem. Eng.* **2017**, *5*, 5608–5616. [CrossRef]
216. Saygili, H.; Güzel, F. High surface area mesoporous activated carbon from tomato processing solid waste by zinc chloride activation: Process optimization, characterization and dyes adsorption. *J. Clean. Prod.* **2016**, *113*, 995–1004. [CrossRef]
217. Acosta, R.; Nabarlatz, D.; Sánchez-Sánchez, A.; Jagiello, J.; Gadonneix, P.; Celzard, A.; Fierro, V. Adsorption of Bisphenol A on KOH-activated tyre pyrolysis char. *J. Environ. Chem. Eng.* **2018**, *6*, 823–833. [CrossRef]
218. Hassan, A.F.; Elhadidy, H. Production of activated carbons from waste carpets and its application in methylene blue adsorption: Kinetic and thermodynamic studies. *J. Environ. Chem. Eng.* **2017**, *5*, 955–963. [CrossRef]
219. Nieto-Márquez, A.; Pinedo-Flores, A.; Picasso, G.; Atanes, E.; Kou, R.S. Selective adsorption of Pb^{2+}, Cr^{3+} and Cd^{2+} mixtures on activated carbons prepared from waste tires. *J. Environ. Chem. Eng.* **2017**, *5*, 1060–1067. [CrossRef]
220. Xiao, W.; Garba, Z.N.; Sun, S.; Lawan, I.; Wang, L.; Lin, M.; Yuan, Z. Preparation and evaluation of an effective activated carbon from white sugar for the adsorption of rhodamine B dye. *J. Clean. Prod.* **2020**, *253*, 119989. [CrossRef]
221. Gebreegziabher, T.B.; Wang, S.; Nam, H. Adsorption of H_2S, NH_3 and TMA from indoor air using porous corncob activated carbon: Isotherm and kinetics study. *J. Environ. Chem. Eng.* **2019**, *7*, 103234. [CrossRef]
222. Taleb, F.; ben Mosbah, M.; Elaloui, E.; Moussaoui, Y. Adsorption of ibuprofen sodium salt onto Amberlite resin IRN-78: Kinetics, isotherm and thermodynamic investigations. *Korean J. Chem. Eng.* **2017**, *34*, 1141–1148. [CrossRef]
223. Saad, M.K.; Khiari, R.; Elaloui, E.; Moussaoui, Y. Adsorption of anthracene using activated carbon and Posidonia oceanica. *Arabian J. Chem.* **2014**, *7*, 109–113. [CrossRef]
224. Saad, M.K.; Mnasri, N.; Mhamdi, M.; Chafik, T.; Elaloui, E.; Moussaoui, Y. Removal of methylene blue onto mineral matrices. *Desalin. Water Treat.* **2015**, *56*, 2773–2780. [CrossRef]
225. Baheti, V.; Naeem, S.; Militky, J.; Okrasa, M.; Tomkova, B. Optimized preparation of activated carbon nanoparticles from acrylic fibrous wastes. *Fibers Polym.* **2015**, *16*, 2193–2201. [CrossRef]

226. Fan, S.; Wang, Y.; Wang, Z.; Tang, J.; Tang, J.; Li, X. Removal of methylene blue from aqueous solution by sewage sludge-derived biochar: Adsorption kinetics, equilibrium, thermodynamics and mechanism. *J. Environ. Chem. Eng.* **2017**, *5*, 601–611. [CrossRef]
227. Abdelhafez, A.A.; Li, J. Removal of Pb(II) from aqueous solution by using biochars derived from sugar cane bagasse and orange peel. *J. Taiwan Inst. Chem. Eng.* **2016**, *61*, 367–375. [CrossRef]
228. Liu, L.; Li, Y.; Fan, S. Preparation of KOH and H_3PO_4 Modified Biochar and Its Application in Methylene Blue Removal from Aqueous Solution. *Processes* **2019**, *7*, 891. [CrossRef]
229. Inyang, M.; Dickenson, E. The potential role of biochar in the removal of organic and microbial contaminants from potable and reuse water: A review. *Chemosphere* **2015**, *134*, 232–240. [CrossRef] [PubMed]
230. Liu, L.; Fan, S.; Li, Y. Removal Behavior of Methylene Blue from Aqueous Solution by Tea Waste: Kinetics, Isotherms and Mechanism. *Int. J. Approx. Res. Public Health* **2018**, *15*, 1321. [CrossRef] [PubMed]
231. Ramesh, T.N.; Kirana, D.V.; Ashwini, A.; Manasa, T.R. Calcium hydroxide as low cost adsorbent for the effective removal of indigo carmine dye in water. *J. Saudi Chem. Soc.* **2017**, *21*, 165–171. [CrossRef]
232. Fan, S.; Tang, J.; Wang, Y.; Li, H.; Zhang, H.; Tang, J.; Wang, Z.; Li, X. Biochar prepared from co-pyrolysis of municipal sewage sludge and tea waste for the adsorption of methylene blue from aqueous solutions: Kinetics, isotherm, thermodynamic and mechanism. *J. Mol. Liq.* **2016**, *220*, 432–441. [CrossRef]

3

Application of Ionizing Radiation in Wastewater Treatment

Rehab O. Abdel Rahman [1,*] **and Yung-Tse Hung** [2]

1 Hot Lab. Center, Atomic Energy Authority of Egypt, Cairo P.O. No. 13759, Egypt
2 Department of Civil and Environmental Engineering, Cleveland State University, Cleveland, OH 44115, USA; yungtsehung@gmail.com
* Correspondence: alaarehab@yahoo.com

Abstract: Technological applications of nuclear science and technology in different sectors have proved their reliabilities and sustainability over decades. These applications have supported various human civilization needs, ranging from power generation to industrial, medical, and environmental applications. Environmental applications of radiation sources are used to support decision making processes in many fields; including the detection and analysis of pollutant transport, water resources management, and treatment of municipal and industrial wastewaters. This work reviewed recent advances in the research and applications of ionizing radiation in treating different wastewater effluents. The main objective of the work is to highlight the role of ionizing radiation technology in the treatment of complex wastewater effluents generated from various human activities and to address its sustainability. Results of both laboratory and industrial scale applications of this treatment technology have been reviewed, and information on operational safety of industrial irradiators, which affect the sustainability of this technology, has been summarized.

Keywords: ionizing radiation; agricultural effluents; dye treatment; pharmaceutical effluents; disinfection

---

1. Introduction

Providing clean water and sanitation is one of the sustainable development goals that were proposed by the United Nation (UN). One of the problems that affect this goal is the reduction of freshwater quality. This reduction is attributed to the continuous increase in untreated wastewaters volumes and poor management practices, which led to the introduction of hazardous materials into freshwater sources [1]. Wastewater is defined, as indicated in UNEP/UN-Habitat, as a combination of one or more of the following effluents: domestic, commercial, industrial, horticultural, aquaculture, and storm water [2]. Recent advances in wastewater management helped in addressing some of the problems in water supply, pollution control, water recycling, and environmental protection. Now wastewaters are proposed as a resource, where many wastewater treatment plants are operated by the biogas generated from the anaerobic digestion of sludge, and the effluents from these plants could be used after appropriate treatment to meet the industrial, agricultural, and potable water requirements. Currently, wastewater management in developing countries is characterized by the discharge of large quantity of wastewater into surface water bodies without proper treatment. It is also challenged by the difficulties to sustain financing, operating and maintaining infrastructure for wastewater treatment.

Conventional wastewater treatment plants (WWTP) aims at reducing the contamination levels to acceptable limits required by the national regulatory agencies and at complying with international guidelines, which allow its safe discharge or reuse. Multi-stages of treatment processes are used, where pre-treatment stage is applied to remove coarse and large solids from the waste stream, using physical treatment technologies such as screens and grit chambers [3]. Primary treatment methods are then applied to remove suspended solids that could be settled using gravity sedimentation with or without coagulation and flocculation [3,4]. The effluent from the primary treatment is directed to the secondary treatment stage to remove residual suspended solids and organic materials by using biological treatment processes [3]. The effluents from secondary stage contain some heavy metals, synthetic bio-refractory organic pollutants, and soluble microbial products derived during biological treatment [5,6]. The synthetic bio-refractory organic pollutants may include emerging micro-pollutants and disinfection by-products. In the tertiary stage of treatment, the effluents from the secondary stage are polished by removing persistence organic containments and heavy metals using advanced wastewater treatment technologies [3,4,7]. These technologies include filtration, sorption, gas stripping, ion-exchange, advanced oxidation processes (AOP), and distillation [7]. Finally, disinfection could be applied depending on the potential use of the treated effluents and the effluents characteristics [3].

Regulations on discharge/reuse indicators varied from country to another. In Denmark, Belgium, Spain, Germany, France, and Netherlands, the regulation covers heavy metals concentrations, total suspended solids (TSS), chemical oxygen demand (COD), 5 days biochemical oxygen demand (BOD_5), total nitrogen (TN), total phosphorus (TP), and quantity of effluent discharge [8]. On the other hand, toxicity is only considered as reuse/discharge indicator in Germany, France, and United Kingdom [8]. Table 1 lists some international guidelines on the maximum limits for the reuse of effluents containing heavy metals, and U.S. regulation for their reuse and discharge [9–12]. Table 2 presents effluent discharge and reuse quality requirements in USA, Canada, and EU [11,13–15]. It should be noted that three classes of effluents are listed in the Canadian regulation based on the degree of treatment processes used in the treatment plants, namely, A, B and C, that refers to effluent from tertiary and disinfection treatment, tertiary treatment, and secondary treatment, respectively. For reuse regulations in EU, the limit of each class is determined based on the processes used in the treatment plants, irrigation methods, and the type of corps. Class A represents effluents from tertiary and disinfection treatment used any irrigation method and to produce all food corps. Classes B and C are effluent from secondary treatment and disinfection, which are used to produce food and processed food and non-food corps using all irrigation methods, and drip irrigation, respectively. Finally class D in EU regulation represents effluents from secondary treatment and disinfection, where all irrigation methods are allowed to produce industrial, energy, and seed corps [13]. Discharges from un-complied conventional WWTP can lead to the introduction of persistent chemicals and eco-toxic micro-pollutants into the aquatic systems [16]. The incomplete removal of these containments (even in 10^{-9}–10^{-6} g/L concentration range) was reported to induce potential long-term detrimental impact on the environment and the human health [17]. Recent research studies reported in the literature supports the application of advanced oxidation processes (AOP) for wastewater treatment to remove these contaminants [16,17].

Table 1. Guidelines of maximum limits for discharge and reuse of some inorganic pollutants in treated wastewater.

	WHO, Reuse, ppb [11]	**FAO, Reuse, ppb [12]**	**US, ppb [9]**			**WHO, Reuse, ppb [11]**	**FAO, Reuse, ppb [12]**	**US, ppb [9]**	
			Discharge	**Reuse**				**Discharge**	**Reuse**
As	50	100	3000	50	Hg	1	NA	2000	2
Cd	5	100	15,000	5	Ni	0	200	12,000	100
Cr	NA	100	10,000	50	Ag	NA	NA	5000	50
Cu	1	200	15,000	NA	Zn	5000	200	25,000	5000
CN	NA	NA	10,000	5.2	pH	6.5–8.5	6.5–8.4	6–9	
Pb	50	5000	40,000	50					

NA: Not Available.

Table 2. Effluent reuse and discharge limits in different regulations [11,13–15].

Indicator	Reuse				Discharge				
	EU [14,15]				EU [14,15]	US [11]	Canada [13]		
	A	B	C	D			A	B	C
BOD_5, ppm	10	25				45	10	10	45
TSS, ppm	10	35				45	10	10	45
COD, ppm	NA	NA	NA	NA	125	NA	NA	NA	NA
E. coli, (CFU/100 mL)	10	10^2	10^3	10^4	NA	2.2	2.2	400	NA
Turbidity (NTU)	5	NA	NA	NA	NA	2	2	NA	NA
TN, ppm	NA	NA	NA	NA	15	10	20	NA	NA

NA: Not Available.

AOP use in-situ generated hydroxyl or sulfate radicals for organic pollutants degradation and heavy metals toxicity reduction [5,18,19]. AOP convert synthetic organic pollutants and soluble microbial products into simple biodegradable and harmless products, which lead to the reduction of COD and BOD in the treated effluents. AOP technologies employ various activation methods, where induced oxidation is achieved via exposure to photochemical or ionizing radiations, or chemicals [18,20,21]. Several studies were conducted to assess the feasibility of using ionizing radiation, in the form of gamma (γ) rays or electrons (e^-), to remove persistence contaminants and to disinfect treated water and sludge. The results of these studies indicated that the ionizing radiation treatment is technically and economically promising. The efficiency of this technology was also proved in the reduction of persistence heavy metals under different treatment conditions. Fifteen pilot plants and several full scale irradiation treatment facilities that employ electron accelerator or γ irradiators were established with varying capacities. International Atomic Energy Agency (IAEA) reported that there is a need to continue research in this area to increase the general awareness of these processes in the environmental engineering community [22]. This awareness could be achieved by presenting integrated reviews on the advances in ionizing radiation applications in industrial effluents treatment from technical, operational safety, and economical aspects. Some review papers were published that reviewed the role of ionizing radiation in the degradation of azo dyes [23], summarized the results of the IAEA coordination research project [24], and have reported the optimized doses and procedures [25]. There is a lack in review papers that summarize recent advances in this field and that provide integrated insights into the role of this technology in eliminating hazardous biodegradation and disinfection products. In this work, the effort is directed to summarize the current understanding of the decomposition and removal mechanisms for organic and inorganic pollutants, respectively. Principles and advances in investigating the scientific basis of the applicability of this technique in the treatment and disinfection of agriculture, dyes, pharmaceutical, and petrochemical effluents will be presented. Operational safety and economical factors that affect the sustainability of this technology will be summarized. Finally, knowledge gaps will be identified and research areas that need to be addressed will be highlighted.

2. Decomposition and Removal Mechanisms

Generally any emitted radiation is characterized by its ability to deposit some of its energy in the surrounding media. This energy excites the media atoms by striping their electrons or break the chemical bonds between its molecules [26]. This particular characteristic leads to the application of ionizing radiation in treating wastewater, where the radiation imparts some of its energy in the radiolysis of water molecules. The effectiveness of the irradiation process is evaluated by calculating the radiation chemical yield (G-value, μmol·J⁻) that quantifies the number of formed species due to the absorption of 100 eV and is given by:

$$G = \frac{6.023 \times 10^{23} C}{D \times 6.24 \times 10^{16}} \quad (1)$$

where C is the formed species concentration (mol/L) and D is the absorbed dose (Gy). Figure 1 illustrates the two stages radiolysis process for water molecules and its corresponding time scale, radiolysis products (primary intermediates), and their G-values. Primary intermediate (PI) then reacts with the pollutants (A) leading to the formation of secondary intermediates (SI) that are less persistence in a bi-molecular reaction to produce the degradation products (C + D), as follows [25,27,28]:

$$A + PI \underset{=}{\rightarrow} SI \rightarrow C + D \quad (2)$$

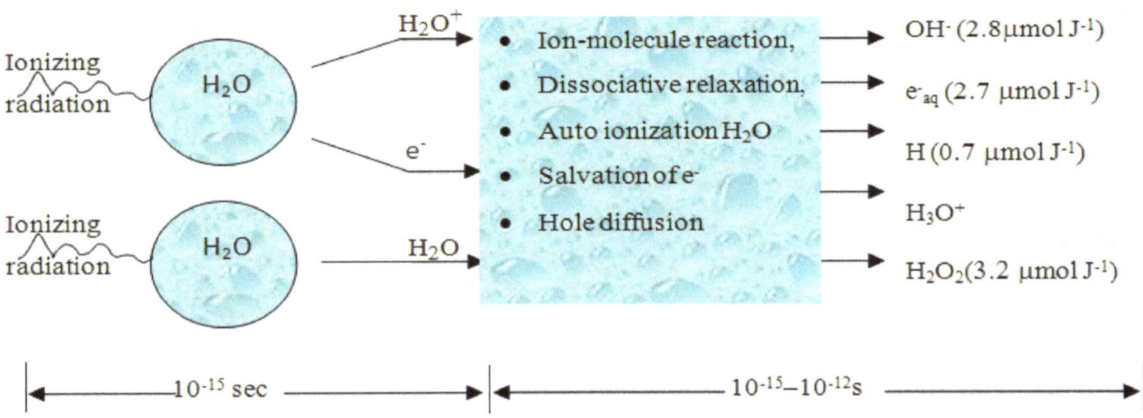

Figure 1. Two stages radiolysis process.

In actual wastewater, there is a competition between the pollutants and the anions in the solution, i.e., Cl^-, $CO_3{}^{2-}$, $HCO_3{}^-$, $SO_4{}^{2-}$, for the reaction with the primary intermediates. This competition can affect the efficiency of the overall treatment process [27]. Gehringer studied the competition kinetics of two pollutants (A, B) in wastewater. The reaction rate was attributed to the bi-molecular rate constant (k_A) between the solute concentration ([A] or [B]) and the primary intermediates concentration [OH] as follow:

$$-\frac{dA}{dt} = k_A[A][OH] \quad (3)$$

Table 3 lists the bi-molecular reaction rate constants for some pollutants [22,25,27,29]. The amount of the specific radicals available for interaction with certain solute could be calculated using the reaction probability (P_A) according to the following equation:

$$P_A = \frac{k_A[A]}{k_A[A] + k_B[B]} \quad (4)$$

Figure 2 illustrates the reactions between the primary intermediates and the available anions in the wastewater effluents. Wojnarovits and Takacs [30] indicated that intermediate reaction with chloride ions is dominant at $pH < 5$, whereas the reaction with carbonate and bicarbonate ions are dominant at neutral or slightly alkaline pH. The produced free radicals will subsequently react with dissolved organic matters in the wastewater via direct electron transfer (outer sphere electron), addition to double bonds (inner sphere electron) or abstraction of H-atoms from C-H bonds. The last reaction takes place between oxidizing radicals and saturated molecules. It should be noted that in some cases, it is very hard to determine if the reaction is electron transfer or radical addition/elimination.

Table 3. Bimolecular reaction rate constants for some pollutants [22,25,27,29].

Compound	Rate Constant, $M^{-1}\ s^{-1}, \times 10^9$			Compound	Rate Constant, $M^{-1}\ s^{-1}, \times 10^9$		
	OH	e^-	H		OH	e^-	H
Perchloroethylene	1.7	13		Bicarbonate	0.0085		4×10^{-3}
Trichloroethylene	2.6	19		Nitrogen		9.7	
Dichloroethylene	7	7.5		Oxygen		19	
Vinylchloride	12	0.25		Methyl t-butyl ether	2	0.018	
1,1,1-Trichloroethane	0.04			Ethyl t-butyl Ether	1.8	0.01	7×10^{-3}
Chloroform	0.005	30		Diisopropyl Ether	2.5	7×10^{-3}	0.067
T-Amyl Methyl Ether	2.4	3×10^{-3}	3×10^{-3}	Acetylenes	0.1–1		
Alcohols	0.1–1			Aldehydes	1		
Alkanes	0.001–1			Aromatics	0.1–100		
Carboxyl Acids	0.01–1			Ketones	1		
Organo-Nitrogen	0.1–100			Olefins	1–10		
Phenols	1–10			Organo-Sulfure	1–10		
Trichloronitromethane (TCNM), Chloropicrin	0.0497 ± 0.28	21.3 ± 0.03		Carbofuran	6.6		
Benzene	7.8			Carbendazim	2.2		

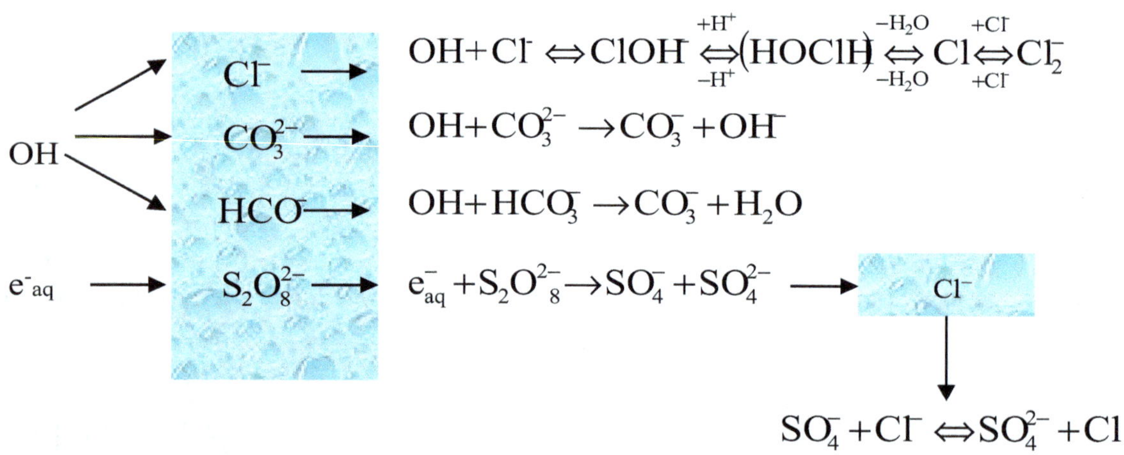

Figure 2. Schematic diagram of intermediate reaction with free radicals.

Ionizing radiation can reduce various forms of mercury via reaction with e^-_{aq} (Equation (5)) and H (Equation (6)) to form unstable compound that dimerized (Equation (7)) to produce insoluble form [31,32].

$$HgCl_2 + e^-_{aq} \rightarrow HgCl + Cl^- \tag{5}$$

$$HgCl_2 + H \rightarrow HgCl + Cl^- + H^+ \tag{6}$$

$$2HgCl_2 \rightarrow Hg_2Cl_2 \tag{7}$$

The presence of hydroxyl radicals inhibited the production of insoluble mercury (Hg_2Cl_2) through re-oxidation of HgCl [31]. To enhance the removal performance of this technology the use of organic radical to act as hydroxyl scavenger was proposed. For example, ethanol may be used (Equations (8) and (9)) with the following reactions [9]:

$$CH_3CH_2OH + OH \rightarrow CH_3CHOH + H_2O \quad (8)$$

$$HgCl_2 + CH_3CHOH \rightarrow HgCl + Cl^- + CH_3CHO + H^+ \quad (9)$$

Successive reduction of Cr species from (VI) state to (III) state is achieved via reaction with H [32]. On the other hand, cadmium (II) and lead (II) are reduced via reaction with H or/and $e^-{}_{aq}$ (Equation (10)) to produce Cd(I) that undergoes disprotonation (Equation (11)) or oxidization through a reaction with OH^- or H_2O_2 (Equation (12)), or react with hydrogen to produce unstable MH^+ species that decay (Equation (13)) [18].

$$M^{2+} + e^-_{aq} \rightarrow M^+ \quad (10)$$

$$2M^+ \rightarrow M + M^{2+} \quad (11)$$

$$M^+ + OH \rightarrow M^{2+} + OH^- \quad (12)$$

$$2H + 2M^+ \rightarrow 2MH^+ \rightarrow H_2 + M^{2+} + M \quad (13)$$

To reduce the oxidation effect and subsequently to enhance cadmium and lead precipitation, an organic OH^- scavenger could be used or the process could be operated in the absence of oxygen [31–33].

3. Advances in Treating Agricultural Wastewaters

Despite chemical pesticides, herbicides, and fungicides are applied according to national agricultural guidelines to enhance the agricultural production efficiency. Residues of these persistence pollutants and their toxic byproducts exist in agricultural wastewater, and could migrate to surface-and ground-waters. Ionizing radiation treatment technology proved its effectiveness in decomposing a varied number of these pollutants using tertiary treatment method. Most of the researches conducted in this area focused on optimizing the irradiation conditions, investigating the effect of other waste components on the degradation process, and the possibility of using combined treatment technologies to enhance the overall efficiency of the treatment process. Table 4 summarizes the performance of ionizing radiation in removing some pollutants in agricultural wastewaters.

Table 4. Performance of agricultural wastewater pollutants degradation using irradiation.

Pollutant	Origin	Degradation Performance	Initial Contaminant Concentration, ppm	Dose, KGy	References
Polychlorinated Biphenyls (PCB)	Pesticide	96%	NA	<0.1	[34]
Alkali Halides	Herbicides, fungicides, insecticide	98%	100	1	[25]
Trihalomethanes	Disinfection by products	87.4%	NA	2	[25]
		95%	145–780	6	[35]
		87.8%	264	8	[35]
Nitrophenols	Degradation products		0.139	5	[25]
Carbofuran, Dimethoate, Imidacloprid, Procloraz	Pesticide	99%	50, (pH 5.5)	5	[36]
Methiocarb	Pesticide	67%			[36]
2,4-Dichlorophenol	Pesticide	Complete degradation	50	10	[22]

NA: Not Available.

The optimal irradiation conditions for six commercial pesticides, i.e., diazinon, dimethoate, procloraz, metiocarb, imidacloprid, and carbofuran, using electron-beam facility, were determined for different wastewater compositions and at different operational conditions [36]. The initial pesticide concentrations (C_i) varied in the range $40 < C_i < 400$ ppm, the initial pH in the range ($4.5 < pH < 8$) and the irradiation doses were varied from 2–10 KGy. The application of 5 KGy was found to be the most efficient for the treatment of procloraz, dimethoate, imidacloprid, and carbofuran (99% degradation), but not efficient for the treatment of metiocarb. Low irradiation doses (<1 KGy) led to the formation of hydroxylated intermediates that build-up in the solution with the progress of the irradiation process. By increasing the irradiation period, a complete removal of the pollutants is achieved as a result of hydroxylation of the accumulated intermediates. The effect of the combined process of aeration and irradiation treatment was evaluated. It was found that the use of aeration can improve the metiocarb removal by 18%. In general, the hydroxylation of the aromatic ring to produce hydroxyl-cyclohexadienyl radicals is reversible [37]. Using aeration to inhibit the formation of these reversible radicals by peroxidation will lead to more efficient degradation [37]. The optimum irradiation conditions for the degradation of three herbicide and one fungicide, i.e., 2,4-dichlorophenoxyacetic acid (2,4-D), 3,6-dichloro-2-methoxy-benzoic acid (dicamba), 4-chloro-2-methylphenoxyacetic acid (MCPA), and carbendazim were determined. It was found that complete radiolytic degradation of dicamba ($C_i = 110$ ppm) is achieved using an irradiation dose of 5 KGy and the process is insensitive to the presence of other waste constituents, i.e., NO_3^-. 2,4-D-degradation was found to be affected by the presence of other waste constituents [35]. Another study addressed the use of ionizing radiation in the treatment of 2,4-D, and MCPA at lower concentration levels (<50 ppm), in the presence Cl^-, Br^- and NO_3^-. The results confirmed the role of oxidative radicals (OH^-) in the degradation of these pollutants and the sensitivity of the irradiation process to the presence of the anions, where the secondary intermediate formed at low doses (<1 KGy) are more toxic than the original pollutants [38,39].

The feasibility of using ionizing radiation in the treatment of agricultural effluents containing chlorinated organic pesticide, i.e., (4-chloro phenoxyacetic acid (4-CPA), 2.4-dichlorophenoxyacetic acid (2,4-D), 2.4-dichlorophenoxyacetic propionic acid (2,4-DP), and 2.4-dichlorophenoxyacetic butanoic acid (2,4-DB) was studied. The results indicated good efficiency of the irradiation process at 1 KGy in decomposing these pollutants. It was observed that chlorine was released as a result of their degradation. To reduce the disinfection byproducts in the treated wastewater prior to its use in chicken and fish livestock, an irradiation of the wastewater stream at 16.2 KGy dose was proposed, and 27 KGy was proposed for sludge treatment [40].

Swine wastewaters, which are alkaline agricultural wastewaters, that contain different pollutants, i.e., carbohydrates, proteins, lipids nitrates, nitrite, phosphate, and ammonia, were treated using a combined process of irradiation and ion-exchange biological treatment method. The use of electron-beam at 75 KGy achieved 85.1% removal efficiency of chemical oxygen demand (COD) at organic loading rate of 1.41 kg/m^3.day and achieved 75% removal efficiency of total nitrogen. The nitrogen removal was found to be sensitive to variation in current density [41].

4. Advance in Dyes Treatments

The presence of organic dyes in industrial effluents can lead to serious health and environmental problems. Ionizing radiation treatment was proposed to remove these contaminants. Research efforts were directed to study model compounds, single polluted solution, simulated wastewater, and real effluents. Table 5 summarizes the kinetic reactions of model compounds with their primary intermediates, their corresponding rate coefficients and their secondary intermediates [23]. Most of these compounds, except Azobenzene, decompose as a result of hydroxylation reaction, where the rate coefficients are slightly varied depending on the molecular structure of the contaminant. On the other hand, reactions with e^- and H have small contribution to the decomposition process [23]. G value for phenol degradation was found to be inversely proportional to the irradiation dose. Degradation reaction is first order and favors neutral pH. Phenol degradation is enhanced with the addition of oxidants, i.e., O_3, or $S_2O_8^{2-}$. $S_2O_8^{2-}$ was more efficient due to the selectivity of SO_4^{2-} radical to the formed by-products (carboxylic acids) [42].

Table 5. Degradationkinetics of model compounds [23].

Model Compound	Primary Intermediate	Rate Coefficient, $\times 10^{10}$ $mol^{-1} \cdot L \cdot s^{-1}$	Secondary Intermediate
Aniline	OH^-	0.86–1	-
	H^+	0.2	Anilino radical (directly and after water elimination.)
Phenol	OH^-	0.66–1.4	Cyclohexadienyl-type radical
	H^+	0.17	
Atrazine	OH^-	0.24	
Azobenzene	e^-	1–3.3	Hydrazyl radicals
	OH^-	2	
4-amino-5-hydroxynaphthalene-2,7-disulphonic acid	OH^-	-	Anilino-type

65% de-coloration of aqueous solution containing Alizarin Yellow GG (AY-GG, C_i = 100 ppm) was achieved using 9 KGy dose. Irradiation pre-biodegradation was proposed to improve the efficiency of treatment by 30% due to the formation of heterocyclic aromatic amines and cyanides. Increasing the irradiation dose can enhance the biodegradation due to the elimination of toxic secondary intermediates [43]. The aqueous solutions of Reactive Blue 15 (RB15) and Reactive Black 5 (RB5) dyes were irradiated with doses 0.1–15 KGy and at 2.87 and 0.14 KGy/h dose rates. Complete de-coloration was observed at 1 and 15 KGy doses for RB5 and RB15, respectively [25]. The de-coloration mechanisms for Apollofix Red (AR) and RB5 in aqueous solutions were investigated using γ-rays irradiation [44]. It was found that the de-coloration that was due to reactions with e^- and H, and increased linearly with increasing the irradiation dose, whereas de-coloration that was due to reaction with OH^- radical increase logarithmically. De-coloration was attributed to the destruction of the color bearing part of the dye as a consequence of OH^- addition to the aromatic ring [25]. Aqueous solutions containing AR and Apollofix Yellow (AY) were irradiated with doses of 1.0–8.0 KGy and at 0.14 KGy/h dose rate. The complete de-coloration was observed using 3.0 and 1.0 KGy doses for AR and AY, respectively [25].

The use of ionizing radiation to decompose wastewater effluents of varying pH ($1.6 < pH < 11.5$) and COD concentration ($650 < C_i < 2210$ ppm) from dye manufacturing factory was evaluated [45]. The use of 15 KGy decreased the optical density by 95% and removed COD by 72%. Use of aeration and H_2O_2 enhanced the degradation. The treatment of simulated wastewater contains four dyes, i.e., direct and reactive azo and anthraquinone dyes, using electron beam and γ radiation at different dyes concentration and irradiation doses in the presence and absence of H_2O_2 was studied [46]. The highest de-coloration performance (≈100%) was obtained via electron beam irradiation using 7 KGy for direct and reactive azo dye (C_i = 1000 ppm). A combined process of electron-beam and biological treatment was used in a pilot plant study in treating two waste streams. The first stream was generated during the operation of dying process and the second stream was from polyester fiber production enriched with ethylene glycol and terephthalic acid in 1998. An industrial facility was later commissioned and operated based on the same technology in 2005 [24,25]. The use of electron-beam technology to de-color and detoxify three effluents that represent chemical, final, and standard textile effluents was conducted. The effluents were de-colored (96, 55, and 90%) using 40, 2.5, and 2.5 KGy, respectively. The use of irradiation pre-biological treatment was found to reduce the toxic effect to the subsequent biological treatment process [47]. The potential use of combined treatment for alkaline and nearly neutral textile effluent of COD (632 < COD < 127 ppm), BOD (311 < BOD < 490 ppm), and turbidity (75–77%) at irradiation doses (<3 KGy) was investigated. It was found that the application of coagulation prior to irradiation enhanced the de-coloration performance [48]. Electron beam irradiation of simulated effluents that contains reactive dye (reactive yellow 15), size (starch), synthetic size (PVA), alkali, color, and pigment (pigment red 139) was studied [49]. Application of 1 KGy irradiation for post-biodegradation treatment enhanced the quality of the treated effluents.

5. Advances in Wastewater and Sludge Disinfection

Disinfection of treated wastewaters using ionizing radiation has been studied extensively. The application of 4 KGy reduced the BOD_5, COD, and total organic carbon (TOC) to acceptable limits [50]. In another study, anaerobic digested sludge irradiation at 1 KGy eliminated 98% of the total and fecal coliform, whereas BOD_5 was reduced to an acceptable limit at 4 KGy and COD was not affected by irradiation up to 20 KGy [44]. The disinfection of acidic industrial effluents and sludge ($2 < pH < 3$) of high concentration of BOD_5 and COD (7093, and 32,664 ppm, respectively) using γ irradiation was investigated [17]. Sludge and treated water disinfection could be achieved using 7 and 4 KGy, respectively, whereas the reduction of the BOD_5 and COD concentration to acceptable limits was obtained at 18 KGy. A comparative study was conducted to evaluate the use of different advanced oxidation methods in the disinfection of the municipal wastewater, re-growth control, and the associated operating costs [51]. The results revealed that UV efficiency was affected by the seasonal variation in the wastewater composition, whereas ionizing radiation efficiency was respectively unaffected by this factor. Ionizing radiation provided high stable disinfection efficiency (95%) for the total colony count and total coliform at radiation doses >0.25 KGy and inhibited the re-growth. From economical point of view, the electric power consumption for UV and ozone is three orders of magnitude higher than that required for ionizing radiation. Electron beam irradiation used to disinfect a municipal wastewater at irradiation doses (<3 KGy) and removed 90% of coliforms [51].

6. Advances in Pharmaceutical and Petrochemical Wastewater Treatments

The possibility of treating pharmaceutical effluents was addressed by studying the effect of ionizing radiation on the biodegradability and toxicity of individual drugs. Changes in biodegradability and toxicity induced in aqueous solutions containing sulfamethoxazole (SMX, C_i = 0.1 mmol/L) using ionizing radiation treatment revealed that SMX biodegradability was improved by applying 0.4 KGy dose. At 2.5 KGy dose, SMX conversion to biologically treatable substances was noted [52]. The degradation of carbamazepine (CBZ) by ionizing radiation was enhanced by the application of oxidant, 10 mM H_2O_2 [20]. The decomposition of mutagenic and carcinogenic secondary intermediates, i.e., acridine (ACIN), was enhanced in the presence of H_2O_2. The ionization of aqueous solutions containing ciprofloxacin (CIR) and norfloxacin (NOR) ($C_i = 10^{-4}$ M) was investigated. The degradation reaction proceeds via OH^- and e^- reactions with comparable rate constant ($\approx 10^9$ $mol^{-1} \cdot L \cdot s^{-1}$). At low irradiation doses, the antibacterial activities of the secondary intermediates vanished. Pollutants hydroxylation during γ irradiations proceeds on the hydroxylated molecules and desethylene derivatives and during pulse radiolysis is attributed to absorbance of hydroxyl-cyclohexadienyl radicals. In hydrated electron reactions, electron adduct is formed then it underwent protonation yielding cyclohexadienyl type radical [53]. The treatment of real pharmaceutical effluent using combined process of coagulation, biological treatment, and γ irradiation was investigated [54]. Two tested neutral effluents, namely, low organic strength (LSW; BOD < 6730 ppm, COD < 12,715 ppm) and high organic strength (HSW; BOD < 27,242, COD < 51,223 ppm). The use of irradiation led to maximum reduction in COD of 45% in acidic media at 50 KGy (LSW) and 30% in acidic media at 100 KGy (HSW). The application of coagulation pre-treatment was found to affect the efficiency, where 55% and 50% could be achieved using 100 KGy, for LSW and HSW, respectively. The use of H_2O_2 led to enhanced COD and TOC removal efficiency, when compared to $S_2O_8{}^{2-}$. The combined treatment led to overall 92.7% ± 2.3% and 90.2% ± 2.9% removal of COD from LSW and HSW, respectively. In a separate study, combined process of coagulation, electron beam treatment, and biological treatment was performed. An overall reduction in COD of 94% and 89% was achieved LSW and HSW, respectively [55]. The radiation doses were varied from 25–100 KGy at different pH that represent acidic, neutral and alkaline media. The slightly improved performance of the electron irradiation process was related to the reaction of $e^{-}{}_{aq}$ with H through parallel reaction with the organic contaminants to generate H^{+}, which subsequently inhabits the recombination of $e^{-}{}_{aq}$ and OH^{-}. The consideration of the total cost of electron beam irradiation facility (20 MeV, 100 kW), its shielding and maintenance, and capital costs was estimated and the cost of the treatment was estimated to be 0.6 USD/m^{-3}.

The decomposition of naphthalene (C_i = 5–32 ppm) in aqueous solution was studied using γ irradiation combined with both H_2O_2 and TiO_2nano-particles. The application of 3 KGy dose led to high naphthalene removal performance (>98%) and TOC reduction (28–31%) due to hydroxylation reaction. This performance is enhanced by 35% due to the presence of 40 ppm of H_2O_2 and 48% due to the presence of 0.8 g/L TiO_2 [56].

7. Sustainability of the Technology

In general, the sustainability of the nuclear industry is governed by its technical competitive performance, economical feasibility, and safe operational practice. As presented in the previous sections, the technical performance of the application of ionizing radiation technology in wastewater treatment is effective in disinfection and reduces the bio-refractory nature of several persistence organic pollutants. Several studies proved the economical feasibility of the e-beam and γ irradiation technologies [21,32,41,48,51,54–63]. For e-beam technology, the capital costs include the accelerator price, building shield, conveyer, cooling and ventilation systems, and monitoring system [57]. The capital cost of the e-beam accelerator is dependent on the power (P, kW) consumed to produce optimum dose (D, KGy) for specified plant flow rate (Q, m^3/h) taking into account the utilization factor (φ) [48,60]:

$$p = \frac{DQ}{\varphi} \tag{14}$$

The cost of the accelerator installation (K_i, k$) is determined using the applied electron energy (E, MeV), power, accelerator type and manufacturer (b, d), and installation coefficient (a) as follows [48,57]:

$$K_i = a \cdot b(1 \pm d)E\sqrt{p} \tag{15}$$

The values of the coefficients b and d varied with time due to the evolution and advances in the manufacturing process [48]. On the other hand, the costs of γ irradiator are not accompanied by power consumption [54], where γ source irradiate spontaneously. The average cost (ATC, k$/$m^3$) of this irradiation technique is determined based on the price of the used radioactive source (R, k$/$C_i$), required activity (I, C_i), irradiation time (t, h) and half life ($t_{0.5}$, h), and the irradiation chamber volume (v, m^3)

$$A\ (C_i) = R \times I \times t/(t_{0.5} \times v) \tag{16}$$

To facilitate the cost comparison between the use of e-beam technology and other treatment technologies the relative treatment costs estimated at different time is presented in Figure 3 [48,64]. In terms of cost, only disinfection process, which uses chlorine, is better than the use of e-beam technology. Similar data are not available for γ irradiation. To compare the cost of this technique alone and cost of this technology combined with coagulation, the data reported by Changotra et al. [54] were plotted for two types of industrial effluents (Figure 4).

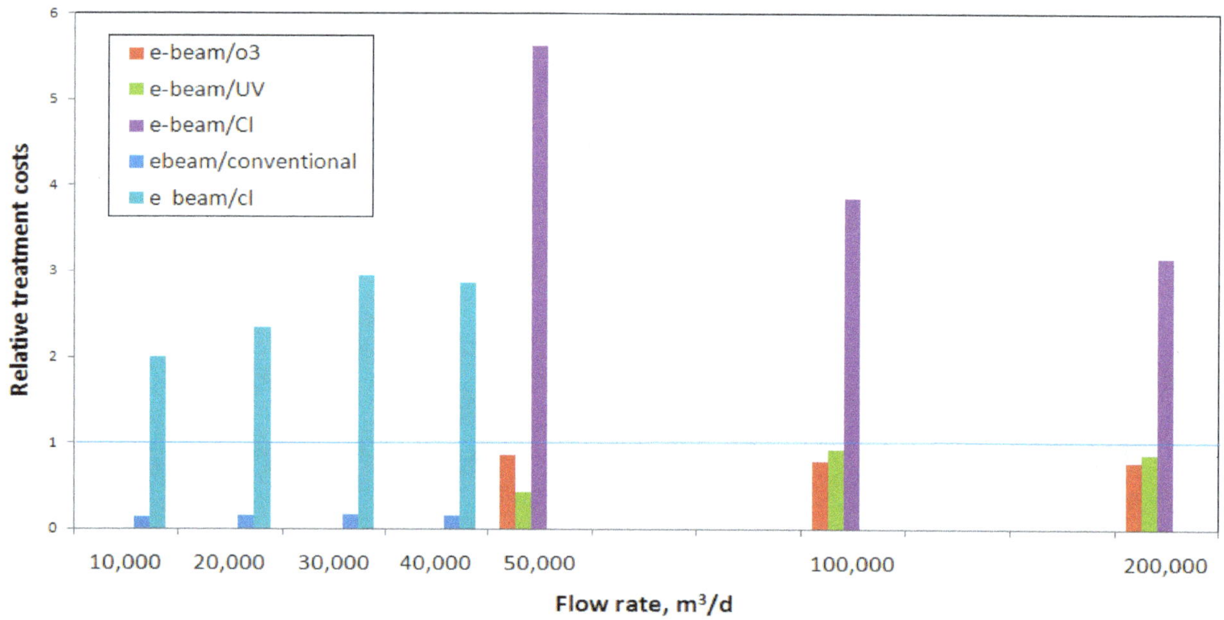

Figure 3. Relative treatment costs for wastewater using e-beam and other disinfection and conventional treatment methods [48,64].

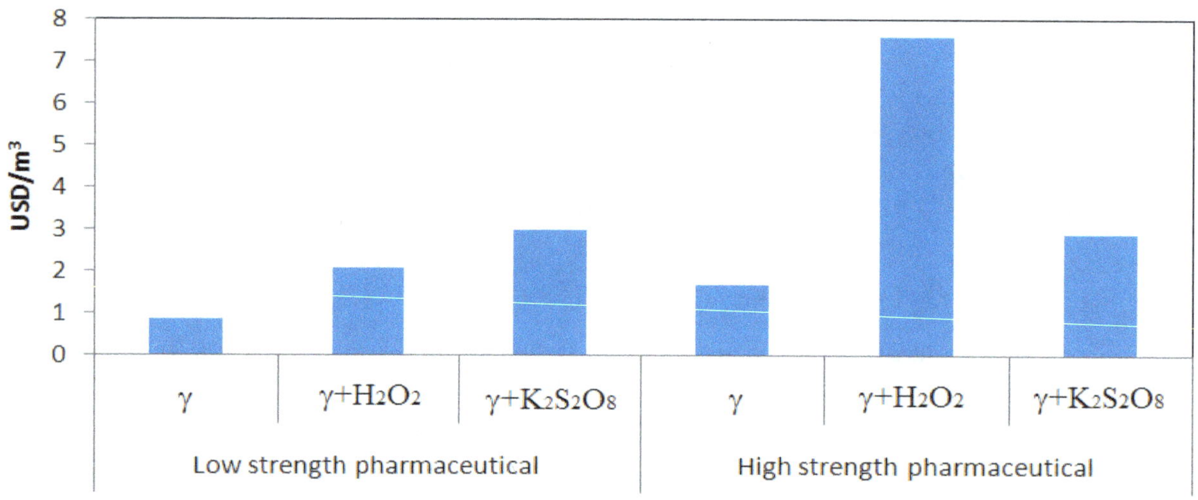

Figure 4. Treatment costs for γ irradiation and combine coagulation-γ irradiation [54].

There are some technical and non-technical issues that need to be resolved to ensure the sustainability of using this technology. These issues are listed below [22]

1. The usable output power of the electron accelerator limits its wide scale use as a safe substitute for chlorination for medium scale wastewater treatment plants.
2. Capital costs of electron accelerators needs to be reduced to ensure economical feasibility, where capital cost of installing an electron accelerator of 1MeV/400KW, could be attributed to half the accelerator price, whereas the design, construction, transportation, and installation of the facility contribute to 37.5% of capital cost.

From safety point of view, radiological accidents associated with different application of industrial irradiator, either γ irradiators or accelerators are limited, and could be classified as level 4 on the international nuclear and radiological event scale (INES). This level is used to describe accident with local consequence; Figure 5 summarizes the radiological exposures in these historical accidents [65]. Except the 1967 accelerator accident, one worker was exposed to radiation in each accident with a probability 0.092 and 0.12 a^{-1} for accelerators and γ irradiators, respectively. It should be noted that due to the more stringent regulatory safety requirements that were issued later, similar accidents were not reported over three decades ago. These regulatory requirements are not only related to safety aspects but also to the security aspects to ensure the application of 3S concept (Safety, Security, and Safeguards) [66,67]. To reduce these accidents, and based on the national regulation, the following measures should be considered [68–70]:

1. Appropriate safety measures, i.e., shielding requirements, operational procedures, provision of safety assessment documents, should be applied to ensure radiological containments.
2. Appropriate security measures should be applied to ensure safety of workers and public.

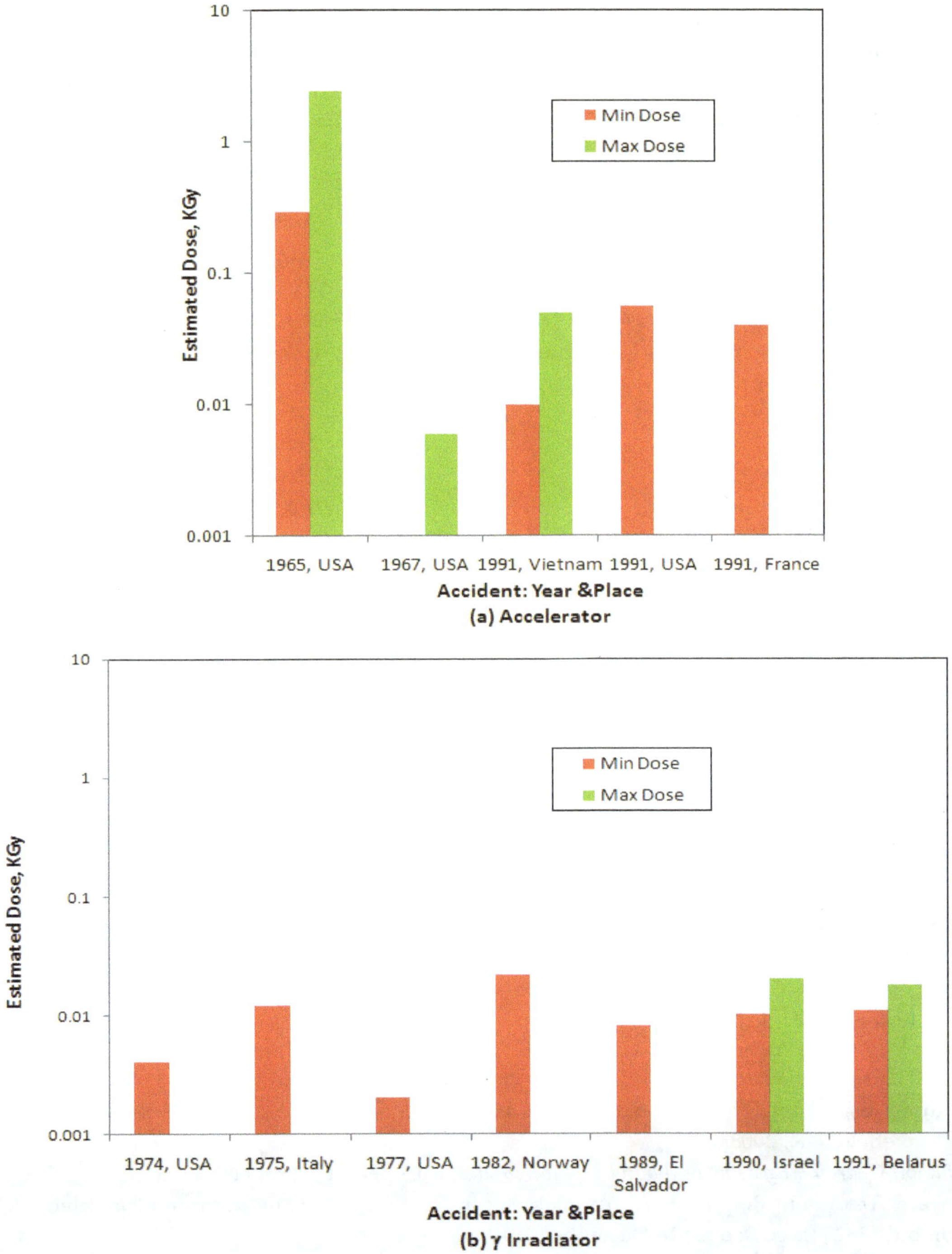

Figure 5. Summary of the historical industrial irradiator accidents; (**a**) accelerators, (**b**) γ irradiator.

Currently, the control strategy in these facilities is based on the combination of physical protection means, i.e., shields, barriers, and interlocks, and operational procedures [65,71]. This strategy ensured an accident probability for industrial irradiator in the order of 2×10^{-4} a^{-1}.

8. Conclusions

The review of ionizing radiation technology in decomposing bio-refractory organic contaminants and disinfecting different wastewater effluents were presented in this study. Factors that affect the sustainability of this technology were summarized. From this review the following conclusions could be drawn.

1. Most of the published work focused on the quantification of ionizing radiation effects on primary pollutants and the secondary intermediates toxicity, and determination of optimum irradiation conditions in different effluents. Research efforts are needed to study the feasibility of ionizing radiation treatment for petrochemical wastewater effluents,
2. Despite real industrial wastewater effluents were studied, there is a need to study the kinetics reactions in these complex systems to enable a better understanding and a better design for combined treatment schemes.
3. Compared to other disinfection technologies, the ionizing radiation technology provides economical, reliable, and safer operations that are not affected by the seasonal variation in the effluent composition, and reduces the generation of secondary toxic intermediates.
4. Operational safety of industrial irradiators has been improved due to the increasing stringent regulatory requirements and the updating the operational procedures, which lead to the reduction of accidents probability from 10^{-2} to 10^{-4} a^{-1}.

Author Contributions: Conceptualization, R.O.A.R. and Y.-T.H.; methodology, R.O.A.R.; resources, R.O.A.R. and Y.-T.H.; writing—original draft preparation, R.O.A.R.; writing—review and editing, R.O.A.R.; supervision, Y.-T.H.; funding acquisition, Y.-T.H. All authors have read and agreed to the published version of the manuscript.

References

1. Allaoui, M.; Schmitz, T.; Campbell, D.; Porte, C.A.D.L. *Good Practices, For Regulating Wastewater Treatment, Legislations, Policies and Standards*; United Nations Environment Programme: Nairobi, Kenya, 2015.
2. UNESCAP. *Policy Guidance Manual on Wastewater Management, with a Special Emphasis on Decentralized Wastewater Treatment Systems*; UNESCAP: Bangkok, Thailand, 2015.
3. FAO. *Wastewater Treatment and Use in Agricultural-FAO Irrigation and Drainage Paper 47*; FAO: Rome, Italy, 1992; Available online: http://www.fao.org/3/t0551e/t0551e00.htm#Contents (accessed on 20 November 2019).
4. Quach-Cu, J.; Herrera-Lynch, B.; Marciniak, C.; Adams, S.; Simmerman, A.; Reinke, R.A. The Effect of Primary, Secondary, and Tertiary Wastewater Treatment Processes on Antibiotic Resistance Gene (ARG) Concentrations in Solid and DissolvedWastewater Fractions. *Water* **2018**, *10*, 37. [CrossRef]
5. Deng, Y.; Zhao, R. Advanced Oxidation Processes (AOPs) in Wastewater Treatment. *Curr. Pollut. Rep.* **2015**, *1*, 167–176. [CrossRef]
6. Shon, H.; Vigneswaran, S.; Snyder, S. Effluent organic matter (EfOM) in wastewater: Constituents, effects, and treatment. *Crit. Rev. Environ. Sci. Technol.* **2006**, *36*, 327–374. [CrossRef]
7. Tchobanoglous, G.; Burton, F.L.; Stensel, H.D. *Waste Water Engineering: Treatment and Reuse*; Metcalf and Eddy: Wakefield, UK, 2003.
8. EU. *Effluent Charging Systems in the EU Member States*; European Parliament: Luxembourg, 2001.
9. Wastewater Treatment Requirement. Available online: https://www.lacsd.org/civicax/filebank/blobdload.aspx?blobid=3325 (accessed on 18 December 2019).

10. Munter, R. Industrial wastewater characteristics. In *Sustainable Water Management in the Baltic Sea Basin: Water Use and Management*; Lundin, L.C., Ed.; The Baltic University Programme, Uppsala University: Uppsala, Sweden, 2000; pp. 185–194.
11. Kumar, M.; Puri, A. A review of permissible limits of drinking water. *Indian J. Occup. Environ. Med.* **2012**, *16*, 40–44. [PubMed]
12. WHO. *A Compendium of Standards for Wastewater Reuse in the Eastern Mediterranean Region*; WHO-EM/CEH/142/E; WHO: Cairo, Egypt, 2006.
13. Environmental Management Act. Municipal Wastewater Regulation B.C. Reg. 87/2012, Consolidation Current to 20 June2018. Available online: http://archives.leg.bc.ca/civix/document/id/crbc/crbc/87_2012 (accessed on 20 August 2019).
14. Councel Directive of 21 May 1991 Concerning Urban Wastewater Treatment (91/271/EEC). Available online: https://eur-lex.europa.eu/LexUriServ/LexUriServ.do?uri=OJ:L:1991:135:0040:0052:EN:PDF%20 (accessed on 18 December 2019).
15. EU. *Minimum Quality Requirements for Water Reuse in Agricultural Irrigation and Aquifer Recharge*; Publications Office of the European Union: Luxembourg, 2017.
16. Luo, Y.; Guo, W.; Ngo, H.H.; Nghiem, L.D.; Hai, F.I.; Zhang, J.; Liang, S.; Wang, X.C. A review on the occurrence of micropollutants in the aquatic environment and their fate and removal during wastewater treatment. *Sci. Total Environ.* **2014**, *473*, 619–641. [CrossRef]
17. Naing, T.T.; Lay, K.K. Utilization of gamma radiation in industrial wastewater treatment. *Int. J. Mech. Prod. Eng.* **2015**, *3*, 1–5.
18. Miklos, D.B.; Remy, C.; Jekel, M.; Linden, K.G.; Drewes, J.E.; Hübner, U. Evaluation of advanced oxidation processes for water and wastewater treatment—A critical review. *Water Res.* **2018**, *139*, 118–131. [CrossRef]
19. Babu, D.S.; Srivastava, V.; Nidheesh, P.V.; Kumar, M.S. Detoxification of water and wastewater by advanced oxidation processes. *Sci. Total Environ.* **2019**, *696*, 133961. [CrossRef]
20. Liu, N.; Lei, Z.D.; Wanga, T.; Wang, J.J.; Zhang, X.D.; Xu, G.; Tang, L. Radiolysis of carbamazepine aqueous solution using electron beam irradiation combining with hydrogen peroxide: Efficiency and mechanism. *Chem. Eng. J.* **2016**, *295*, 484–493. [CrossRef]
21. Meng, M.; Pellizzari, F.; Boukari, S.O.B.; Leitner, N.K.V.; Teychene, B. Impact of e-beam irradiation of municipal secondary effluent on MF and RO membranes performances. *J. Membr. Sci.* **2014**, *471*, 1–8. [CrossRef]
22. IAEA. *Radiation Treatment of Polluted Water and Wastewater*; IAEA: Vienna, Austria, 2008.
23. Wojnarovits, L.; Takacs, E. Irradiation treatment of azo dye containing wastewater: An overview. *Radiat. Phys. Chem.* **2008**, *77*, 225–244. [CrossRef]
24. Sampa, M.H.O.; Takacs, E.; Gehringer, P.; Rela, P.R.; Ramirez, T.; Amro, H.; Trojanowicz, M.; Botelho, M.L.; Han, B.; Solpan, D.; et al. Remediation of polluted waters and wastewater by radiation processing. *Nukleonika* **2007**, *52*, 137–144.
25. Jan, S.; Kamili, A.N.; Parween, T.; Hami, R.; Parray, J.A.; Siddiqi, T.O.; Ahmad, M.P. Feasibility of radiation technology for wastewater Treatment. *Desalin. Water Treat.* **2015**, *55*, 2053–2068. [CrossRef]
26. Abdel Rahman, R.O.; Kozak, M.W.; Hung, Y.T. Radioactive pollution and control. In *Handbook of Environment and Waste Management*; Hung, Y.T., Wang, L.K., Shammas, N.K., Eds.; World Scientific Publishing Co: Singapore, 2014; pp. 949–1027. [CrossRef]
27. Gehringer, P. Advances in radiation processing of wastewater—Basics of the process. In *Status of Industrial Scale Radiation Treatment of Wastewater and Its Future*; IAEA-TECDOC-1407; IAEA: Vienna, Austria, 2003; pp. 7–17.
28. Shojaie, F. Comparison of the rate constants of a bimolecular reaction using two methods. *Arab. J. Chem.* **2017**, *10*, S148–S156. [CrossRef]
29. IAEA. *Status of Industrial Scale Radiation Treatment of Wastewater and Its Future*; IAEA-TECDOC-1407; IAEA: Vienna, Austria, 2003.
30. Wojnárovits, L.; Takacs, E. Wastewater treatment with ionizing radiation. *J. Radioanal. Nucl. Chem.* **2017**, *311*, 973–981. [CrossRef]
31. Chaychian, M.; Al-Sheikhly, M.; Silverman, J.; McLaughlin, W.L. The mechanisms of removal of heavy metals from water by ionizing radiation. *Radiat. Phys. Chem.* **1998**, *53*, 145–150. [CrossRef]

32. Chmielewski, A.G. Application of ionizing radiation to environment protection. *Nukleonika* **2005**, *50*, s17–s24. [CrossRef]
33. Dalalia, N.; Kazeraninejada, M.; Akhavan, A. Removal of heavy metal ions from wastewater samples using electron beam radiation in the presence of TiO_2. *Desalin. Water Treat.* **2017**, *71*, 136–144. [CrossRef]
34. El-Motaium, R.A. Application of nuclear techniques in Environmental studies and pollution Control. In Proceedings of the 2nd Environmental Physics Conference, Alexandria, Egypt, 18–22 February 2006; pp. 169–182.
35. Rela, P.R.; Sampa, M.H.O.; Duarte, C.L.; Costa, F.E.D.; Rela, C.S.; Borrely, S.I.; Mori, M.N.; Somessari, E.S.R.; Silveira, C.G. The status of radiation process to treat industrial effluents in Brazil. In *Radiation Treatment of Polluted Water and Waste Water*; IAEA TECDOC-1598; IAEA: Vienna, Austria, 2008; pp. 27–41.
36. Ramirez, T.; Armas, M.; Uzcategui, M. Effect of accelerated electron beam on pesticides removal of effluents from flower plantations. In *Radiation Treatment of Polluted Water and Waste Water*; IAEA TECDOC-1598; IAEA: Vienna, Austria, 2008; pp. 43–58.
37. Emmi, S.S.; De Paoli, G.; Takács, E.; Caminati, S.; Pálfi, T. The e-induced decomposition of pesticides in water: A gamma and pulse radiolysis investigation on carbofuran. In *Radiation Treatment of Polluted Water and Waste Water*; IAEA TECDOC-1598; IAEA: Vienna, Austria, 2008; pp. 77–88.
38. Şolpan, D. The degradation of some pesticides in aqueous solutions by gamma radiation. In *Radiation Treatment of Polluted Water and Waste Water*; IAEA TECDOC-1598; IAEA: Vienna, Austria, 2008; pp. 177–199.
39. Utrill, J.R.; Abdel Daiem, M.M.; Polo, M.S.; Pérez, R.O.; Peñalver, J.J.L.; Gala, I.V.; Mota, A.J. Removal of compounds used as plasticizers and herbicides from water by means of gamma irradiation. *Sci. Total Environ.* **2016**, *569*, 518–526. [CrossRef]
40. Sabbagh, S.; El Mahmoudi, A.S.; Al-Dakheel, Y.Y. Waste Water Sterilization by Cobalt Co-60 for the Agricultural Irrigation: A Case Study. *Int. J. Water Resour. Arid Environ.* **2014**, *3*, 11–18.
41. Lim, S.J.; Kim, T.H. Combined treatment of swine wastewater by electron beam irradiation and ion-exchange biological reactor system. *Sep. Purif. Technol.* **2015**, *146*, 42–49. [CrossRef]
42. Alkhuraiji, T.S.; Boukari, S.O.B.; Alfadhl, F.S. Gamma irradiation-induced complete degradation and mineralization of phenol in aqueous solution: Effects of reagent. *J. Hazard. Mater.* **2017**, *328*, 29–36. [CrossRef] [PubMed]
43. Sun, W.; Chen, L.; Tian, J.; Wang, J.; He, S. Degradation of a monoazo dye Alizarin Yellow GG in aqueous solutions by gamma irradiation: Decolorization and biodegradability enhancement. *Radiat. Phys. Chem.* **2013**, *83*, 86–89. [CrossRef]
44. Takacs, E.; Wojnarovits, L.; Palfi, T. Azo dye degradation by high-energy irradiation: Kinetics and mechanism of destruction. *Nukleonika* **2007**, *52*, 69–75.
45. Muneer, M.; Bhatti, I.A.; Bhatti, H.N.; Rehman, K.U. Treatment of Dyes Industrial Effluents by Ionizing Radiation. *Asian J. Chem.* **2011**, *23*, 2392–2394.
46. Abdou, L.A.W.; Hakeim, O.A.; Mahmoud, M.S.; El-Naggar, A.M. Comparative study between the efficiency of electron beam and gamma irradiation for treatment of dye solutions. *Chem. Eng. J.* **2011**, *168*, 752–758. [CrossRef]
47. Borrely, S.I.; Morais, A.V.; Rosa, J.M.; Pedroso, C.B.; Pereira, M.D.C.; Hig, M.C. Decoloration and detoxification of effluents by ionizing radiation. *Radiat. Phys. Chem.* **2016**, *124*, 98–202. [CrossRef]
48. Meibodi, M.E.; Parsaeian, M.R.; Amraei, R.; Banaei, M.; Anvari, F.; Tahami, S.M.R.; Vakhshoor, B.; Mehdizadeh, A.; FallahNejad, N.; Shirmardi, S.P.; et al. An experimental investigation of wastewater treatment using electron beam irradiation. *Radiat. Phys. Chem.* **2016**, *125*, 82–87. [CrossRef]
49. Deogaonkar, S.C.; Wakode, P.; Rawat, K.P. Electron beam irradiation post treatment for degradation of non biodegradable contaminants in textile wastewater. *Radiat. Phys. Chem.* **2019**, *165*, 108377. [CrossRef]
50. Basfar, A.A.; Abdel Rehim, F. Disinfection of wastewater from a Riyadh Wastewater Treatment Plant with ionizing radiation. *Radiat. Phys. Chem.* **2002**, *65*, 527–532. [CrossRef]
51. Lee, O.M.; Kim, H.Y.; Park, W.; Kim, T.H.; Yu, S. A comparative study of disinfection efficiency and regrowth control of microorganism in secondary wastewater effluent using UV, ozone, and ionizing irradiation process. *J. Hazard. Mater.* **2015**, *295*, 201–208. [CrossRef] [PubMed]
52. Sági, G.; Kovács, K.; Bezsenyi, A.; Csay, T.; Takács, E.; Wojnárovits, L. Enhancing the biological degradability of sulfamethoxazole by ionizing radiation treatment in aqueous solution. *Radiat. Phys. Chem.* **2016**, *124*, 179–183. [CrossRef]

53. Tegze, A.; Sági, G.; Kovács, K.; Tótha, T.; Takács, E.; Wojnárovits, L. Radiation induced degradation of ciprofloxacin and norfloxacin: Kinetics and product analysis. *Radiat. Phys. Chem.* **2019**, *158*, 68–75. [CrossRef]
54. Changotra, R.; Rajput, H.; Guin, J.P.; Varshney, L.; Dhir, A. Hybrid coagulation, gamma irradiation and biological treatment of real pharmaceutical wastewater. *Chem. Eng. J.* **2019**, *370*, 595–605. [CrossRef]
55. Changotra, R.; Rajput, H.; Guin, J.P.; Khader, S.A.; Dhir, A. Techno-Economical Evaluation of Coupling Ionizing Radiation and Biological Treatment Process for the Remediation of Real Pharmaceutical Wastewater. *J. Clean. Prod.* **2020**, *242*, 118544. [CrossRef]
56. Chu, L.; Yu, S.; Wang, J. Gamma radiolytic degradation of naphthalene in aqueous solution. *Radiat. Phys. Chem.* **2016**, *123*, 97–102. [CrossRef]
57. Economical Aspects of Radiation Sterilization with Electron Beam. Available online: https://www.osti.gov/etdeweb/servlets/purl/644045 (accessed on 18 December 2019).
58. Maruthi1, Y.A.; Das, N.L.; Hossain, K.; Rawat, K.P.; Sarma, K.S.S.; Sabharwal, S. Application of electron beam technology in improving sewage water quality: An advance technique. *Eur. J. Sustain. Dev.* **2013**, *2*, 1–18.
59. Hossain, K.Y.; Maruthi, A.; Das, N.L.; Rawat, K.P.; Sarma1, K.S.S. Irradiation of wastewater with electron beam is a key to sustainable smart/green cities: A review. *Appl. Water Sci.* **2018**, *8*, 6. [CrossRef]
60. Gala, I.V.; Peñalver, J.J.L.; Polo, M.S.; Utrilla, J.R. Degradation of X-ray contrast media diatrizoate in different water matrices by gamma irradiation. *Chem. Technol. Biotechnol.* **2013**, *88*, 1336–1343. [CrossRef]
61. Maruthi, A.Y.; Das, N.L.; Hossain, K.; Rawat, K.P.; Sarma, K.S.S.; Sabharwal, S. Disinfection and reduction of organic load of sewage water by electron beam radiation. *Appl. Water Sci.* **2011**, *1*, 49–56. [CrossRef]
62. Ndong, J.E.; Uribe, R.M.; Gregory, R.; Gangoda, M.; Nickelsen, M.G.; Loar, P. Effect of electron beam irradiation on bacterial and Ascaris ova loads and volatile organiccompounds in municipal sewage sludge. *Radiat. Phys. Chem.* **2015**, *112*, 6–12. [CrossRef]
63. IAEA. *Radiation Technology for Cleaner Products and Processes*; IAEA-TECDOC-1786; IAEA: Vienna, Austria, 2016.
64. Pereira, L.S.; Cordery, I.; Iacovides, I. *Coping with Water Scarcity*; IHP VI, Technical Documents in Hydrology 58; UNSCO: Paris, France, 2002.
65. Ortiz, P.; Oresegun, M.; Weatley, J. Lessons from Major Radiation Accidents. In Proceedings of the 10th International Congress of the International Radiation Protection Association—IRPA 10, Hiroshima, Japan, 14–19 May 2000.
66. IAEA. *Management for the Prevention of Accidents from Disused Sealed Radioactive Sources*; IAEA, TECDOC-1205; IAEA: Vienna, Austria, 2001.
67. Abdel Rahman, R.O. Introductory Chapter: Introduction to Current Trends in Nuclear Material Research and Technology. In *Nuclear Material Performance*; Abdel Rahman, R.O., Saleh, H.E.M., Eds.; Intech: Rijeka, Croatia, 2016; pp. 3–14. Available online: http://www.intechopen.com/books/nuclear-material-performance/introductory-chapter-introduction-to-current-trends-in-nuclear-material-research-and-technology (accessed on 18 December 2019).
68. USDOE. *Safety of Accelerator Facilities*; DOE O440.C; USDOE: Washington, DC, USA, 2011.
69. IAEA. *Directory of Gamma Processing Facilities in Member States*; IAEA-DGPF/CD; IAEA: Vienna, Austria, 2004.
70. IAEA. *Radiation Safety of Gamma, Electron and X Ray Irradiation Facilities*; SSG No. 8; IAEA: Vienna, Austria, 1996.
71. Janžekovič, H. Overview of Radiation Accidents at Industrial Accelerator Facilities. In Proceedings of the 26th International Conference on Nuclear Energy for New Europe, NENE, Bled, Slovenia, 11–14 September 2017; p. 7.

4

Adsorption of Chromium (VI) on Calcium Phosphate: Mechanisms and Stability Constants of Surface Complexes

Ahmed Elyahyaoui [1,*], Kawtar Ellouzi [1], Hamzeh Al Zabadi [2], Brahim Razzouki [3], Saidati Bouhlassa [1], Khalil Azzaoui [4], El Miloud Mejdoubi [4], Othman Hamed [5], Shehdeh Jodeh [5,*] and Abdellatif Lamhamdi [4,6]

1. Laboratory of Radiochemistry, Department of Chemistry, Faculty of Sciences, Mohamed V-Agdal, B.P 1014 Rabat, Morocco; ellouzi.kawtar@gmail.com (K.E.); bouhlass@fsr.ac.ma (S.B.)
2. Public Health Department, An-Najah National University, P.O. Box 7, Nablus 44830, Palestine; halzabadi@gmail.com
3. Laboratory of Spectroscopy, Molecular Modeling, Materials and Environment, Department of Chemistry, Faculty of Sciences, Mohamed V, B.P 1014 Rabat, Morocco; bramrwk@gmail.com
4. Laboratory LMSAC, Faculty of Sciences, Mohamed 1st University, P.O. Box 717, Oujda 60000, Morocco; k.azzaoui@yahoo.com (K.A.); ee.mejdoubi@gmail.com (E.M.M.); abdellatiflamhamdi@hotmail.com (A.L.)
5. Department of Chemistry, An-Najah National University, P.O. Box 7, Nablus 44830, Palestine; ohamed@najah.edu
6. National School of Applied Sciences Al Hoceima, Mohamed 1st University, P.O. Box 717, Oujda 60000, Morocco

* Correspondence: yahyaoui@fsr.ac.ma (A.E.); sjodeh@hotmail.com (S.J.).

Academic Editor: Faisal Ibney Hai

Abstract: The adsorption of chromate on octacalcium phosphate (OCP) was investigated as a function of contact time, surface coverage, and solution pH. The ion exchange method was adapted to establish the interaction mechanism. Stoichiometry exchange of H^+/OH^- was evaluated at a pH range of 3–10, and obtained values ranged between 0.0 and 1.0. The surface complexes formed between chromate and OCP were found to be $\overline{> S(HCrO_4)}$ and $\overline{> S(CrO_4)}$. The logarithmic stability constant $\log K_{1\text{-}1}$, and the $\log K_{10}$ values of the complexes, were 6.0 in acidic medium and 0.1 in alkaline medium, respectively. At low pH and low surface coverage, the bidentate species $\overline{> S(HCrO_4)_2}$ with $\log K_{10.5}$ of about 2.9, was favored at a hydration time of less than 150 min. The contribution of an electrostatic effect to the chromium uptake by the OCP sorbent, was also evaluated. The results indicate that the adsorption of chromate on OCP is of an electrostatic nature at a pH $\leq$ 5.6, and of a chemical nature at a pH > 5.6.

Keywords: octacalcium phosphate; chromium (VI); adsorption; environmental; ion exchange

---

1. Introduction

The hexavalent chromium Cr (VI) generated from various industrial processes, such as metallurgy, dyes, paints, inks, and plastics, is a major global concern, due to its harmful effects on humans and nature [1].

As a result, the presence of this metal cation in nature is well controlled. The maximal concentration level of Cr (VI) allowed in drinking water, as determined by the US-Environmental Protection Agency (EPA), is 0.05 mg/L [2]. Compliance with the EPA's chromium rule requires additional industrial monitoring. Designing a treatment process which reduces the number of chromium ions in industrial effluent to an acceptable level, has become crucial. A number of methods for this purpose have already been developed, among which are reduction and precipitation [3], ion exchange [4], solvent extraction [5], adsorption [6], and electrochemical precipitation [7]. When considering these methods, adsorption is the most promising technique [8]. This process is usually performed using conventional adsorbents, such as silica, zeolites, iron(III) (hydr)oxides, and activated carbon, or nonconventional adsorbents, such as red mud, sewage sludge, and bone char [9–11]. Phosphate materials (synthetics and minerals) have also been used as effective adsorbents for heavy metals from wastewaters and polluted soils. They have an excellent stabilization efficiency for several metal ions. For this reason, they are highly efficient metal adsorbents [12–14], particularly calcium phosphate. This material has a large specific area, high thermal and geochemical stability, low solubility, high ionic exchange capacity, and high stability towards ionization by radiation. For these reasons, it has been used as a backfill material for geological repositories for nuclear waste [15], and as an adsorbent in engineered barriers for environmental restoration [16]. It was also reported that phosphates with an amorphous structure are more efficient adsorbents of lead, uranium, and plutonium [17].

Although several adsorbents have been developed, and the retention of the metals hasbeen extensively studied, the sorption mechanisms are still rather difficult to identify. This could be because several phenomena, such as iso-morphous substitutions, surface complexation, and dissolution–precipitation, can occur simultaneously during the sorption process [18–21]. In addition, there is a lack of data on the adsorption of many metals on amorphous or poorly-crystallized phosphates.

Published studies show that the uptake of hexavalent chromium by calcium phosphate, exhibits typical anionic (such as $HCrO_4^-$ and CrO_4^{2-}) sorption behavior, and that the adsorption decreases by increasing the pH [22]. They also show that Cr (VI) adsorption is favored on phosphate that ispositively charged, at a low to neutral pH level (i.e., high point of zero charge (PZC)) [23]. The findings suggest that the retention of Cr (VI) by phosphates occurs through an electrostatic attraction and via binding to the surface functional groups OH_2^+ and OH^-. In other studies, it was found that hydroxapatite (HAp) and tricalcium phosphate (TCP) composite are able to adsorb a significant amount of chromium (VI), at a pH level of about 5. So, the composites showed a lower PZC than HAp (6.2–8.5) [22–24].

Moreover, there is no systematic understanding of the mechanism of chromium immobilization that involves the protonation/deprotonation of surface hydroxyl groups, and their interaction with the metal oxyanion. The combined effect of both pH and contact time on the adsorption mechanism requires more investigation (Figure 1).

Taking these considerations into account, the present study aims to investigate the complexation of hexavalent chromium with low-crystallized octacalcium phosphate (OCP). To achieve the aim of this study, the removal of Cr (VI) from aqueous solution was studied through batch experiments, as a function of contact time, the amount of adsorbent, and the equilibrium pH of chromate (10^{-4} M) solution. The ion exchange method, which has already been successfully implemented, especially in solvent extraction, has been chosen by the authors, in order to study the behavior of chromium (VI) on the surface of OCP. Another aim of this work is to investigate the surface complexation of OCP.

2. Experiment

2.1. Materials and Methods

Octacalcium phosphate (OCP) was synthesized in our laboratory. Phosphoric acid (99%), Chromium (VI), Calcium hydroxide (99%), Potassium hydroxide (KOH) (99%), and Nitric acid (HNO_3) (99%), were purchased from Sigma Aldrich, and were used in the same form as they were received. High-purity distilled water was used for all of the experiments.

2.2. Synthesis and Characterization of OCP

Calcium phosphate was synthesized using the microwave-hydrothermal method. In this method, phosphoric acid (0.3 M) and calcium hydroxide solution (0.5 M) are used as the starting materials. The preparation method has been described in [22]. The obtained mixture was heated at 150 °C for 1 h, and then irradiated in a microwave oven (800 w) for 5 min. The resulting gel was filtered off and dried overnight in an air oven, at 80 °C. The obtained solid was repeatedly washed with hot distilled water, and was identified as OCP by the associated XRD patterns and FT-IR. The characteristics of the diffraction line observed in the XRD patterns at $2\theta = 4.7°$, is evidence for the formation of OCP [22].

This result was confirmed by the FTIR characteristic peaks of OCP at 1089 and 1033, and by the Ca/P ratio of 1.34, which was close to the theoretical OCP ratio of 1.33 [25,26].

2.3. Surface Properties and Adsorption Experiments

The adsorption experiments were carried out by a batch method. A stock solution of Cr (VI) (10^{-4} M) was prepared from potassium dichromate. The pH level of the solution was measured using a Hanna combined electrode (Hanna pH 210). Nitric acid and KOH were used to adjust the starting acidity of the aqueous solutions of Cr (VI), with a concentration of 10^{-4} M in all cases. Sorption experiments were conducted at room temperature, as a function of the pH, contact time, and amount of adsorbent (m). The supernatants (5.0 mL) were filtered and analyzed for aqueous chromium using the 1,5-diphenylcarbazide (EPA 7196A) spectrophotometry method. The adsorbed chromium was calculated from the difference between the concentrations before and after equilibrium with calcium phosphate. The ratio of Cr (VI) concentrations in solid and aqueous phases led to the distribution coefficient D.

3. Results and Discussion

3.1. Effect of pH on Chromium Adsorption

The logarithmic variation of D with pH at different contact times, for 0.5, 1.0, and 1.5 g/L of calcium phosphate solutions, is plotted in Figure 1.

As shown in Figure 1, all cases exhibit similar behavior; log D increased by increasing the pH, to reach the maximum at pH_{max} of 4.0 to 5.0, before decreasing as the pH continued to increase. The maximum adsorption efficiencies were found to increase with the amount of adsorbent (m). The results also show that the pH_{max} is dependent on the contact time; when the contact time increased from 5 to 150 min, the pH_{max} rose from about 4 to 5.

The variation in the extraction efficiency with the solution's pH could be related to the protonation/deprotonating of both surface groups, and to the acidity of H_2CrO_4 ($pK_{a1} = 0.2$ and $pK_{a2} = 6.5$) [27–29]. According to the chromium speciation pH-diagram, the chromium (VI) was adsorbed as hydrogen chromate ($HCrO_4^-$) at pH ≤ 5.0 ($\geq$95%), as chromate (CrO_4^{2-}) at pH ≥ 7.6 ($\geq$95%), and as a mixture of these species between pH 5.0 and 7.6. In this case, both electrostatic and chemical sorption mechanisms could occur, and generally, it is not possible to distinguish between these two mechanisms [30]. It was assumed that the maximum adsorption occurred when the combination of a high positive surface charge and a high concentration of anionic chromium species, are achieved. Thus, the uptake of weak acid was maximal at a pH value around its dissociation constant, or near the PZC of surface sorbent materials [31]. In general, anion adsorption is strongly dependent on the pH of the medium, exhibiting the greatest removal in acidic to neutral solution. As has been demonstrated in previous studies, optimal Cr (VI) adsorption occurs at a pH lower than 4 for various sorbents, such as some metallic (oxy)hydroxides [32–34] and natural bio-sorbents, for example, larch bark [35], cooked tea dust [36], papaya seeds [37], raw Bagasse [38], and activated carbon [39]. In the case where iron and aluminum oxides are used as adsorbents, an adsorption efficiency of higher than 80% was reached at a pH of 4 to 6. The adsorption was seen to be highly dependent on the pH of the medium; the Cr (VI) uptake increased by increasing the pH values from 1.0 to 7.0, after which the uptake decreased.

Similar results were reported for the variation of hexavalent chromium adsorption with pH solution, with other adsorbents such as clay minerals [40], oxide-coated sand [40], aluminosilicate [41], zeolite [42], chitosan [43], and hydroxyapatite [44], which all exhibited a maximum uptake at a pH of 5 to 7. It has been suggested that, at a low pH, the adsorption is also low, due to the competition between the metal ions and protons for the adsorption sites. In these cases, the uptake of Cr (VI) followed the ion exchange mechanism. From these results, it could be concluded that, at a pH higher than pH_{max}, Cr (VI) exhibited typical anionic sorption behavior, with adsorption decreasing when the pH was increased. Similar results were reported for the retention of similar anions on oxide surfaces [45,46]. This adsorption pattern is the result of the protonation of surface hydroxyl sites and of Cr (VI) hydrolysis [47]. Thus, at a pH higher than pH_{max}, the retention of Cr (VI) was due to the interaction of $HCrO_4^-$ and/or CrO_4^{2-} with OH surface groups, rather than with OH_2^+ groups, which were predominant at a pH lower than pH_{max} [44]. The significant influence on the adsorption of CrO_4^{2-} was found at the slope of the pH adsorption edges. Distinct pH regions with different slopes characterized the various adsorbed species. This result reflected the change in the sorption mechanism [48].

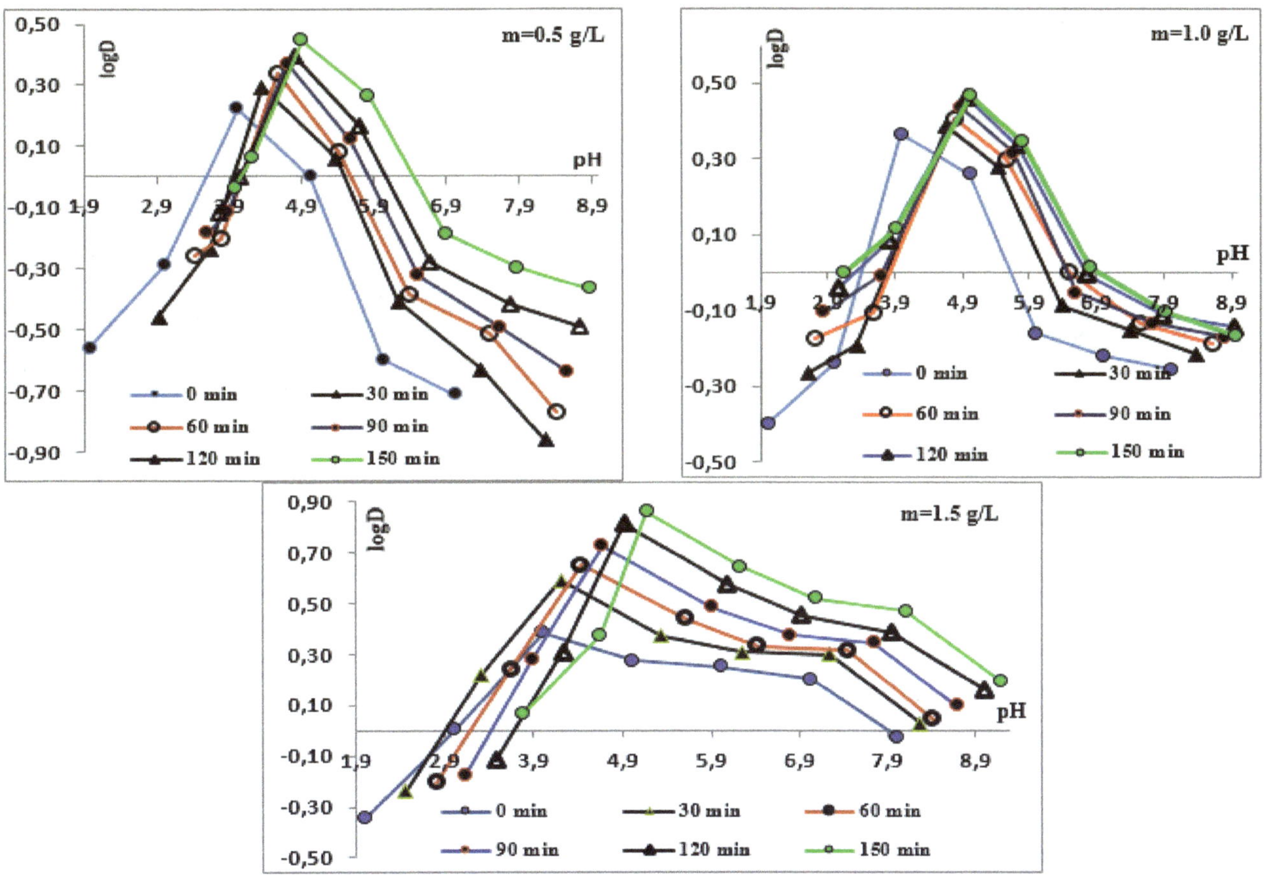

Figure 1. Log(D) versus pH for different amounts of OCP and different contact times.

3.2. Effect of Contact Time

The effect of contact time on the adsorption efficiency was examined, and the results are shown in Figure 2. At a pH lower than pH_{max}, a distinct difference was observed for adsorption envelopes in relation to the sorbent dose, or sorbet/sorbent ratio (surface coverage). At a sorbent dose of 0.5 and 1.0 g/L, the pH adsorption edges followed a similar trend for $t \geq 15$ min. At a pH of 50% of adsorption (pH_{50}), a negligible variation occurred at pH 3.9–4.1. When the sorbent dose was1.5 g/L, the adsorption process was dependent on the hydration time. Therefore, the adsorption envelopes shifted to higher pH values, resulting in an increase in pH_{50} from 2.9 to 3.9, as the time increased from 0 to 150 min. The mechanism for the oxyanion adsorption was dependent on the surface coverage and hydration of the surface sites, involving the formation of distinct surface complexes. This was in agreement with the spectroscopic study results, which indicated that chromium (VI) resulted in the formation of both monodentate and bidentate surface complexes on iron oxides [49,50]. The proportion of these species was dependent on the metallic ion concentration. Recently, through the use of a spectroscopic study, it has been shown that the monodentate chromate complexes on ferrihydrite were predominant at a low surface coverage and a pH $\geq$ 6.5. In contrast, bidentate surface complexes were formed at a high surface coverage and pH $\leq$ 6 [51].

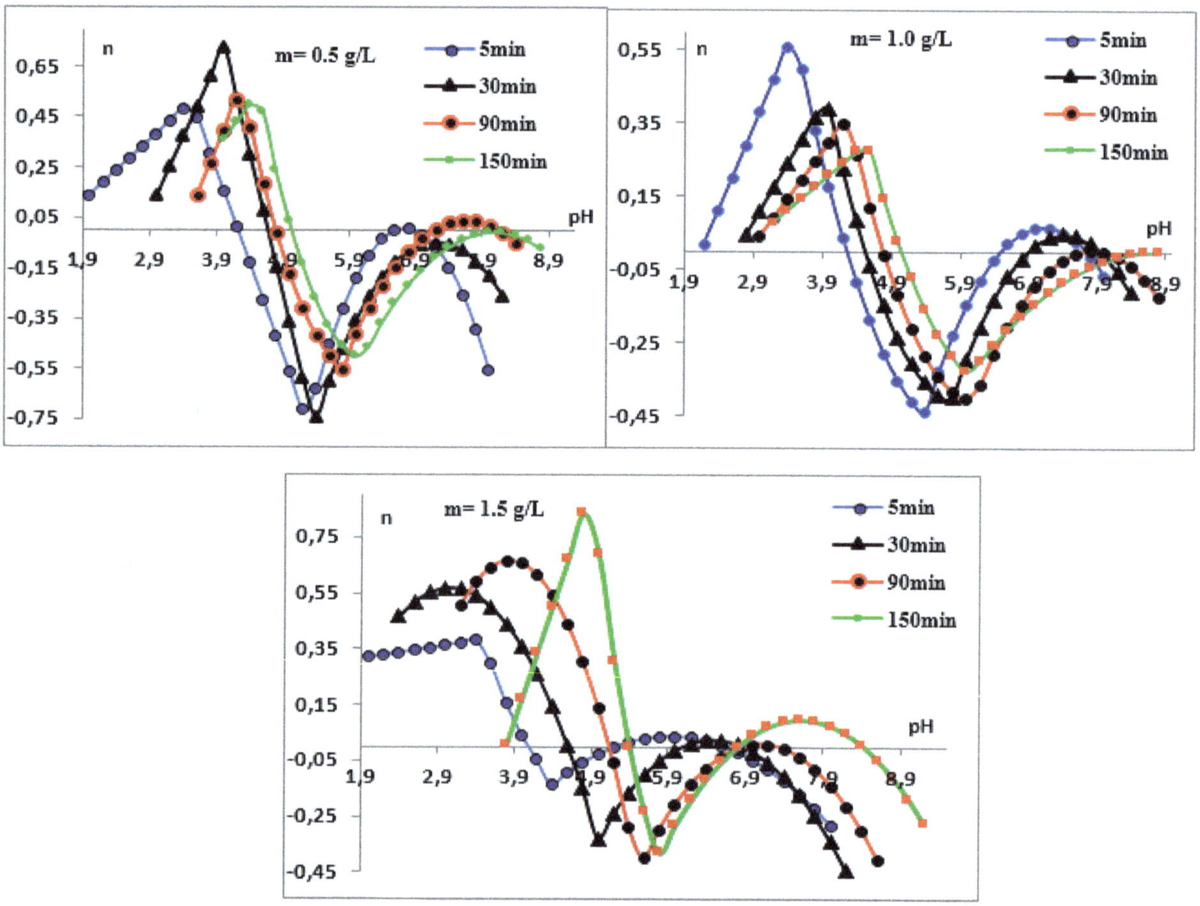

Figure 2. Variation of n = f(pH) obtained at various contact time and at different OCP sorbent amounts of m = 0.5, 1.0 and 1.5 g/L: 5 min (•); 30 min (▲); 90 min (•); and 150 min (■).

Taking into account the previous results, it could be deduced that, at a low adsorbent dose, monodentate surface complexes prevailed for all of the examined experimental conditions. In contrast, at a higher adsorbent dose, the bidentate surface complexes became predominant at a low hydration level of sorbent materials at pH $\leq pH_{max}$, and were subsequently converted to monodentate species, when hydration equilibrium was reached.

At a pH higher than pH_{max}, adsorption envelopes become more alkaline. The shift to this alkalinity increased by increasing the hydration time. Due to electrostatic repulsion with the negative surface charges, the chromium uptake initially involved the adsorption of a $HCrO_4^-$ anion, which was followed by the subsequent slower, and less important, uptake of CrO_4^{2-} in the alkaline region. These electrostatic factors could influence both the kinetics and equilibrium of chromate ions, as observed for the adsorption of a similar oxyanion on ferrihydrite [52]. When the contact time increased, the adsorption of the chromate anion also increased, and the Log(D) = f(pH) curves show a pronounced difference in the slopes at t = 150 min. As previously discussed, this phenomenon could be due to a change in the adsorbed fraction of the Cr (VI)-predominant surface complexes.

3.3. Chromium (VI) Adsorption Reactions and Stability Constants

The acid-based properties of calcium phosphate were described by the protonation and deprotonation reactions of the phosphate surface functional groups, as shown in Equations (1) and (2), $\overline{> SOH}$ [53,54]:

$$\overline{> SOH} + H^+ \leftrightarrow \overline{> S(OH_2)^+} \quad K^+ \tag{1}$$

$$\overline{> SOH} \quad \leftrightarrow \overline{> SO^-} + H^+ \quad K^- \tag{2}$$

K^+ and K^- were the surface stability constants, and the on lined species referred to the solid phase.

The adsorption reaction of chromium (VI) on calcium phosphate can be expressed as follows:

$$(\overline{> SOH)_l(H_iCrO_4)H_{-n}} + nH^+; \quad l = 1;2, \quad i = 1;2 \tag{3}$$

where n is the number of protons, which ranges from −1 to +1.

The initial ionization of H_2CrO_4 is relatively strong, so $HCrO^-{}_4$ was the main species found at pH > 5, and i = 1 at equilibrium (Equation (3)), under these conditions.

The symbol $H_{-n,}$ stands for both hydrogen atoms (n < 0) and for OH groups (n > 0). Taking into account that $H_2O \equiv H_{-1} + H_l$, the surface complexes $\overline{> (SOH)_l(H_iCrO_4)H_{-n}}$ noted thereafter $C_{ln,}$ represented a general formulation of species, differentiated by water composition. So $\overline{> (SOH)_l(H_2CrO_4)H_{-n}}$ could be $\overline{> (SOH)_l(HCrO_4)H_{-n+1}}$ + H_2O or $\overline{> (SOH)_l(CrO_4)H_{2-n}}$ + $2H_2O$, and even $\overline{> (S)_l(H_2CrO_4)H_{-n-l}}$ or $\overline{> (S)_l(HCrO_4)H_{-n-l+1}}$ + H_2O or $\overline{> (S)_l(H_2CrO_4)H_{-n-l+2}}$ + $2H_2O$.

The surface complexation constant for the relationship (3) is:

$$K_{ln} = \frac{\left[(\overline{> SOH)_l(H_iCrO_4)H_{-n}}\right]\left[H^+\right]^n}{\left[\overline{> SOH}\right]^l\left[H_iCrO_4{}^{(i-2)+}\right]} \tag{4}$$

The distribution coefficient being:

$$D = \frac{\left[(\overline{> SOH)_l(H_iCrO_4)H_{-n}}\right]}{\left[H_iCrO_4{}^{(i-2)+}\right]} \tag{5}$$

where $[H_iCrO_4{}^{(i-2)+}]$ represents the equilibrium concentration of Cr (VI) in solution, it was obtained that:

$$\log D = \log K_{ln} + \log m + npH \tag{6}$$

where $\overline{> SOH}$ = m, was the concentration of sorbent used in g/L.

Assuming that the first approximation is the essential formation of mononuclear (l = 1), the surface complexes of Equation (6) become:

$$\log D = \log K_{1n} + \log m + npH \tag{7}$$

The variation of the distribution coefficient with pH allowed us to define the nature of the C_{ln}-adsorbed species. The values of (l,n) were obtained according to:

$$\left(\frac{\delta \log D}{\delta pH}\right)_m = n \tag{8}$$

So, the surface complexes of Cr (VI) with calcium phosphate could be well described from the experimental data logD = f(pH), shown in Figure 1. The analysis of the obtained results showed that the plots of logD = f(pH) were linear at various pH ranges. The straight lines of the slope correspond to the mean values of n, and varied between −1 and 1. The value of the surface complexes (1,n), involved in this case were (l,0), (l,1), (l,2), and (l,−1). In this case, the co-precipitation/adsorption process of chromium (VI), and the species distribution diagram as a function of pH, are needed. It is worth noting that the interaction of Cr (VI) with the OCP surface could be described as an n = f(pH) variation. For this purpose, fitting of the data into a polynomial equation was carried out for the three pH regions. The obtained results show that, when in acidic solution (pH < 5), a second-degree equation fitted (R^2 > 99%) the ascendant curve, whereas in low acidic to alkaline media, a cubic polynomial fitted (R^2 > 99%) two distinct segments of the descendant curve, at pH regions of about 4 to 6 and 6 to 10.

3.3.1. Surface Complexes and Effect of pH and Contact Time on H_3O^+/OH^- Exchange

The variations of n = f(pH) are illustrated in Figure 2. As shown in Figure 2, the protonation/deprotonating reaction followed a similar trend, with respect to the pH value. In all cases, the maximum H_3O^+ and OH^- exchange of $|n|$ = 0.8 occurred at a pH range of 4.0–6.0. When considering the obtained results, it becomes evident that the H_3O^+/OH^- stoichiometry was not an integer, as might be expected from the theoretical single reaction. Similar results were observed for other adsorbed elements on iron (oxy) hydroxides [55,56].

It was assumed that the chromium adsorption occurred by different reactions, and consequently, resulted in a combination of at least two predominant surface complexes. In the case l = 1, the predominant complexes would be different for the sorbent amounts of 0.5 and 1.0 g/L, while a different adsorption behavior was observed for 1.50 g/L. Indeed, a low value of around 0.5 was found at low pH and at *t* < 150 min, indicating that C_{10} is not the predominant species under these conditions.

Taking into account these considerations, the uptake of Cr (VI), characterized by n = 0.5, and the general adsorption reaction, can be expressed by:

$$\overline{>\mathrm{SOH}} + 2\mathrm{H_2CrO_4} \leftrightarrow \overline{>\mathrm{S(OH_2)^+(HCrO_4)_2H^{-n}}} + 1\mathrm{H^+};\quad 1 = 1 \quad n = 2 \tag{9}$$

$$\overline{>\mathrm{SOH}} + 2\mathrm{H_2CrO_4} \leftrightarrow \overline{>\mathrm{SH(HCrO_4)_2H^{-n}}} + \mathrm{H_2O} + \mathrm{nH^+};\quad 1 = 1 \quad n = 2 \tag{10}$$

The adsorption constant $K_{10.5}$ for equilibrium (10) was given by:

$$K'_{1n} = \frac{\left[\overline{>\mathrm{SH(HCrO_4)_2H^{-n}}}\right]\left[\mathrm{H^+}\right]^{\mathrm{n}}}{\left[\overline{>\mathrm{SOH}}\right]\left[\mathrm{H_2CrO_4}\right]^2} = \frac{\mathrm{D}\ \left[\mathrm{H^+}\right]^{\mathrm{n}}}{\left[\overline{>\mathrm{SOH}}\right]\left[\mathrm{H_2CrO_4}\right]} \tag{11}$$

Based on this adsorption mechanism, the following relationship could be obtained:

$$\log(\mathrm{D(D+1)}) = \log K'_{1n} + \log \mathrm{m} + \log[\mathrm{Cr}]_0 + \mathrm{n\,pH} \tag{12}$$

For $[Cr]_0 = 10^{-4}$ M, which was the chromium analytical concentration used, the equation becomes:

$$\log(\mathrm{D(D+1)}) = \log K'_{1n} + \log \mathrm{m} - 4 + \mathrm{n\,pH} \tag{13}$$

At a pH range of 2.5 to 5.0, the relation log(D(D + 1)) = f(pH) exhibited a linear variation, with slopes increasing from 0.5 to 1.0, as the time increased from 0 to 150 min.

Accordingly, C_{11}, $C_{10.5}$ and C_{10} were the predominant surface species in the case of l = 1. A non-protonated C_{11} ($\overline{>\mathrm{SOH(H_2CrO_4)H_{-1}}} \equiv \overline{>\mathrm{S(HCrO_4)H_{-1}}} \equiv \overline{>\mathrm{SCrO_4}}$) complex was always formed at low acidity (pH ~4.4), combined with the protonated C_{10} ($\overline{>\mathrm{SOH(H_2CrO_4)H_0}} \equiv \overline{>\mathrm{S(HCrO_4)H_0}} \equiv \overline{>\mathrm{SHCrO_4}}$), complex during hydration equilibrium conditions. Nevertheless, when this equilibrium was not reached at the 1.5 g/L sorbent amount, the C_{10} complex was not favored, and disappeared, to the benefit of the bidentate: $C_{10.5}$ ($\overline{> SOH(H_2CrO_4)_2H_{-1}} \equiv \overline{> SH(HCrO_4)_2H_{-1}} \equiv > S(HCrO_4)_2$) species. As a result, in this condition, $C_{10.5}$ and C_{11} were the prevailing complexes. The $C_{10.5}$ surface complex can also be formulated as the $\overline{> S(H_2Cr_2O_7)}$ species. However, the bichromate surface complexes were not found at 0.5 and 1.0 g/L sorbent doses and were neglected, as previously reported [57].

The results show that, as the sorbent amount increased, the number of surface sites also increased. At high-hydration equilibrium, the hydrogen chromate anions ($HCrO_4^-$) and water molecules were competing for the active surface sites. At a low-hydration level of the sorbent material's surface site, the $HCrO_4^-$ ions displaced the H_2O molecules, in the hydration sphere of the C_{11} complex. Bidentate surface species were then formed at a high sorbent dose and short contact time. These results were in accordance with previous studies, suggesting that high-chemisorbed water molecules prevented the bidentate complexes from forming, as noted for the complexation of chromium with ferrihydrite [48]. A similar substitution of water molecules by $HCrO_4^-$ anions could take place at a higher chromium concentration, even at hydration equilibrium, explaining the formation of bidentate surface species when there is a high surface coverage, observed for the adsorption of Cr (VI) or a similar anion on ferrihydrite [29].

It should be noted that the $C_{10} \equiv \overline{> S(HCrO_4)H_0}$ complex could also be expressed as an outer-sphere $\overline{> SOH_2^+ - HCrO_4^-}$ species, since we could not distinguish between the different complexes' formulations or structures, based on one H_2O molecule. Generally, when at a low pH value, the formation of such outer-sphere Cr (VI) surface complexes is favored, similar to that previously obtained on amorphous aluminum oxides [58], and supported by the formation of SOH_2^+ in acidic media, as was indicated in a titration experiment [59].

At a pH higher than 4.5, n varied between -1 and 0, in all investigated conditions. The predominant surface species formed in these conditions, were C_{10} and $C_{1\text{-}1}$. Thus, Cr (VI) was adsorbed via the formation of $\overline{> S(OH)(HCrO_4)H_0}$ (C_{10}) and $\overline{> S(OH)(HCrO_4)H_1}$ ($C_{1\text{-}1}$) complexes, that could be expressed as $\overline{> SCrO_4}$ and $\overline{> SHCrO_4}$, respectively. As the contact time increased, the optimal pH formation of protonated $C_{1\text{-}1} \equiv \overline{> SHCrO_4} \equiv > S(OH_2^+)(HCrO_4^-)$ and the un-protonated $C_{10} \equiv \overline{> SCrO_4} \equiv \overline{> S(OH)(HCrO_4)}$ species, shifted from 5 to 6 and 7 to 8, respectively. The adsorption of Cr (VI) increased with t, and at a higher hydration time, the adsorption reaction was likely to comply to OH^- surface exchange. Thus, for a pH lower than PZC, the negative surface charge of the phosphate material reacts in acidic media to form $>S(OH_2)^+$, which adsorbs chromium as $HCrO_4^-$. Maximal OH^- exchange was observed at a pH range of 5.2–6.0, which approximately coincided with the pH range of zero charge and the iso-electric points of unloaded material. Whilst higher than PZC, the repulsion between chromium anions and negatively sorbent surface charge, increased with pH. The contribution of the columbic effect to the overall uptake of Cr (VI) could be a more important process. Nevertheless, at a pH value around the iso-electric point, the electrostatic repulsion reached a minimum value, and then the intrinsic process (with n = 0) became the major adsorption mechanism. As shown from the obtained results, the pH value became higher with contact time, reaching 8.2 at $t \approx 150$ min.

Consequently, the overall adsorption equilibrium could be obtained by the intrinsic reaction with 1 mole of H^+ (n = -1), or 1 mole of OH^- (n = 1) exchange reactions per mole of $HCrO_4^-$.

3.3.2. Equilibrium Constants

In the Log(D)= f(pH),Figure 1 plots the various contact times and sorbent amounts, and the straight lines show slopes ranging from -0.4 to 0.6. In the cases when the surface complexes are mononuclear (l = 1), apparent equilibrium constants, $K_{ap} = K_{1n}$ or K'_{1n}, were obtained from the origin ordinates A = LogK_{1n} + Log(m) or logK'_{1n} + Log(m) $-$ 4. From the obtained results, it can be that the variations of logK_{1n} = f(n) (Figure 3) and logK'_{1n} = f(n) (not shown), were linear under all experimental conditions.

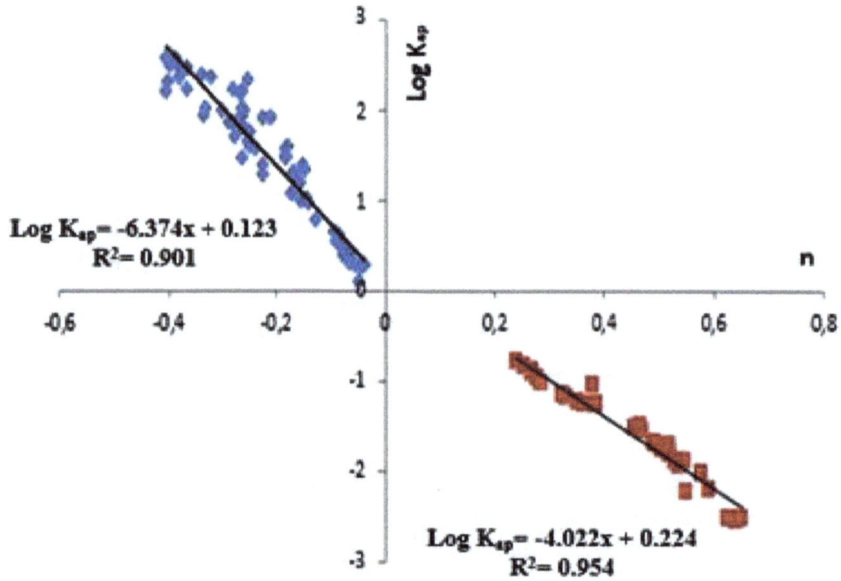

Figure 3. Variations of log Kap = f(n).

Taking into account that |n| also represented the fraction (x) of the predominant $1H^+$ or $1OH^-$ exchange reactions contributing to the overall adsorption equilibrium, the apparent constants (K_{ap}) of the overall equilibrium were given by the expressions shown in Equations (14) and (15):

$$K_{ap} = (K_{10})^{(1-|n|)} (K_{1\pm1})^{|n|} \quad (14)$$

$$\log K_{ap} = (1 - |n|)\log K_{10} + |n|\log K_{1\pm1} \quad (15)$$

with $K_{1\pm1} = K_{11}$ or K_{1-1}.

As an example, the overall partition equilibrium prevailing for pH > 5.5, and involving successive exchange reactions $n = 0$ and -1, can be summarized in the following reactions:

$$(1-x)(\overline{>SOH} + HCrO_4{}^- \leftrightarrow \overline{>S(OH)(HCrO_4)H_0} \equiv \overline{>S(CrO_4)} + H_2O): \quad (K_{10})^{(1-x)},\ n = 0 \quad (16)$$

Overall adsorption reactions are:

$$(x)(\overline{>SOH} + HCrO_4{}^- + H_2O \leftrightarrow (\overline{>SOH(HCrO_4)H_1} \equiv \overline{>S(HCrO_4)} + H_2O) + OH^-: \quad (K_{1-1})^{x},\ n = -1 \quad (17)$$

$$\overline{>SOH} + HCrO_4{}^- + nH_2O \leftrightarrow (\overline{>SOH(HCrO_4)H_n} \equiv \overline{>S(HCrO_4)} + H_2O) + nOH^-: \quad (K_{1n}),\ -1 \le n \le 0 \quad (18)$$

Where x is the molar fraction of the sorption mechanism with 1 OH^- exchange. The apparent constant, $K_{ap} = K_{1n}$, is then given by:

$$\log K_{1n} = (1 - x)\log K_{10} + x\log K_{11} = (1 - |n|)\log K_{10} + |n| \log K_{10} \quad (19)$$

As shown above, the K_{ap} value could be determined experimentally, and was variable, depending on the surface charge and pH. Therefore, the constants K_{10} and $K_{1\pm1}$ were the intersect and the slope of $\log K_{ap} = f(n)$, respectively, whereas $K_{1\text{-}1}$ was the opposite of the slope. From this information, the intrinsic constants K_{10}, for H^+ (formation of $\overline{>S(OH_2^+)(HCrO_4^-)H_0} \equiv \overline{>S(HCrO_4)}$) and OH^- (formation of $\overline{>S(OH)(HCrO_4)H_0} \equiv \overline{>S(CrO_4)}$) adsorption reaction exchange, were determined, with corresponding logarithmic values of $\log K_{10} = 0.2$ and 0.1, respectively. The intrinsic constants ($K_{int} = K_{10}$) were obtained in general, by extrapolating the apparent constants to a zero surface charge [57,58].

As discussed previously, the variations of $\log K_{ap} = f(n)$ could be due to the contribution of the electrostatic effect (K_{col}), related to K_{ap} [60,61] according to:

$$K_{int} = K_{ap}K_{col} = K_{ap}\exp(-nF\Psi_0/RT) \quad (20)$$

$$K_{ap} = K_{int}\left(K_{col}\right)^{-1} = \left(K_{10}\right)^{(1-|n|)}\left(K_{1\pm1}\right)^{|n|} \quad (21)$$

In the particular case of $|n| = 1$, Equation (22) was obtained:

$$K_{ap} = K_{int}\left(K_{col}\right)^{-1} = K_{1\pm1} \quad (22)$$

where ψ_0 was the surface potential, and R, T, and F were the gas constant, absolute temperature, and Faraday constant, respectively.

Since $K_{int} = K_{10}$, the columbic effect was determined according to

$$\log K_{col} = \log K_{10} - \log K_{1\pm1} \quad (23)$$

While $\log K'_{ap}$ for equilibrium (10) was:

$$\log K_{col} = \log K'_{10} - \log K_{10.5} \quad (24)$$

The origin ordinate of the $\log K'_{ln} = f(n)$ plots resulted in $\log K'_{10} = 5.3$. These results are summarized in Table 1.

Table 1. Surface complexation constants for Cr (VI) sorption onto Octacalcium phosphate.

Species	n	Adsorption Reaction	$\log K_{1n}$	$\log K_{col}$
		Acidic medium (pH < 5)		
$\overline{>SOH_2^+} - \overline{HCrO_4^-}$	0	$(\overline{>SOH}) + H_2CrO_4 \leftrightarrow \overline{>SOH_2^+} - \overline{HCrO_4^-}$	0.1	
$\overline{>SHCrO_4}$	0	$(\overline{>SOH}) + H_2CrO_4 \leftrightarrow \overline{>SHCrO_4} + H_2O$	0.1	
$\overline{>SCrO_4}$	+1	$(\overline{>SOH}) + H_2CrO_4 \leftrightarrow \overline{>SCrO_4} + H_2O + 1H^+$	−4.0	4.1
$\overline{>S(HCrO_4)_2}$	0.5	$(\overline{>SOH}) + 2H_2CrO_4 \leftrightarrow \overline{>SH(SCrO_4)_2H^-} + H_2O + 1H^+$	3.0	2.4
		Lower acidic to alkaline medium (pH > 5)		
$\overline{>SHCrO_4^-}$	0	$(\overline{>SOH}) + HCrO_4^- \leftrightarrow \overline{>SCrO_4^-} + H_2O$	0.2	
$\overline{>SHCrO_4}$	−1	$(\overline{>SOH}) + HCrO_4^- \leftrightarrow \overline{>SHCrO_4} + 1OH^-$	−6.7	6.1

It is worth noting that the interaction of chromate anions with phosphate materials was of an essentially electrostatic nature in acidic media (pH < 5), and of a chemical character at a lower acidity (pH > 5) to alkaline solution. Two surface species were always formed when hydration equilibrium was reached; deprotonated ($\overline{>SCrO_4}$) and protonated ($\overline{>SHCrO_4}$) complexes, which were more stable in near neutral, than acidic, solution. This was consistent with modeling adsorption data, indicating a mixture of both monodentate and bidentate chromate surface complexes on goethite [62,63].

4. Conclusions

The adsorption of hexavalent chromium on OCP material was thoroughly investigated. One goal of this study was to develop a method for studying the surface complexation of OCP. The obtained distribution coefficient (D) was dependent on the contact time, pH, and surface coverage. The treatment of Log(D) = f(pH) experimental data were used to evaluate H^+/OH^- exchange stoichiometry in adsorption reactions, and the results were used to specify the predominant Cr (VI) surface species. At hydration equilibrium, protonated hexavalent chromium formed $\overline{>SHCrO_4}$ and unprotonated $\overline{>SCrO_4}$ complexes, under all explored conditions. When the hydration equilibrium was not reached at a low surface coverage, the protonated species disappeared, to the benefit of the bidentate ($\overline{>S(HCrO_4)_2}$) complex. The stability constants were $\log K_{10} = 0.123$, for $\overline{>SHCrO_4}$, which could be formulated as $\overline{>S(OH_2^+)(HCrO_4^-)}$ and $\log K_{11} = -4.0$ for $\overline{>SCrO_4}$, in acidic media.

In alkaline media, the log $K_{1\text{-}0}$ = 0.2 for $\overline{> SCrO_4)}$ and log $K_{1\text{-}1} = -6.4$ for $\overline{> SHCrO_4}$. Whilst, for the bidentate $\overline{> S(HCrO_4)_2}$ surface species, the log$K_{10.5}$ = 2.9.The electrostatic effect was evaluated for the predominant adsorption reactions. The obtained results suggested that Cr (VI) adsorption on OCP was of an electrostatic nature in acidic solutions, and of a chemical nature in lower acidic to alkaline solutions. The results could have practical and promising applications in the fields of environmental health, for the removal of hazardous chromium from industrial wastewater before dumping it into the environment (water and soil). This could enhance the environmental risk management process and will play a major role in preventing future coastal contamination.

Abbreviations

OCP	Octacalcium Phosphate ($Ca_8H_2(PO_4)_{6.5}H_2O$)
XRD	X-ray Diffraction
FTIR	Fourier transform infrared spectroscopy
D	distribution coefficient (the ratio of concentration of adsorbed Cr (VI) to its concentration in aqueous phase)
m	OCP sorbent amounts in g/L
t	contact time.
K^+, K	surface stability of active site > SOH
l	number of functional surface group involved in adsorption reaction
H n	hydrogen atoms (n < 0) or OH groups (n > 0)
K_{ln}	adsorption constant
Ψ_0	surface potential
R	universal gas constant (8.31 J/mol·K)
T	temperature (K)

References

1. Dhal, B.; Thatoi, H.N.; Das, N.N.; Pandey, B.D. Chemical and microbial remediation of hexavalent chromium from contaminated soil and mining/metallurgical solid waste: A review. *Hazard J. Mater.* **2013**, *250–251*, 272–291. [CrossRef] [PubMed]
2. Zachara, J.M.; Ainsworth, C.; Brown, G.E., Jr.; Catalano, J.G.; McKinley, J.P.; Qafoku, O.; Smith, S.C.; Szecsody, J.E.; Traina, S.J.; Warner, J.A. Chromium speciation and mobility in a high level nuclear waste vadose zone plume. *Geochim. Cosmochim. Acta* **2004**, *68*, 13–30. [CrossRef]
3. Zhou, X.; Korenaga, T.; Takahashi, T.; Moriwake, T.; Shinoda, S. A process monitoring/controlling system for the treatment of wastewater containing Chromium (VI). *Water Res.* **1993**, *27*, 1049–1054. [CrossRef]
4. Tiravanti, G.; Petruzzelli, D.; Passino, R. Pretreatment of Industrial Wastewaters II, Selected Proceedings of the Second IAWQ International Conference on Pretreatment of Industrial Wastewaters. *Water Sci. Technol.* **1997**, *36*, 197–207. [CrossRef]
5. Pagilla, K.R.; Canter, L.W. Biosorption of Hexavalent Chromium from Aqueous Medium with Opuntia BiomassAthens, Greece. *Environ. J. Eng.* **1999**, *125*, 243–248. [CrossRef]
6. Bruce Manning, A.; Goldberg, S. Modeling Competitive Adsorption of Arsenate with Phosphate and Molybdate on Oxide Minerals. *Soil. Sci. Soc. Am. J.* **1996**, *60*, 121–131. [CrossRef]
7. Kongsricharoern, N.; Polprasert, C. Removal of Heavy Metals from Wastewater by Adsorption and Membrane Processes: A Comparative Study. *Water Sci. Technol.* **1996**, *34*, 109–116. [CrossRef]
8. Hu, J.; Chen, G.; Lo, I.M.C. Removal and recovery of Cr (VI) from wastewater by maghemite nanoparticles. *Water Res.* **2005**, *39*, 4528–4536. [CrossRef] [PubMed]
9. Dahbi, S.; Azzi, M.; de la Guardia, M. Removal of hexavalent chromium from wastewaters by bone charcoal. *Fresenius J. Anal. Chem.* **1999**, *363*, 404–407. [CrossRef]
10. Dahbi, S.; Azzi, M.; Saib, N.; de la Guardia, M.; Faure, R.; Durand, R. Removal of trivalent chromium from tannery waste waters using bone charcoal. *Anal. Bioanal. Chem.* **2002**, *374*, 540–546. [PubMed]
11. Hyder, A.H.M.G.; Shamim, A.B.; Nosa, O.E. Adsorption isotherm and kinetic studies of hexavalent chromium removal from aqueous solution onto bone char. *J. Environ. Chem. Eng.* **2015**, *3*, 1329–1336. [CrossRef]
12. Yoon, J. Phosohate-Induced Lead Immobilization in Contaminated Soil. Master's Thesis, University of Florida, Gainesville, FL, USA, May 2005.

13. Chowdhury, S.R.; Ernest Yanful, K. Arsenic and chromium removal by mixed magnetite-maghemite nanoparticles and the effect of phosphate on removal. *J. Environ. Manag.* **2010**, *91*, 2238–2247. [CrossRef] [PubMed]
14. Lee, J.Y.; Elzinga, E.J.; Reeder, R. Lead and zinc removal with storage period in porous asphalt pavement. *J. Environ. J. Sci. Technol.* **2005**, *39*, 4042–4048. [CrossRef]
15. Qureshi, M.; Varshney, K.G. *Inorganic Ion Exchangers*; Chemical Analysis CRC Press: Boca Raton, FL, USA, 2002; p. 282.
16. Correa, F.G.; Martínez, J.B.; Gómez, J.S. Synthesis and characterization of calcium phosphate and its relation to Cr (VI) adsorption properties. *Rev. Int. Contam. Ambient.* **2010**, *26*, 129–134.
17. Conca, J.L.; Lu, N.; Parker, G.; Moore, B.; Adams, A.; Wright, J.V.; Heller, P. Remediation of Chlorinated and Recalcitrant Compounds. *J. Environ. Qual.* **2000**, *7*, 319–326.
18. Saxena, S.; D'Souza, S.F. Heavy metal pollution abatement using rock phosphate mineral. *Environ. Int.* **2006**, *32*, 199–202. [CrossRef] [PubMed]
19. Aklil, A.; Mouflih, M.; Sebti, S. Removal of Cadmium from Water Using Natural Phosphate as Adsorbent. *Hazard J. Mater. A* **2004**, *112*, 183–190. [CrossRef] [PubMed]
20. Corami, A.; D'Acapito, F.; Mignardia, S.; Ferrini, V. Removal of Cu from Aqueous Solutions by Synthetic Hydroxyapatite: EXAFS Investigation. *Mater. Sci. Eng. B* **2008**, *149*, 209–213. [CrossRef]
21. Zhao, X.-Y.; Zhu, Y.-J.; Zhao, J.; Lu, B.-Q.; Chen, F.; Qi, C.; Wu, J. Hydroxyapatite nanosheet-assembled microspheres: Hemoglobin-templated synthesis and adsorption for heavy metal ions. *J. Colloid Interface Sci.* **2014**, *416*, 11–18. [CrossRef] [PubMed]
22. Serrano-Gómez, J.; Ramírez-Sandoval, J.; Bonifacio-Martínez, J.; Granados-Correa, F.; Badillo-Almaraz, V.E. Uptake of CrO42-Ions by Fe-Treated Tri-Calcium Phosphate. *J. Mex. Chem. Soc.* **2010**, *54*, 34–39.
23. Razzouki, B.; El Hajjaji, S.; ElYahyaoui, A.; Lamhamdi, A.; Jaafar, A.; Azzaoui, K.; Boussaoud, A.; Zarrouk, A. Kinetic investigation on arsenic (III) adsorption onto iron hydroxide (III). *Pharm. Lett.* **2015**, *7*, 53–59.
24. Soundarrajan, M.; Gomathi, T.; Sudha, P.N. Understanding the Adsorption Efficiency of Chitosan Coated Carbon on Heavy Metal Removal. *Int. J. Sci. Res. Publ.* **2013**, *3*, 2250–3153.
25. Sakai, S.; Anada, T.; Tsuchiya, K.; Ymazaki, H.; Margouli, H.C.; Suzuki, O. Comparative study on the resorbability and dissolution behavior of octacalcium phosphate, β-tricalcium phosphate, and hydroxyapatite under physiological conditions. *Dent. Mater. J.* **2016**, *35*, 216–224. [CrossRef] [PubMed]
26. Osamu, S. Octacalcium phosphate (OCP)-based bone substitute materials. *Jpn. Dent. Sci. Rev.* **2013**, *49*, 58–71.
27. Samake, D. Treatment of Tannery Wastewater Using Clay Materials. Ph.D. Thesis, Joseph Fourier-Grenoble University, Bamako, Mali, December 2008.
28. Zachara, J.M.; Cowan, C.E.; Schmidt, R.L.; Ainsworth, C.C. Chromate Adsorption by Kaolinite. *Clays Clay Miner.* **1988**, *36*, 317–326. [CrossRef]
29. Grossl, V.; Eick, M.; Sparks, D.; Goldberg, S.; Ainsworth, C.C. Arsenate and Chromate Retention Mechanisms on Goethite. 2. Kinetic Evaluation Using a Pressure-Jump Relaxation Technique. *Environ. Sci. Technol.* **1997**, *31*, 321–326. [CrossRef]
30. Lorphensri, O.; Intravijit, J.; Sabatini, D.A.; Kibbey, T.C.G.; Osathaphan, K.; Saiwan, C. Characterization of Commercial Ceramic Adsorbents and its Application on Naphthenic Acids Removal of Petroleum Distillates. *Water Res.* **2006**, *40*, 1481–1491. [CrossRef] [PubMed]
31. Harding, I.S.; Rashid, N.; Hing, K.A. urface charge and the effect of excess calcium ions on the hydroxyapatite surface. *Biomaterials* **2005**, *26*, 6818–6826. [CrossRef] [PubMed]
32. Zachara, J.M.; Girvin, D.C.; Scmidt, R.L.; Resch, C.T. Chromate adsorption on amorphous iron oxyhydroxide in presence of major ground water ion. *Environ. Sci. Technol.* **1987**, *21*, 589–594. [CrossRef] [PubMed]
33. Demetriou, A.; Pashalidis, I. Spectophotometric studies on the competitive adsorption of boric acid (B(iii)) and chromate (Cr(Vi)) onto iron (oxy) hydroxide (Fe(O)OH). *Glob. Nest J.* **2012**, *14*, 32–39.
34. Hsia, T.H.; Lo, S.L.; Lin, C.F.; Lee, D.Y. Chemica and spectroscopie evidence for specifie adsorption of chromate on hydrous iron oxide. *Chemosphere* **1993**, *26*, 1897–1904. [CrossRef]
35. Aoyama, M.; Tsuda, M. Removal of Cr(VI) from aqueous solutions by larch bark. *Wood Sci. Technol.* **2001**, *35*, 425–434. [CrossRef]

36. Odeh, L.; Odeh, I.; Khamis, M.; Khatib, M.; Qurie, M.; Shakhsher, Z.; Qutob, M. Hexavalent Chromium Removal and Reduction to Cr (III) by Polystyrene Tris(2-aminoethyl)amine. *Sci. Res. Acad. Publ.* **2016**, *6*, 26–37. [CrossRef]
37. Chang, L. Metal Ions Removal from polluted Waters by Sorption onto Exhausted Coffee Waste. Application to Metal Finishing Industries Wastewater Treatment. Ph.D. Thesis, University of Girona, Girona, Spain, September 2014.
38. Stankiewicz, A.I.; Moulijn, J.A. Process Intensification: Transforming Chemical Engineering. *Chem. Eng. Prog.* **2000**, *96*, 22–34. [CrossRef]
39. Oldham, D.J.; Mohsen, E.A. A technique for predicting the performance of self-protecting buildings with respect to traffic noise. *Noise Control Eng. J.* **1980**, *15*, 11–19. [CrossRef]
40. Bhattacharjee, S.; Swain, S.K.; Sengupta, D.K.; Singh, B.P. Effect of heat treatment of hydroxyapatite on its dispersibility in aqueous medium. *J. Colloids Surf.* **2006**, *277*, 164–170. [CrossRef]
41. Wu, X.W.; Ma, H.W.; Zhang, Y.R. Adsorption of chromium (VI) from aqueous solution by a mesoporous aluminosilicate synthesized from microcline. *Appl. Clay Sci.* **2010**, *48*, 538–541. [CrossRef]
42. Pandey, P.K.; Sharma, S.K.; Sambi, S.S. Kinetics and equilibrium study of chromium adsorption on zeoliteNaX. *Int. J. Environ. Sci. Technol.* **2010**, *7*, 395–404. [CrossRef]
43. Schmuhl, R.; Krieg, H.M.; Keizer, K. Adsorption of Cu(II) and Cr(VI) ions by chitosan: Kinetics and equilibrium studies. *Water SA* **2001**, *27*, 1–8. [CrossRef]
44. Asgari, G.; Rahmani, A.R. Preparattion of an Adsorbent from Pumice Stone and Its Adsorption Potential for Removal of Toxic Recalcitrant Contaminants. *J. Res. Health Sci.* **2013**, *13*, 53–57. [PubMed]
45. Dzombak, D.A.; Morel, F.M.M. *Adsorption of Inorganic Contaminants in Ponded Effluents from Coal-Fired Power Plants*; Massachusetts Institute of Technology Cambridge: Cambridge, MA, USA, 1985.
46. Farley, K.J.; Dzombak, D.A.; Morel, F.M.M. Distinguishing Adsorption from Surface Precipitation. *J. Colloid Surf. Sci.* **1985**, *106*, 226–242. [CrossRef]
47. Correa, F.G.; Gómez, J.S.; Martínez, J.B. Synthesis and Characterization of Inorganic Materials to Be Used as Adsorbents of Toxic Metals and of Nuclear Interest. In *Contributions of the National Institute of Nuclear Research to the Advance of Science and Technology in Mexico. Commemorative Edition 2010*; Instituto Nacional de Investigaciones Nucleares: Ocoyoacac, Mexico, 2010; pp. 195–210.
48. Müller, A.; Duffek, A. Similar Adsorption Parameters for Trace Metals with Different Aquatic Particles. *J. Aquat. Geochem.* **2001**, *7*, 107–126. [CrossRef]
49. Fendorf, S.; Eick, M.J.; Grossl, P.; Sparks, D. Arsenate and Chromate Retention Mechanisms on Goethite. 1. Surface Structure. *Environ. Sci. Technol.* **1997**, *31*, 315–320. [CrossRef]
50. Waychunas, G.A.; Fuller, C.C.; Davis, J.A. Surface complexation and precipitate geometry for aqueous Zn (II) sorption on ferrihydrite. *Geochim. Cosmochim. Acta* **2002**, *66*, 1119–1137. [CrossRef]
51. Mario, V.; Maya, A.T.; James, O.L. Surface Complexation Modeling of Carbonate Effects on the Adsorption of Cr(VI), Pb(II), and U(VI) on Goethite. *Environ. Sci. Technol.* **2001**, *35*, 3849–3856.
52. Raven, K.P.; Jain, A.; Loeppert, R.H. rsenite and Arsenate Adsorption on Ferrihydrite: Kinetics, Equilibrium, and Adsorption Envelopes. *Environ. Sci. Technol.* **1998**, *32*, 344–349. [CrossRef]
53. Perrone, J.; Fourest, B.; Giffaut, E. Surface and Physicochemical Characterization of Phosphates Vivianite, $Fe_2(PO_4)_3$ and Hydroxyapatite, $Ca_5(PO_4)_3OH$. *J. Colloid Interface Sci.* **2002**, *249*, 441–452. [CrossRef] [PubMed]
54. Smičiklas, I.; Dimović, S.; Plećaš, I.; Mitrić, M. Adsorption and removal of strontium in aqueous solution by synthetic hydroxyapatite. *Water Res.* **2006**, *40*, 2267–2274. [CrossRef] [PubMed]
55. Antelo, J.; Fiol, S.; Gondar, D.; López, R.; Arce, F. Comparison of arsenate, chromate and molybdate binding on schwertmannite: Surface adsorption vs. anion-exchange. *J. Colloid Interface Sci.* **2012**, *386*, 338–343. [CrossRef] [PubMed]
56. Gunnarsson, M. *Surface Complexation at the Iron Oxide/Water Interface Experimental Investigations and Theoretical Developments*; Institution för kemi Göteborgs University: Göteborgs, Germany, 2002.
57. Jain, A.; Raven, K.P.; Loeppert, R.H. Understanding Arsenate Reaction Kinetics with Ferric Hydroxides. *Environ. Sci. Technol.* **1999**, *33*, 1179–1184. [CrossRef]
58. Álvarez-Ayuşço, E.; García-Sánchez, A.; Querol, X. Adsorption of Heavy Metals from Aqueous Solutions on Synthetic Zeolite. *Hazard J. Mater.* **2007**, *142*, 191–198.

59. Ellouzi, K.; Elyahyaoui, A.; Bouhlassa, S. Octacalcium phosphate: Microwave-assisted hydrothermal synthesisand potentiometric determination of the Point of Zero Charge (PZC) and isoelectric point (IEP). *Pharm. Lett.* **2015**, *7*, 152–159.
60. Jin-Wook, K. *The Modeling of Arsenic Removal from Contaminated Water Using Coagulation and Sorption*; Texas A&M University: College Station, TX, USA, 2005.
61. Singh, S.P.N.; Mattigod, S.V. Modelling boron adsorption on kaolinite. *Clays Clay Miner.* **1992**, *40*, 192–205. [CrossRef]
62. Eick, M.J.; Peak, J.D.; Brady, W.D. The effect of oxyanions on the oxalatepromoted dissolution of goethite. *Soil. Sci. Soc. Am. J.* **1999**, *63*, 1133–1141. [CrossRef]
63. Tanuma, Y.; Anada, T.; Honda, Y.; Kawai, T.; Kamakura, S.; Echigo, S.; Suzuki, O. Granule Size–Dependent Bone Regenerative Capacity of Octacalcium Phosphate in Collagen Matrix. *Tissue Eng. A* **2011**, *18*, 546–557. [CrossRef] [PubMed]

5

Filtration of Uncharged Solutes: An Assessment of Steric Effect by Transport and Adsorption Modelling

Simona M. Miron [1,2], Patrick Dutournié [1,2,*] and Arnaud Ponche [1,2]

[1] Institut de Science des Matériaux de Mulhouse IS2M (UMR CNRS 7228), Université de Haute Alsace, 68100 Mulhouse, France; simona-melania.miron@uha.fr (S.M.M.); arnaud.ponche@uha.fr (A.P.)
[2] Université de Strasbourg, 3 bis rue A. Werner, 68093 Mulhouse CEDEX, France
* Correspondence: patrick.dutournie@uha.fr.

Abstract: The major aim of this work was to understand and estimate the evolution of the membrane selectivity of neutral solutes after the filtration of protein or amino acid solutions. Classical methodologies led to the estimation of the mean pore radius, different for each filtrated neutral solute. The use of pore size distribution from nitrogen adsorption/desorption experiments enabled a good description of hydraulic and selectivity performances. The modification of the membrane hydraulic properties after the successive filtration of protein solutions revealed that the decrease is quasi linear, the same for all the studied membranes and independent of prior tests. According to the experimental observations, an adsorption model was developed, considering a layer by layer adsorption in the larger pores of the membrane. The predictive obtained results are in good agreement with the experimental rejection rates, validating the assumptions.

Keywords: protein adsorption; neutral solute; ultrafiltration; selectivity modelling; pore size distribution

1. Introduction

Membrane-based technology offers a reliable option in a growing market to the standard separation processes (especially in terms of power consumption, addition of chemical reagents, and operating convenience) [1]. Currently, membrane separation technology is widely used to separate, concentrate, and rectify solutions for various purposes (for example, waste treatment, water desalination, purification of pharmaceuticals) in many industrial sectors (food, medicine, waste decontamination, chemical, textile industries) [2–5]. Specifically, they are used for the separation of neutral solutes from aqueous solutions (proteins for food industry, vitamins, peptides, drugs in general for the medicine industry) [6,7].

Alternatively, neutral solute filtrations are mainly used for characterizing membranes (steric effect, cut-off). For this purpose, a transport model is used. For example, Schönherr et al. [8] filtrated raffinose to estimate the mean pore radius by using the Paganelli and Solomon model [9]. Schaerp et al. [10] used the Spiegler–Kedem equation [11] derived from irreversible thermodynamics to numerically approximate the experimental rejection curves of galactose–, maltose–, and raffinose–water solutions. The reflection coefficient is expressed by the steric hindrance pore model. Ito et al. [12] filtrated five dextran samples (molecular weights ranging from 10.5 kg mol^{-1} to 2000 kg mol^{-1}) to investigate the modification of pore size as a function of the Ba^{2+} concentration. Velicangil and Howell [13] estimated the steric properties of ultrafiltration membranes by using the orifice model with rejection curves of three protein solutions (papain, Bovine Serum Albumin (BSA), and ovalbumin). In recent years, several works investigated the steric effect by using the Nernst–Planck model for uncharged solutes [14]. They studied the rejection rates of vitamin B12, raffinose, sucrose, and glucose and compared the estimated mean pore radius with the results from Atomic Force Microscopy (AFM). Significant differences were observed between the estimated mean pore radius, especially regarding pore size distribution estimated via AFM investigations. This methodology was used in many current works [15–17].

One major observation of these works is that results obtained with several molecules are different. Indeed, the filtration of different uncharged solutes leads to various estimations of mean pore radius. Another major problem, mainly related in the literature, is the adsorption of uncharged solutes in the membrane pore, and so, the partial clogging of the active layer. Proteins are indeed known to adsorb on various surfaces due to their high propensity of deformability and conformation modification, as demonstrated by Karlson et al. [18] and Norde [19]. Even if the amount of protein on hydrophobic substrates is generally high, the outer layer of membranes constituting of oxides can also absorb a non-negligible quantity of protein. Thereby, it reduces the apparent pore size and also modifies the surface properties of the active layer, as demonstrated by Huang et al. [20] and Ponche et al. [21].

For example, Robertson and Zydney [22] observed protein adsorption in ultrafiltration membrane pores, reducing the hydraulic permeability and increasing the selectivity. They estimated the decrease in pore size due to adsorption, a decrease compatible with a monolayer adsorption. A.M. Comerton et al. [23] studied the membrane adsorption of pharmaceutically active compounds. They observed that ultrafiltration (UF) membranes are more subject to adsorption than nanofiltration and reverse osmosis membranes. This phenomenon can be easily explained by the pore size of UF membranes. In most cases, the adsorption of uncharged solutes increases the membrane selectivity and decreases the hydraulic performances and, in some cases, can extend to complete pore clogging [24]. If studies referring to molecule adsorption in membrane pore are abundant [25], fewer studies deal with adsorption mechanisms in the pore, except the classical theoretical models (monolayer) [26] or for microfiltration membranes [27]. In any case, there has not been a study that dealt with the numerical estimation of solute rejection rate (predictive) by taking into account pore size distribution and pore adsorption.

In this study, the filtration of neutral solutes is performed with a ceramic ultrafiltration membrane to investigate size selectivity. To this end, the experimental pore size distribution estimated from nitrogen adsorption/desorption experiments is used as an input of the model. Adsorption phenomena are taken into account to understand the change in hydraulic performances and the membrane selectivity.

2. Materials and Methods

Experiments were performed with tubular ceramic membranes provided by TAMI Industries (Nyons, France). The active layer in TiO_2 was deposited on the internal surface of an alumina macroporous tube (1 kDa, length = 25 cm, inner diameter = 8 mm). The TiO_2 layer was observed by scanning electron microscopy (Philips XL30 First FEG, SEMTech Solutions, North Billerica, MA, USA) to estimate layer thickness.

Nitrogen gas adsorption–desorption isotherms were performed with a Micromeritics ASAP 2420 (Micrometrics, Ottawa, Canada, apparatus at T = 77 K. Prior to experiments, each sample (broken pieces of TiO_2 membrane and alumina tube) was out gassed to a residual pressure lower than 0.8 Pa at 350 °C for 15 h. The micropore volume was calculated using the *t*-plot method. The cumulative volume and the pore size diameter (distribution) were calculated using the density functional theory method (DFT).

Filtration tests were performed in a laboratory pilot plant (stainless steel), previously described in literature [28]. The studied solution was stored in a 5 L tank and a volumetric pump provided solution circulation for tangential flow filtration (retentate). The flow was controlled by an analog flow rate sensor. The experiments were performed at 700 L/h, a value corresponding to a mean fluid velocity higher than 5 m/s to avoid concentration polarization at the surface of the active layer. After filtration, the permeate flow was sampled for analysis. Both retentate and permeate returned into the solution tank. The applied pressure was adjusted by a manual valve (4 to 12 bar) and was measured by two sensors upstream and downstream the membrane carter. A cooling unit maintained the fluid temperature at 25 °C. Between each filtration test, the experimental set-up was rinsed with demineralized water (conductivity < 0.1 µS/cm).

Before performing filtration tests, a conditioning step of the membrane was required. This step consisted of the filtration of pure water until hydraulic performances were in steady state.

Filtration tests of uncharged solute-water solution ware performed with vitamin B12 (Alfa Aesar, purity 98%), *L*-phenylalanine (Fluka, purity 99%), *L*-tyrosine (Fluka, purity 99%), and lysozyme (Sigma-Aldrich, from chicken egg white, activity > 70,000 U/mg). The concentrations of retentate (C_r) and permeate (C_p) samples were investigated by absorbance measurements with a UV-visible spectrophotometer (Lambda 35, Perkin Elmer Instrument). Information about the studied molecules are given in Table 1.

Table 1. Information on the molecules used for experimental tests.

Products	Concentration (mol m^{-3})	Molar Mass (g mol^{-1})	Stokes Radius (Å)	Wavelength (nm)
L-phenylalanine	5.45	165.19	3.7	257
L-tyrosine	2.48	181.19	3.8	275
Vitamin B12	9.22×10^{-3}	1355.38	7.4	362
Lysozyme	0.025	14,300	19.0	281

The observed rejection rate R was calculated using Equation (1):

$$R = \frac{C_r - C_p}{C_r}. \tag{1}$$

Several tests of analytical measurement uncertainties were performed (repeated eight times). The maximal error was 1.9% for retentate and 2.9% for permeate samples. The rejection rate uncertainty was $\Delta R = (1 - R)\left[\frac{1}{C_p}\Delta C_p + \frac{1}{C_r}\Delta C_r\right]$ and, in all cases, was inferior to 4.8% (result for a rejection rate close to 0).

During the experimental tests, the hydraulic properties were monitored. To do this, between each experiment test, a pure water filtration test was performed. The permeate flux (Jv) was measured and plotted for different applied pressures (ΔP). The hydraulic permeability (L_p) was obtained from the slope of the linear curve via Equation (2):

$$Jv = \frac{L_p}{\mu}\Delta P. \tag{2}$$

Three series of tests were performed with the same membrane. Between each series the membrane was regenerated. The regeneration consisted of a hydrothermal treatment (five days in water at 105 °C) to recover its original hydraulic properties.

Filtration tests of neutral solutes were used to investigate size selectivity. Two models were studied: model A assumed a uniform one size distribution (average pore radius) and the second (model B) used an experimental size distribution determined from nitrogen adsorption/desorption experiment as an input.

In model A, the mean pore radius was estimated by numerically approximating rejection rates of the studied solutes. The equation used to approximate the experimental results was the solution of the mass balance (differential equation) in the membrane pore. The differential equation expressing solute mass balance is described by the Nernst–Planck approach for uncharged solutes [29,30], assuming uniform dispersed, one size cylindrical pores. This differential equation was solved with the equality of chemical potentials on both sides of the active layer [31] at the pore/solution interface. Starting from these assumptions, the rejection rate of an uncharged solute [32] can be calculated with Equation (3):

$$R = 1 - \frac{\varphi K_c}{1-(1-\varphi K_c)\exp\left(-\frac{K_c r_p{}^2 \Delta P}{8\mu K_d D_\infty}\right)}. \quad (3)$$

With $\varphi = \left(1 - \frac{r}{r_p}\right)^2$ the steric partitioning coefficient.

The second model (model B) implied that we had an experimental distribution of pore size. In this case, the permeation flow rate was assumed to be the sum of the flow rate of each pore (Equation (4)). The flow rate in the pore was calculated using Hagen Poiseuille's law:

$$Q = J_v S_m = \frac{Lp}{\mu}\Delta P \times S_m = \sum_i np(i) \times q(i) = \sum_i np(i) \times \frac{\pi r_p(i)^4}{8\mu \Delta x}\Delta P. \quad (4)$$

The rejection rate R was calculated using the Nernst–Planck approach for uncharged solutes (Equation (3)). The calculation was carried out for each pore size, summing all the contribution weighting by the pore number and the relative flow rate (Equation (5)) as follows:

$$R = \frac{1}{Q}\sum_i np(i).q(i).R(i). \quad (5)$$

3. Results

3.1. Experimental Results and Estimation of Steric Effect

Experimental tests were performed to understand the physical phenomena that act on the mass transfer mechanisms in a porous media. For this purpose, the monitoring of hydraulic and selectivity properties was required. Three series of tests were performed with the same membrane. Between each series of tests, the membrane was regenerated according to the protocol previously described in the Materials and Methods section. This regeneration aimed to recover the initial membrane properties. Table 2 provides experimental results obtained for three series in terms of maximal rejection rate (selectivity property) and hydraulic permeability (hydraulic performance). For each series, the results are in chronological order. These series were chosen as the first molecule filtrated was different (*L*-phenylalanine for series 1, vitamin B12 and Lysozyme for series 2, and *L*-Tyrosine for the last).

Table 2. Filtration experiments (chronologic order) for the three series of tests.

Molecule	R_{Max} (%)	r_p (nm)	$Lp \times 10^{14}$ (m^3 m^{-2})	R_{Max} (%)	$Lp \times 10^{14}$ (m^3 m^{-2})	R_{Max} (%)	$Lp \times 10^{14}$ (m^3 m^{-2})
After conditioning step			6.7		6.7		6.6
L-phenylalanine	5	3.3 ± 0.30	5.5				
L-tyrosine	5	3.5 ± 0.30	4.7			5	4.8
VB12	60	1.7 ± 0.15	4.4			40	4.9
Lysozyme	93	2.5 ± 0.05	3.6			96	4.3
Lysozyme	95	2.4 ± 0.05	3.3			99	3.7
VB12	75	1.4 ± 0.10	3.4	20	6.7		
Lysozyme	98	2.2 ± 0.04	3.1	65	4.0	100	2.9
Lysozyme	99	2.0 ± 0.03	2.8	98	2.5		
VB12	81	1.2 ± 0.05	2.8			85	2.8
Lysozyme	100	<2.0	2.2	100	2.0		
VB12	87	1.0 ± 0.04	2.3	90	2.0		
VB12	86	1.0 ± 0.04	2.2				
L-phenylalanine	16	1.6 ± 0.10	2.3				
L-tyrosine						17	2.8

The first experiments (left-hand column) were performed with a new membrane. After conditioning, the hydraulic permeability was 6.7×10^{-14} m^3 m^{-2}. The four studied molecules were filtrated one after another, from the smallest to the largest. The rejection rate increased as filtration tests were performed and, at the same time, the hydraulic permeability significantly decreased, suggesting that a part of the protein was adsorbed at the membrane surface (or in the pore), reducing mass transfer. Nevertheless, filtration of VB12 does not seem to modify the membrane hydraulic performances, and, thus, was chosen as the model solute to follow the membrane selectivity properties. The solute rejection rates obtained with the membrane still having its initial properties (i.e., membrane that had only filtrated water and vitamin B12 after regeneration) were 5% for *L*-phenylalanine (test 1 series 1), 5% for *L*-Tyrosine (test 1—series 3), 20% for Vitamin B12 (test 1 series 2), and 65% for Lysozyme (test 2—series 2). These rejection rates are in good agreement with the size of the studied molecules, but much lower than expected for a membrane with a cut-off of 1 kDa.

After these first experimental tests, the membrane permeability and the solute transmission decreased. This behavior was illustrated by the rejection rate of vitamin B12, which increased as protein or amino acid solutions were filtrated. For example, it increased from 20% to 90% after several filtration tests of lysozyme solutions (series 2), indicating that steric effect increased significantly. This can be explained by protein or amino acid adsorption at the membrane surface and or in the pore, thereby restricting the transfer of solute through the porous medium. These adsorption phenomena are not completely irreversible, since after the regeneration treatment the membrane recovered its initial properties.

The estimated mean pore radii are given in Table 2 for series 1 (from Equation (3)). The results (calculated average pore radii) corroborate previous observations. Indeed, the average pore radius decreased as protein filtration was performed. The results obtained with the different solutes show that this current model (model A) remains unsatisfactory for estimating steric effect because the results are dependent on the studied molecule. The average pore radius estimated with vitamin B12 was systematically lower than for lysozyme and *L*-phenylalanine. Bowen et al. [14] obtained the same results for the filtration of four uncharged solutes (vitamin B12, raffinose, sucrose, and glucose). Indeed, the estimated mean pore radius can vary by a ratio of 2:1. They also observed a modification of water permeability after filtration of each solute.

Additional tests were performed with mixtures of uncharged solutes (vitamin B12 and lysozyme) in water or salted water. These tests were performed following the tests of series 1.

The results (Table 3) show that the rejection rate of the solute was not modified by any another solute in the solution or by salt (NaCl—5 mM) in dilute solution. In the present case, it seems that the solutes in solution did not interact between them. However, in the literature, several studies [33–35] showed an increase in neutral solute transmission in saline solutions. This phenomenon is explained by a partial dehydratation of the molecule in the membrane pore, facilitating the transport through the membrane active layer. This behavior is specifically observed in nanofiltration. In the present study, the filtrated solutes, the porous material (membrane active layer), and the pore size were different, which could explain the observed differences.

Table 3. Filtration tests of solute mixture.

Filtrated Solutions	R_{Max} (%)	Average r_p (nm)	$Lp \times 10^{14}$ (m^3 m^{-2})
Lysozyme	99	2.0	2.3
VB12 + **Lysozyme**	86 + 99	1.0/2.0	2,2
VB12 + **Lysozyme** + NaCl	87 + 100	1.0/<2.0	2,4
VB12	86	1.0	2,3

Moreover, the difference in estimated mean pore radius according to the studied molecule was confirmed. There are two possible explanations. First, the one size cylindrical pore hypothesis is a poor and inadequate representation of the porous medium. Second, the solute interacts with the membrane surface according to the chemical groups of both the solute and the active layer (van der Waals forces or acid-base interactions). To investigate the first possibility, a pore size distribution was studied. To this end, nitrogen adsorption/desorption experiments were used to measure the porous volume according to the pore size.

Figure 1 shows the cumulative porous volume obtained during the nitrogen adsorption step according to pore size. This curve can be suitably approximated by a log-normal distribution. These results were substantiated by Scanning Electron Microscopy (SEM) investigations. Indeed, the SEM images (Figure 2c) show a top view of the active layer. The surface constituted sintered titania aggregates separated by nanometric cracks or channels. The larger ones, measuring about 4–10 nm, were of a lower amount than the smaller ones (less than 1 nm wide).

From these experimental results, we assumed that the porous medium was constituted of cylindrical and unidirectional pores according to the previous size distribution (Figure 1) and calculated the pore number as a function of the pore size.

Assuming that the flow rate in each pore can be described by the Hagen–Poiseuille law, the permeation flux can be estimated with Equation (4). From these results, it is possible to estimate the active layer thickness by equaling the observed permeation flux with the flow rate calculated by Equation (4). For this purpose, the studied range of pore diameter (0–14 nm) was divided into sub-ranges of pore size from $i = 0$ to 1400. The active layer thickness was around 2.9 μm. Previous observations [28] performed by microscopy (SEM) showed that the active layer was about 1–2 μm thick. Figure 2a shows a cross section of the membrane obtained after breaking it. The different alumina layers (different porosities) are visible below the titania active layer (full grey layer). The apparent thickness (Figure 2c) of this layer is in agreement with previous investigation.

Calculations were performed for modelling selectivity performances of the new or regenerated membranes for the four studied molecules (i.e., for membranes, which have only been in contact with water). These calculated results were compared with the experimental observed rejection rates of the four studied solutes (Table 4). The results are very close to the experimental ones for the four studied molecules. This comparison indicates that taking into account pore size distribution provides a good way to estimate rejection rates of neutral solutes.

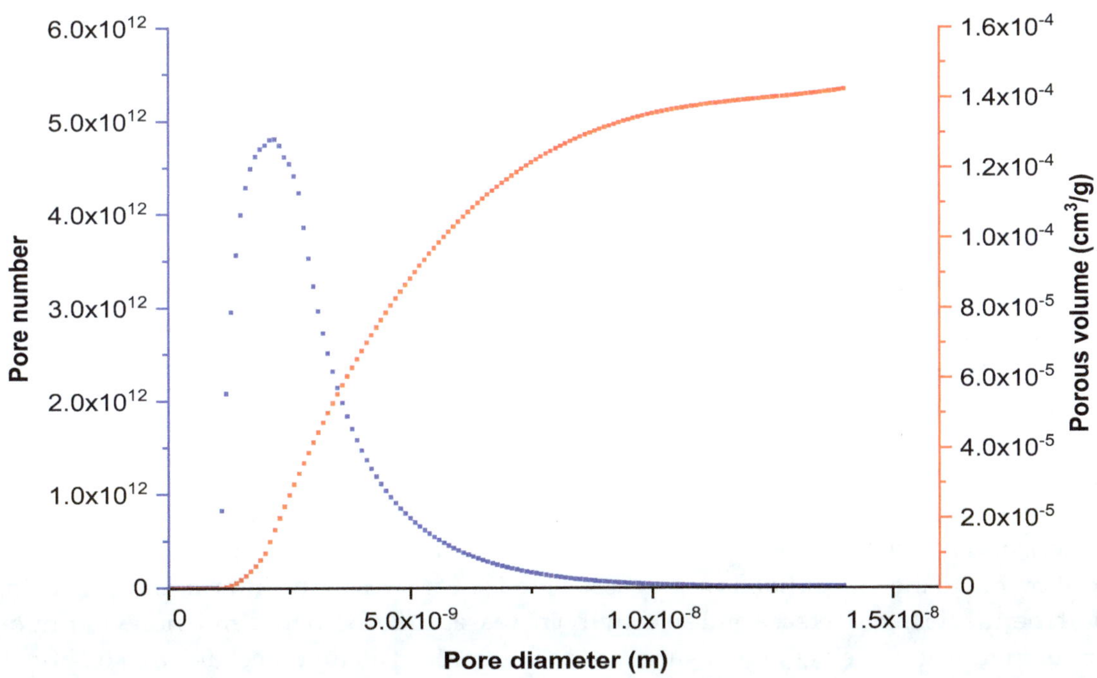

Figure 1. Cumulative porous volume (red curve) versus pore diameter and pore size distribution (blue curve).

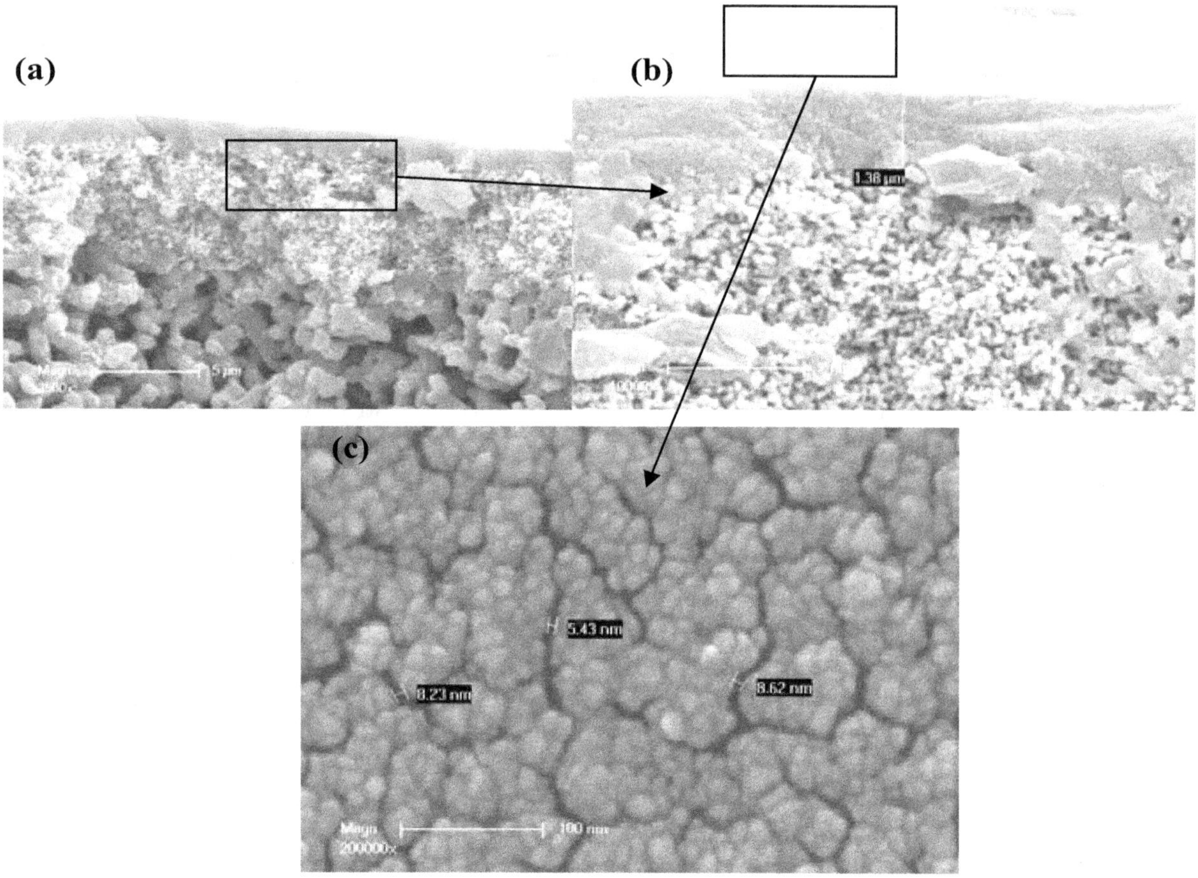

Figure 2. SEM photographs of (**a**,**b**) cross section at different enlargements and (**c**) top view of the active layer.

Table 4. Experimental and calculated (pore size distribution) rejection rate of solutes (membrane with initial properties).

Solutes	*R* Exp (%)	*R* Calc (%)
L-phénylalanine	5	6
Vitamin B12	20	21
Lysozyme	65	66
L-Tyrosine	5	7

3.2. Hydraulic Performance Loss and Adsorption in the Pore

The hydraulic performances of the membranes decreased as filtration tests of neutral solutes were performed (except for Vitamin B12). At the same time, selectivity performances increased, indicating an increase in the steric effect limiting the mass transfer. As shown in Table 2 for the first series, the average pore radius (calculation relative to each filtrated molecule) decreased, step by step, until a minimum value. Four clogging/blocking mechanisms are classically reported in the literature: complete blocking (the filtrated molecule blocks the pore inlet), standard blocking (adsorption of the molecules in the pore), intermediate blocking (formation of a non-continuous layer, blocking partially the flow inlet), and a cake filtration. This last one often occurs in dead-end filtration. In our case, it can be ruled out owing to the tangential flow and associated shear stress. The intermediate blocking cannot explain both permeate flux decrease and selectivity increase. Indeed, to that end, non-continuous layers should preferentially grow, clogging the larger pore inlet only. The first mechanism, i.e., pore blocking, cannot be a plausible hypothesis, because during filtration of amino acids, the pores that clog should be the smallest, which would result in a decrease in rejection rate. So, the only plausible explanation is protein or amino acid adsorption in the larger pores, reducing the flow rate and increasing the selectivity. F. Wand et al. [36] observed that standard blocking occurred first during ultrafiltration of colloid–water solutions (dead-end filtration experiments). K. Katsoufidou et al. [37] observed a rapid irreversible membrane fouling during ultrafiltration of humic acid due to internal pore adsorption.

Figure 3 shows the hydraulic permeability of the membrane divided by the one obtained just before the first filtration test of lysozyme for series 1 and 3 (presented in Table 2). These results were compared with two test series performed in the same operating conditions with another membrane (series 4 and 5). The hydraulic permeability decreased as filtration tests were carried out (quasi linear behavior), indicating the same adsorption kinetics, regardless of the studied membrane and its past experiments. Assuming one size pore distribution, the hydraulic permeability was proportional to the fourth power of the pore radius (Poiseuille flow). In these conditions, the pore radius estimated by the filtration tests of a neutral solute should vary according to the fourth-root dependence of the number of tests.

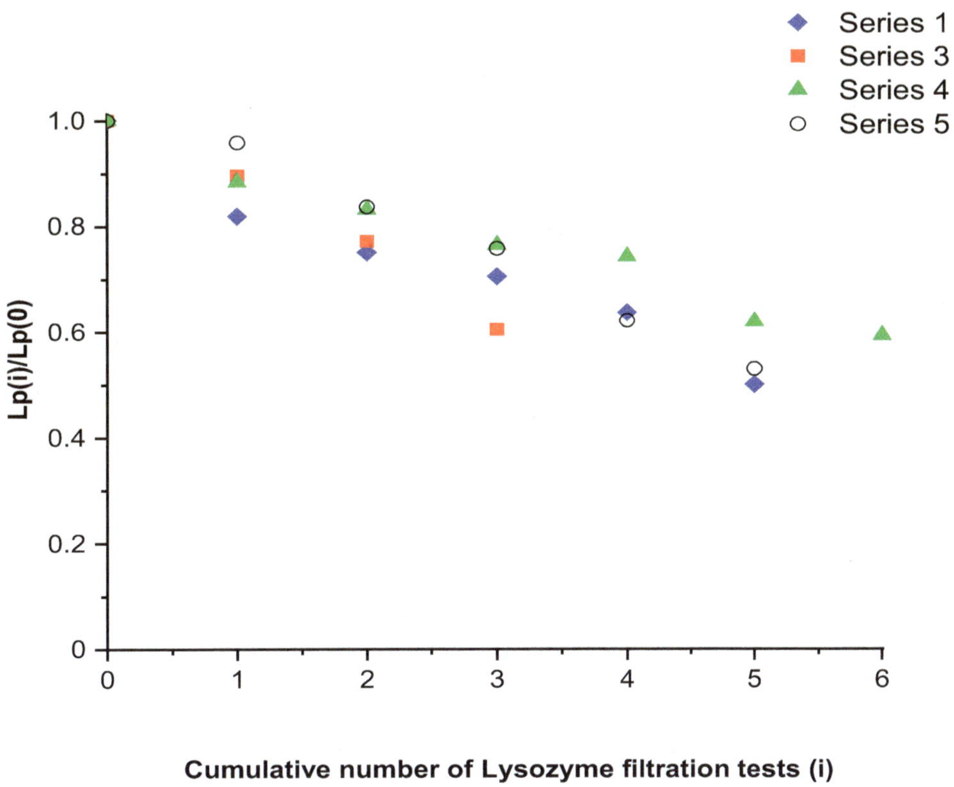

Figure 3. Relative hydraulic permeability versus the number of filtration tests of lysozyme performed.

Figure 4 compares the estimated mean pore radius for vitamin B12 and Lysozyme filtration tests (series 1) with the mean pore radius calculated using a linear function of the fourth-root of the cumulative number of lysozyme test. For the two cases, the estimated pore radius can be approximated by $r_p(i) = r_p(0) \times (1 - 0.165i)^{0.25}$.

The same investigation with pore size distribution was not possible owing to the number of adjustable parameters (i.e., $Np(i)$ and $r_p(i)$).

Steric hindrance after protein adsorption is required in order to predict the selectivity performances of the membrane. Classical models of adsorption or adsorption kinetics do not provide information about flow restriction in pores.

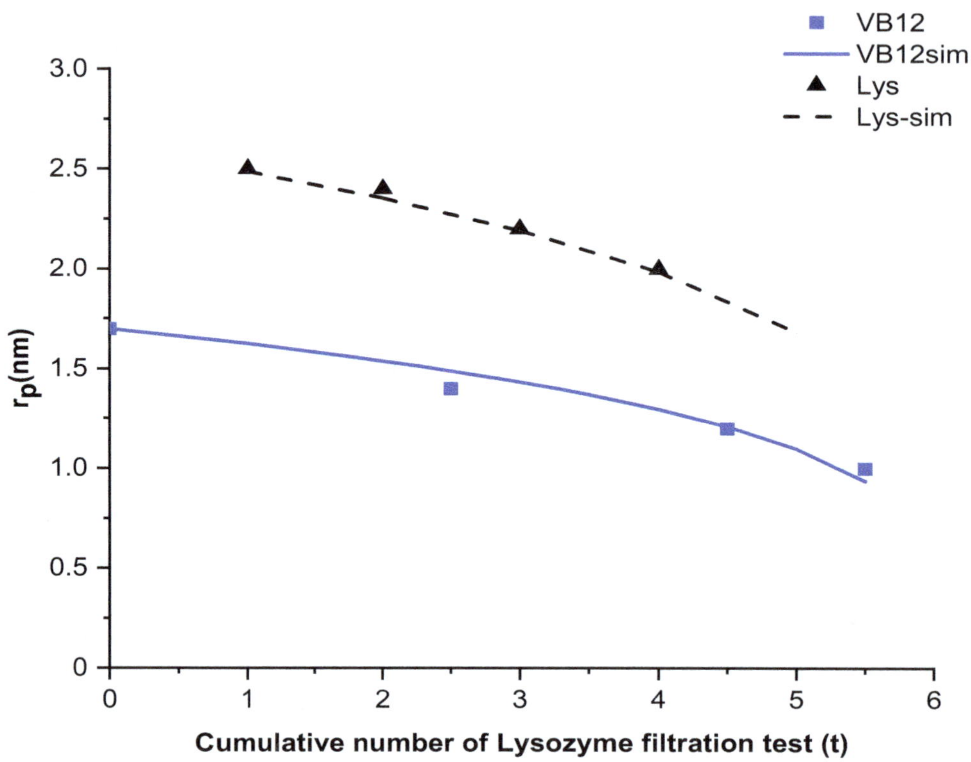

Figure 4. Mean pore radius estimated from filtration experiments and calculation of neutral solute (series 1: Vitamin B12 and Lysozyme).

In these conditions, to model these adsorption phenomena, we considered the adsorption of spherical molecules to be by uniform layers in the pore, reducing its radius of two stokes radius of the adsorbed molecule. This assumption is an arbitrary but required hypothesis for calculating the pore size distribution after molecule adsorption. The active layer thickness and the total number of pores were assumed to be unchanged (only their sizes can be modified). The calculation was carried out in all the pores with diameters twice larger than the molecule size. This procedure was numerically repeated until the calculated mass flow rate equaled the experimental permeation flux. When the equality of experimental and numerical hydraulic performances was reached, the model became predictive and capable of calculating the rejection rates of other filtrated molecules (i.e., Vitamin B12, *L*-Tyrosine, *L*-Phenylalanine, and Lysozyme).

Figure 5 shows the pore size distribution of the membrane after several filtration tests of lysozyme solution (series 1) compared to the initial one. This pore size distribution was obtained by equaling the experimental and calculated permeation fluxes. The results show that adsorption phenomena took place in the larger pores, partially clogging them and reducing their apparent diameters. Consequently, the pores with an initial size higher than 3.9 nm (corresponding to pores larger than adsorbed solute) completely disappeared and the pore quantity with a diameter in between the 1.8–2.8 nm range significantly increased.

To validate this modeling, rejection rates of the four filtrated molecules were calculated before and after lysozyme adsorption.

Experimental and calculated rejection rates of the studied molecules are given in Table 5 for the membrane before (initial properties) and after several filtration tests of lysozyme solutions. Taking into account that a pore size distribution provides a good description of membrane selectivity performances and its modification over time, from this new size distribution, the rejection rates of *L*-phenylalanine, *L*-Tyrosine, and Lysozyme increased and are in good agreement with the experimental results. Nevertheless, the rejection rate of vitamin B12 increased up to 60%, which is rather different regarding the experimental results (between 85% and 90%). These results are basically prior image and need to be further investigated and confirmed by other membranes and molecules.

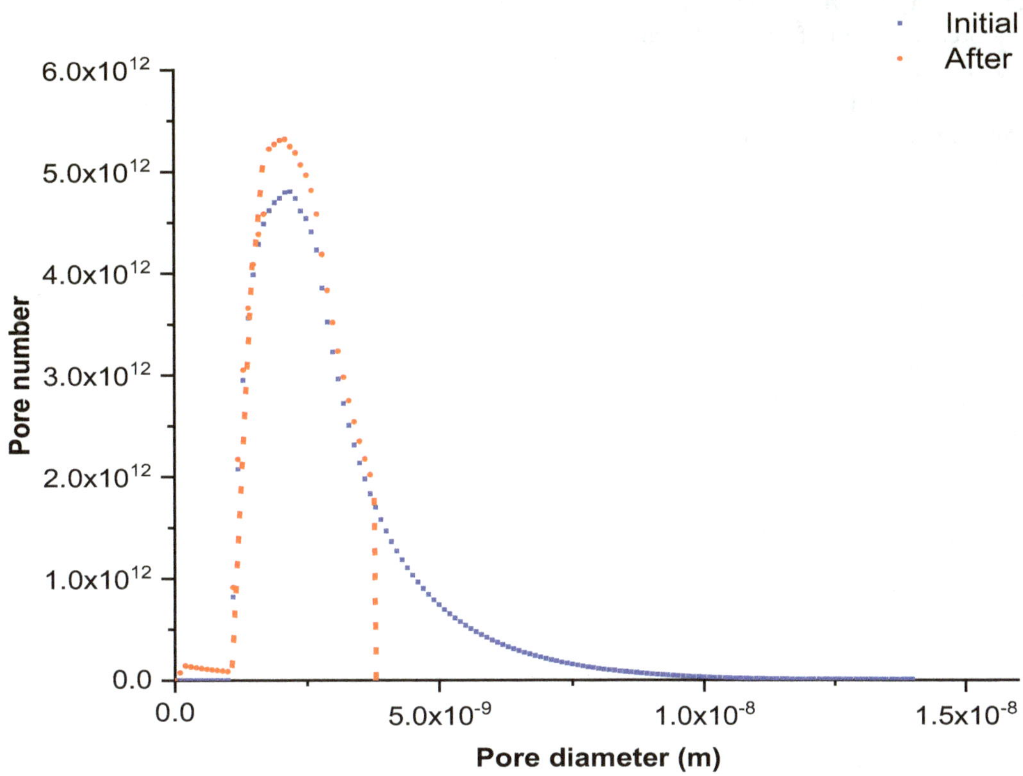

Figure 5. Initial and calculated pore size distribution after lysozyme filtration tests (series 1).

Table 5. Experimental and calculated rejection rates before and after lysozyme filtration.

Molecules	Initial Properties *R*/*R*calc	After Lysozyme Filtration *R*/*R*calc
L-phenylalanine	5%/6%	16%/13%
VB12	20%/21%	90%/60%
Lysozyme	65%/66%	100%/98%
L-tyrosine	5%/7%	17%/13%

4. Conclusions

In this work, filtration tests of four uncharged solutes were performed in order to investigate the steric exclusion effects. These tests were performed with new or regenerated membrane to investigate the pore size distribution and selectivity of the membrane regarding each solute. Taking into account a pore size distribution experimentally determined with nitrogen adsorption/desorption techniques made it possible to accurately describe the membrane selectivity performances. The simulated and experimental rejection rates of the four studied molecules are in good agreement for new membranes. Successive filtration tests showed that proteins and amino acids interact with the membrane (adsorption in the larger pore), reducing the hydraulic performances and increasing the selectivity. To describe this phenomenon, the pore size distribution was recalculated by considering molecule adsorption in the larger pore. In this way, the calculation was performed by equalizing experimental and numerical permeation flux. The simulated results provide a good perspective of trends in membrane selectivity for neutral solute filtration.

Author Contributions: Conceptualization, P.D. and A.P.; Methodology, S.M.M., P.D. and A.P.; Validation, S.M.M., P.D. and A.P.; Modelling P.D.; Investigation, S.M.M., P.D. and A.P.; Writing—Original Draft Preparation, S.M.M., P.D. and A.P.; Supervision, P.D. and A.P.

Glossary

C_p	Permeate concentration of solute (mol m^{-3})
C_r	Solute concentration in the feed solution (mol m^{-3})
D_∞	Diffusion coefficient of solute at infinite dilution (m^2 s^{-1})
J_v	Permeation flux (m^3 m^{-2} s^{-1})
K_c	Hindrance factor for convection (dimensionless) $K_c = 1 - 2.3\lambda + 1.154\lambda^2 + 0.224\lambda^3$
K_d	Hindrance factor for diffusion (dimensionless) $K_d = (2 - \varphi)\times(1 + 0.054\lambda - 0.988\lambda^2 - 0.441\lambda^3)$
L_p	Hydraulic permeability (m^3 $m^{-2}{}_{memb}$)
$np(i)$	Pore number with a size $r_p(i)$
Q	Permeation flow rate (m^3 s^{-1})
$q(i)$	Permeation flow rate in pores with a size $r_p(i)$ (m^3 s^{-1})
r	Stokes radius of solute (m)
R	Observed rejection rate (dimensionless)
$Rcalc$	Calculated rejection rate
$R(i)$	Observed rejection rate of solute in pores with a size $r_p(i)$ (dimensionless)
r_p	Average pore radius (m)
$r_p(i)$	Pore radius
S_m	Membrane surface
t	Number of successive tests

Greek Letters

ΔP	Transmembrane pressure (Pa)
Δx	Active layer thickness (m)
λ	Radii ratio (dimensionless) $\lambda = r/r_p$
μ	dynamic viscosity of the filtrated solution (Pa s)
φ_i	steric partition coefficient relative to pores with a size $r_p(i)$ (dimensionless)

References

1. Baker, R.W. *Membrane Technology and Applications*, 3rd ed.; Wiley: Hoboken, NJ, USA, 2012.
2. Ho, W.S.W.; Sirkar, K.K. *Membrane Handbook*; Springer: New York, NY, USA, 1992.
3. Li, N.N.; Fane, A.G.; Ho, W.S.W.; Matsuura, T. *Advanced Membrane Technology and Applications*; Wiley: Hoboken, NJ, USA, 2008.
4. Pabby, A.K.; Rizvi, S.S.H.; Requena, A.M.S. *Handbook of Membrane Separations: Chemical, Pharmaceutical, Food and Biotechnological Applications*, 2nd ed.; CRC Press: Boca Raton, FL, USA, 2015.
5. Vandevivere, P.C.; Bianchi, R.; Verstraete, W. Review: Treatment and reuse of wastewater from the textile wet-processing industry: Review of emerging technology. *J. Chem. Technol. Biotechnol.* **1999**, *72*, 289–302. [CrossRef]
6. Arise, A.K.; Alashi, A.M.; Nwachukwu, I.D.; Ijabadeniyi, O.A.; Aluko, R.E.; Amonsou, E.O. Antioxidant activities of bambara groundnut (*Vigna subterranea*) protein hydrolysates and their membrane ultrafiltration fractions. *Food Funct.* **2016**, *5*, 2431–2437. [CrossRef] [PubMed]
7. Wang, X.; Li, C.; Du, C.; Gao, J.; Zhao, K.X.; Shi, R.; Jiang, Y. Plasma protein binding monitoring of therapeutic drugs in patients using single set of hollow fiber centrifugal ultrafiltration. *Bioanalysis* **2017**, *9*, 579–592. [CrossRef] [PubMed]
8. Schönherr, J. Water permeability of isolated cuticular membranes: The effect of pH and cations on diffusion, hydrodynamic permeability and size of polar pores in the cutin matrix. *Planta* **1976**, *128*, 113–126. [CrossRef]
9. Paganelli, C.V.; Solomon, A.K. The rate of exchange of tritiated water across the human red cell membrane. *J. Gen. Physiol.* **1957**, *41*, 259–277. [CrossRef]
10. Schaerp, J.; van der Bruggen, B.; Vandecasteerle, C.; Wilms, D. Influence of ion size and charge in nanofiltration. *Sep. Purif. Technol.* **1998**, *14*, 155–162. [CrossRef]

11. Spiegler, K.S.; Kedem, O. Thermodynamics of hyperfiltration (reverse osmosis): Criteria for efficient membrane. *Desalination* **1966**, *1*, 311–326. [CrossRef]
12. Ito, T.; Hioko, T.; Yamaguchi, T.; Shinbo, T.; Nakao, S.I.; Kimura, S. Development of a molecular recognition ion gating membrane and estimation of its pore size control. *J. Am. Chem. Soc.* **2002**, *124*, 7840–7846. [CrossRef]
13. Velicangil, O.; Howell, J.A. Estimation of the properties of high-flux ultrafiltration membranes. *J. Phys. Chem.* **1980**, *84*, 2991–2992. [CrossRef]
14. Bowen, W.R.; Mohammad, A.W.; Hilal, N. Characterisation of nanofiltration membranes for predictive purposes. Use of salts, uncharged solutes and atomic force microscopy. *J. Memb. Sci.* **1997**, *126*, 91–105. [CrossRef]
15. Déon, S.; Lam, B.; Fievet, P. Application of a new dynamic transport model to predict the evolution of performances throughout the nanofiltration of single salt solutions in concentration and diafiltration modes. *Water Res.* **2018**, *136*, 22–33. [CrossRef] [PubMed]
16. Dutournié, P.; Limousy, L.; Blel, W.; Déon, S.; Fievet, P. Understanding of ion transport in a Na-Mordenite membrane: Use of numerical modeling to estimate surface-solute interactions in the pore. *Ind. Eng. Chem. Res.* **2014**, *53*, 8221–8227. [CrossRef]
17. Bandini, S.; Morelli, V. Mass transfer in 1812 spiral wound modules: Experimental study in dextrose-water nanofiltration. *Sep. Pur. Technol.* **2018**, *199*, 84–96. [CrossRef]
18. Karlsson, M.; Ekeroth, J.; Elwing, H.; Carlsson, U. Reduction of irreversible protein adsorption on solid surfaces by protein engineering for increased stability. *J. Biol. Chem.* **2005**, *280*, 25558–25564. [CrossRef]
19. Norde, W. Driving forces for protein adsorption at solid surfaces. *Macromol. Symp.* **1996**, *103*, 5–18. [CrossRef]
20. Huang, T.; Anselme, K.; Sarrailh, S.; Ponche, A. High-performance liquid chromatography as a technique to determine protein adsorption onto hydrophilic/hydrophobic surfaces. *Int. J. Pharm.* **2016**, *497*, 54–61. [CrossRef]
21. Ponche, A.; Ploux, L.; Anselme, K. Protein/Material Interfaces: Investigation on Model Surfaces. *J. Adhes. Sci. Technol.* **2010**, *24*, 2141–2164. [CrossRef]
22. Robertson, B.C.; Zydney, A.L. Protein adsorption in asymmetric ultrafiltration membranes with highly constricted pores. *J. Colloid Interface Sci.* **1990**, *134*, 563–575. [CrossRef]
23. Comerton, A.M.; Andrews, R.C.; Bagley, D.M.; Yang, P. Membrane adsorption of endocrine disrupting compounds and pharmaceutically active compounds. *J. Membr. Sci.* **2007**, *303*, 267–277. [CrossRef]
24. Jepsen, K.L.; Bram, M.; Pedersen, S.; Yang, Z. Membrane fouling for produced water treatment: A review study from a process control perspective. *Water* **2018**, *10*, 847. [CrossRef]
25. Bu, F.; Gao, B.; Yue, Q.; Liu, C.; Wang, W.; Shen, X. The Combination of Coagulation and Adsorption for Controlling Ultra-Filtration Membrane Fouling in Water Treatment. *Water* **2019**, *11*, 90. [CrossRef]
26. Zhang, R.; Liu, Y.; He, M.; Su, Y.; Zhao, X.; Elimelech, M.; Jiang, Z. Antifouling membranes for sustainable water purification: Strategies and mechanisms. *Chem. Soc. Rev.* **2016**, *45*, 5888–5924. [CrossRef] [PubMed]
27. Tien, C.; Ramarao, B.V.; Yasarla, R. A blocking model of membrane filtration. *Chem. Eng. Sci.* **2014**, *111*, 421–431. [CrossRef]
28. Dutournié, P.; Limousy, L.; Anquetil, J.; Déon, S. Modification of the selectivity properties of tubular ceramic membranes after alkaline treatment. *Membranes* **2017**, *7*, 65. [CrossRef]
29. Bowen, W.R.; Welfoot, J.S.; Williams, P.M. Linearized transport model for nanofiltration: Development and assessment. *AIChE J.* **2002**, *48*, 760–773. [CrossRef]
30. Déon, S.; Dutournié, P.; Limousy, L.; Bourseau, P. The two-dimensional pore and polarization transport model to describe mixtures separation by nanofiltration: Model validation. *AIChE J.* **2011**, *57*, 985–995. [CrossRef]
31. Donnan, F.G. The theory of membrane equilibria. *Chem. Rev.* **1924**, *1*, 73–90. [CrossRef]
32. Déon, S.; Dutournié, P.; Bourseau, P. Modelling nanofiltration with Nernst-Planck approach and polarization layer. *AIChE J.* **2007**, *53*, 1952–1969. [CrossRef]
33. Bargeman, G.; Vollen broek, J.M.; Straatsma, J.; Schroën, C.G.P.H.; Boom, R.M. Nanofiltration of multi-component feeds. Interactions between neutral and charged components and their effect on retention. *J. Membr. Sci.* **2005**, *247*, 11–20. [CrossRef]
34. Bellona, C.; Drewes, J.E.; Xu, P.; Amy, G. Factors affecting the rejection of organic solutes during NF/RO treatment—A literature review. *Water Res.* **2004**, *38*, 2795–2809. [CrossRef]

35. Escoda, A.; Fievet, P.; Lakard, S.; Szymczyk, A.; Déon, S. Influence of salts on rejection of polyethyleneglycol by an NF organic membrane: Pore swelling and salting-out effects. *J. Membr. Sci.* **2010**, *347*, 174–182. [CrossRef]
36. Wang, F.; Tarabara, V.V. Pore blocking mechanisms during early stages of membrane fouling by colloids. *J. Colloid Interface Sci.* **2008**, *328*, 464–469. [CrossRef] [PubMed]
37. Katsoufidou, K.; Yiantsios, S.G.; Karabelas, A.J. A study of ultrafiltration membrane fouling bu humic acids and flux recovery by backwashing: Experiments and modelling. *J. Membr. Sci.* **2005**, *266*, 40–50. [CrossRef]

6

Longitudinal Removal of Bisphenol-A and Nonylphenols from Pretreated Domestic Wastewater by Tropical Horizontal Sub-Surface Constructed Wetlands

Andrés Toro-Vélez [1,2,*], Carlos Madera-Parra [3], Miguel Peña-Varón [1], Hector García-Hernández [4], Wen Yee Lee [5], Shane Walker [6] and Piet Lens [4]

1 Grupo de Saneamiento Ambiental, Instituto Cinara, Unversidad del Valle, Cali 100-00, Colombia; miguel.pena@correounivalle.edu.co
2 Doctorado en Ciencias Ambientales, Universidad del Cauca, Popayán 190001, Colombia
3 Escuela EIDENAR-Facultad de Ingeniería, Universidad del Valle, Cali 100-00, Colombia; carlos.a.madera@correounivalle.edu.co
4 UNESCO-IHE Institute for Water Education, 2611 AX Delft, The Netherlands; h.garcia@un-ihe.org (H.G.-H.); p.lens@un-ihe.org (P.L.)
5 Department of Chemistry, University of Texas at El Paso, El Paso, TX 79968, USA; wylee@utep.edu
6 Department of Civil Engineering, University of Texas at El Paso, El Paso, TX 79968, USA; wswalker2@utep.edu
* Correspondence: andres.toro@correounivalle.edu.co

Abstract: Bisphenol A (BPA) and nonylphenols (NPs), with a high potential to cause endocrine disruption, have been identified at levels of nanograms per liter and even micrograms per liter in effluents from wastewater treatment plants. Constructed wetlands (CWs) are a cost-effective wastewater treatment alternative due to the low operational cost, reduced energy consumption, and lower sludge production, and have shown promising performance for treating these compounds. A CW pilot study was undertaken todetermine its potential to remove BPA and NP from municipal wastewater. Three CWs were used: the first CW was planted with *Heliconia* sp., a second CW was planted with *Phragmites* sp., and the third CW was an unplanted control. The removal efficiency of the *Heliconia*-CW was 73 ± 19% for BPA and 63 ± 20% for NP, which was more efficient than the *Phragmites*-CW (BPA 70 ± 28% and NP 52 ± 23%) and the unplanted-CW (BPA 62 ± 33% and NP 25 ± 37%). The higher capacity of the *Heliconia*-CW for BPA and NP removal suggests that a native plant from the tropics can contribute to a better performance of CW for removing these compounds.

Keywords: municipal wastewater; constructed wetlands; Bisphenol A; nonylphenol; biodegradation; tropical environment

---

1. Introduction

Exposure to trace concentrations of certain synthetic and natural chemicals compounds, e.g., pharmaceuticals and personal care products (PPCPs), may induce negative environmental and health effects. The United States Geological Survey found 13 compounds related to organic wastewater contaminants in samples from the water supplies that ranged from 0.0120 to 0.480 $\mu g \cdot L^{-1}$, and concentrations of pharmaceutical and personal care compounds ranged from 0.0037 to 0.0576 $\mu g \cdot L^{-1}$ [1]. The main source of these compounds in the water cycle is their discharge by sewage systems. Several synthetic compounds with a high potential to cause endocrine disruption at low concentrations (e.g., micrograms per liter, $\mu g \cdot L^{-1}$, or even nanograms per liter, $ng \cdot L^{-1}$) have been identified to be present in the effluents of wastewater treatment plants (WWTPs) [2,3]. Furthermore, due to their chemical and recalcitrant characteristics, some of these PPCP compounds pass through conventional wastewater treatment processes without undergoing any transformation, resulting in their direct discharge to the receiving waters [4,5].

Compounds such as Bisphenol A (BPA) are widely used in industrial processes as a primary raw material in the manufacturing of many products such as plastics for engineering applications (e.g., epoxy resins and polycarbonate plastics), electronic devices, food cans, bottles, and dental sealants [6]. There is evidence that relates BPA appearance directly to adverse reproductive and carcinogenic effects in mice with a dose of 25 $ng \cdot kg^{-1}$ per day and 1 $\mu g \cdot kg^{-1}$ per day, respectively [7]. Nonylphenols (NPs) are used in the manufacturing of anionic detergents, lubricants, agrochemicals, tanneries, and lubricant oil additives. The main source of NP in municipal wastewater is due to the intermediate degradation products of soaps and detergents. NPs are found to be endocrine disruptive compounds (EDCs), and have effects in the reproductive system of some mammals, including the reduction in testis and ovaries weight, and the appearance of an irregular estrous cycle [8].

Different treatment systems for the reduction of EDCs from wastewater are being evaluated, such as membrane bioreactors, activated sludge systems, ozonation, photocatalysis, and sequencing batch reactors. Most of these processes require a high economic investment with high environmental costs related to their operation and maintenance which makes their implementation difficult in developing countries. Some studies focus on sustainable wastewater treatment by decreasing electricity consumption and mitigating its greenhouse gas footprint [9]. In this sense, constructed wetlands (CWs) are natural wastewater treatment systems that offer cost-effective treatment for small to medium-sized systems [10]. Recently, the removal of pharmaceuticals in horizontal and vertical subsurface flow constructed wetlandswas evaluated [9] with removal efficiencies for ibuprofen of 51–54% in winter and 85–96% in summer and for carbamazepine of 24–36% in winter and 48% in summer.

The behavior of EDC removal in CW wastewater treatment systems is not fully understood. Data display a high variation in removal efficiency, ranging from 20% to 99%, depending on the chemical compound characteristics, plant type, flow conditions (regime), and geographic location [11]. Particularly under tropical climate conditions, some research has observedthat CW are capable of removing phenolic compounds in a range from 60% to 77% [12]. However, more research is required to understand the potential of CWs in the removal of EDCs, specifically regarding the effect of variables such as plant type and hydraulic retention time. Pilot-scale research is essential to enable extrapolation to full-scale CW design for effective removal of EDCs.

2. Materials and Methods

2.1. Location and Description of Horizontal Sub-Surface Constructed Wetlands (HSSF-CWs)

This research was performed at a test site (3°43′50″ N and 76°16′20″ O) located approximately 1.1 km away from the urban area of Ginebra, Colombia, a small city of Valle del Cauca, located approximately 50 km northwest of Cali (Figure 1a). The test site is a research and technology transfer station of the domestic wastewater treatment and reuse research center of the Universidad del Valle (Cali, Colombia).

The study was carried out in a module consisting of three pilot scale horizontal subsurface flow constructed wetlands (HSSF-CWs), as shown in Figure 1b. The influent to the CWs comes from the effluent of an anaerobic pond as primary treatment of the domestic wastewater of the city. Each CW was designed to treat a flow of 3.5 $m^3 \cdot day^{-1}$ with an effective volume of 6.35 m^3 (9 × 3 × 0.6 m and 40% porosity). This design corresponded to a nominal hydraulic retention time of 1.8 days, a nominal surface loading rate of 0.13 $m \cdot day^{-1}$. One CW was planted with *Heliconias* sp. (a native flowering and marketable plant) and a second CW was planted with *Phragmites* sp. (a native perennial wetland grass). The third CW (in between the other two CWs) was a control without plants.

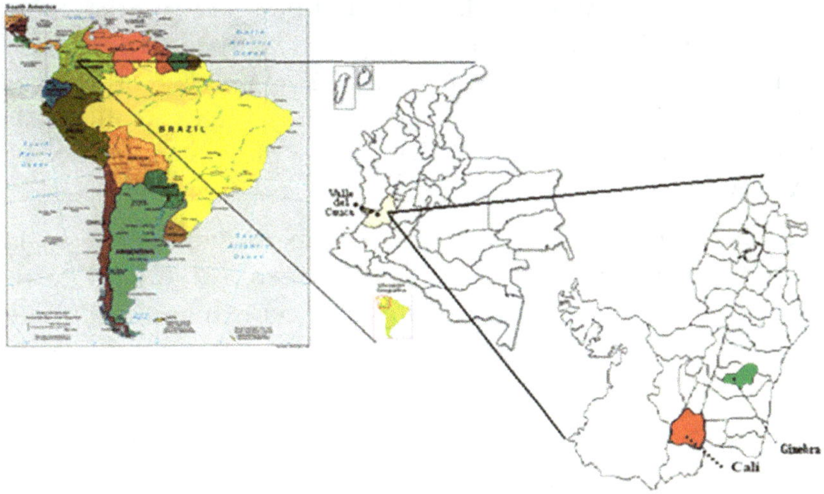

(a)

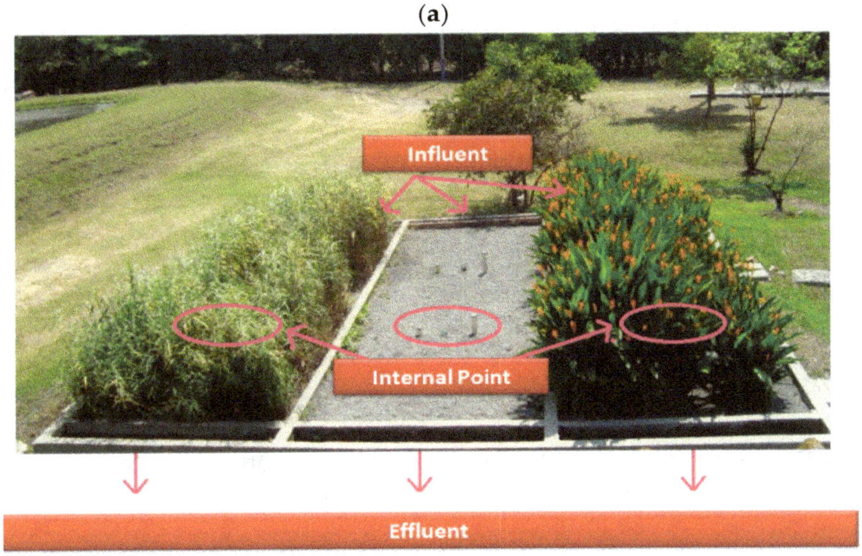

(b)

Figure 1. (**a**) Location of the pilot scale; (**b**) pilot scale horizontal subsurface flow constructed wetlands used in this study.

The EDC removal efficiency throughout each HSSF-CW was measured and two sampling points were provided to quantify the longitudinal concentration changes in BPA and NP concentration. The first was an internal sampling point, located at six meters from the inlet, which represents two-thirds of the length of the CW. These points were labeled H1 for *Heliconias*-CW, P1 for *Phragmites*-CW and C1 for control-CW. The other sample point was located at the outlet and was labeled H2, P2, and C2 respectively. The main response variable was the removal efficiency of BPA and NP. One sample was collected once per week over a seven-week periodat each sampling point, including the beginning of the first week (i.e., eight samples were collected at each sample point). Unfortunately, the third effluent *Phragmites* sample was broken in shipping, and the entire fourth set of samples was lost in shipping.

2.2. *Endocrine Disruptive Compounds*

Bisphenol A was spiked into the wastewater to ensure detection, whereas NP was detected at higher concentrations naturally in the wastewater samples and did not required an external injection. For BPA spiking, a feeding system was built to supplement the anaerobic pond effluent with an average mass loading of 2.08 mg·day^{-1} BPA (i.e., a daily average concentration of 0.59 μg·L^{-1} BPA in the influent to the CW units). The feeding system consisted of a 50 L storage tank with piping to the CWs. From the BPA stock solution (200 mg·L^{-1}, prepared with analytical grade (>98%) BPA from Sigma-Aldrich (St. Louis, MO, USA) and stored at 4 °C), a 25 mg·L^{-1} solution was prepared separately using distilled water. Every day, at the same time, the storage tank was filled with 50 L of anaerobic pond effluent and 83 mL of 25 mg·L^{-1} BPA, (after completing the research, it was observed that spiking the influent with BPA was unnecessary).

Samples from the CWs were collected in 100 mL pre-cleaned amber bottles, labeled and stored in a dark room at <4 °C until shipment to the University of Texas (UTEP), El Paso, USA. To minimize the effect of microbiological degradation, each sample was centrifuged and filtered through 0.45 and 0.12 μm cellulose membrane filter before the shipping to UTEP. The shipping time was between 1.5 and 2 days. Upon arrival, the samples were kept in a refrigerator at 4 °C until chemical analysis.

To determine the EDC concentrations, samples were analyzed at UTEP using stir bar sorptive extraction (SBSE) with in-line thermal-desorption and gas chromatography/mass spectrometry (TDU-GC–MS). Briefly, (i) twenty milliliters of the filteredsample were transferred to a 20-mL screw cap vial, (ii) sodium carbonate (200 mg) was added to adjust the pHto 11.5, and (iii) acetic acid anhydride (200 μL) was added as the derivatization reagent. A pre-conditioned stir bar (three hours at 300 °C in a flow of nitrogen) was placed in each vial, and the samples were stirred at 1000 rpm for four hours. After the extraction, the stir bar (Gerstel, Linthicum, MD, USA) was removed with forceps, rinsed with purified water, and dried with lint-free tissue paper. The stir bar was thermally desorbed in a thermal desorption (TDU) system at the sample introduction inlet of a GC-MS system (Agilent, Santa Clara, CA, USA).

2.3. *Influent Concentration of HSSF-CW*

Table 1 shows the physical and chemical composition of the influent supplied to the HSSF-CW units, including the EDC concentrations. The influent composition varied throughout the study period due to variations in the background concentrations present in the domestic wastewater.

Table 1. Composition of the influent supplied to the horizontal subsurface flow constructed wetlands (HSSF-CWs).

Statistical Results	Parameters					
	BPA (μg·L^{-1})	NPs (μg·L^{-1})	DOC * (mg·L^{-1})	COD * (mg·L^{-1})	COD$_f$ * (mg·L^{-1})	TSS * (mg·L^{-1})
$\hat{y}$	8.80	1671	17.6	252	134	63.7
σ	6.40	838	4.23	48.6	28.8	26.2
C.V.	0.73	0.50	0.24	0.19	0.21	0.41

Notes: DOC: Dissolved Organic Carbon; COD: Chemical Oxygen demand; COD$_f$: Filtered COD; TSS: Total Suspended Solids; $\hat{y}$: mean value; σ: standard deviation; and C.V.: Coefficient of variation. * These parameters were measured in accordance with the Standard methods 21th Ed.

Of the three locations sampled in the CWs, the influent had the lowest redox potential (ORP) value (−420 ± 189 mV), which was understandable given that the influent was from an anaerobic pond. Likewise, the influent dissolved oxygen (DO) concentration was low (<0.15 ± 0.3 mg·L^{-1}). The unplanted (control) CW had higher ORP and DO (−253 mV and 0.6 mg·L^{-1}, respectively) than the influent, but the effluents of the planted CWs (*Phragmites* and *Heliconia*) had the highest average ORP values (−158 and −127 mV, respectively) and DO concentrations (0.9 and 0.8 mg·L^{-1}, respectively). The higher ORP and DO in the planted CWs is likely due to oxygen translocation by plants through its roots system [13].

2.4. Statistical Analysis

Data were recorded and analyzedwith Microsoft Excel, and Minitab 15 software was used for the Friedman Two-Way Analysis of Variance by Ranks combined with a non-parametric post hoc analysis (Wilcoxon signed-rank test).

3. Results and Discussion

3.1. Longitudinal Removal of BPA in HSSF-CWs

Figure 2 shows the influent and effluent BPA concentrations for the (a) *Heliconia*-CW, (b) unplanted (control), and (c) *Phragmites*-CW, as well as the longitudinal removal efficiencies for each (parts (d), (e), and (f), respectively). A significant reduction in the BPA concentrations was observed at the internal sampling points H1, C1 and P1—each at two-thirds of the horizontal length of the CW—compared to the influent (Figure 2). This implies that BPA was transformed, sorbed, or consumed in the first two-thirds of each CW. From this point until the effluent discharge point, the BPA concentrations were fairly consistent, and decreased only marginally. The differences in the partial and total BPA removal for the internal sample point and effluent were established by a Wilcoxon signed rank test in each CW and displayed significant differences for the *Heliconia*-CW and unplanted-CW. Regarding the *Phragmites*-CW, the final third of the CW did not contribute to the improvement of the total BPA removal.

From the literature, the main removal mechanism of BPA in a CW is sorption [14–16], due to a higher octanol–water partition coefficient of the hydrophobic BPA molecule (log K_{ow} is 3.4), as well as a large surface area in the CW, which enhances sorption onto the biofilm, suspended solids, support media, and rhizosphere [17]. The sorption capacity of a compound can be better expressed in terms of the organic carbon partition coefficient (K_{oc} or K_d) related to the quantity of the sorbed compound in the solid phase with respect to the concentration in the aqueous phase [18]. This is important because BPA has two hydroxyphenyl groups, which tend to promote sorption in soils (or support media) and sediments [19]. The research of this manuscript did not include BPA measurements on the sediments or support media, but based on other research of BPA partitioning in wastewater sediment [20], a log K_d ($L \cdot kg^{-1}$) value of 4.37 was calculated for BPA at the average concentration reported in Table 1. If the log $K_d < 2.47$, sorption can be neglected, while for values higher than 4.0, sorption onto the solid phase is one of the major removal processes [18,20]. The calculated log K_d value of BPA suggests that it is likely that sorption of BPA was one of the significant removal processes in each of the three CW types investigated.

Sorption onto the support media or onto the biofilm and sediments on the media implies a longer EDC residence time for BPA in the CW. This may favor bioremediation by increasing exposure to plant uptake or microbial degradation. Indeed, phytoremediation can also be an important removal pathway, where the EDC may possibly be degraded by phytostimulation or rhizodegradation, phytodegradation, phythoextraction, sequestration or volatilization [13,21]. In some cases, the plants growing in the CW may play a significant role in the direct uptake of many organic pollutants from wastewater. For instance, the presence of *Phragmites australis* in a CW improved the removal efficiency of BPA compared to an unplanted CW [12,22,23]. In a study by Dodgen et al. [24], the plant uptake of BPA, 4-NP, diclofenac (DCL), and naproxen (NPX) during the hydroponic cultivation of *Lactuca sativa* (lettuce) and *Brassica oleracea* (collard) was investigated, and EDC accumulation was observed in both plant species with a trend in descending order of BPA > NP > DCL > NPX [24].

The Friedman Test indicates significant differences (p-value: 0.03) for the removal of BPA (Figure 2) at the intermediate sampling point (H1, C1, and P1). The post hoc test revealedthat P1 obtained the highestaverage removal efficiency (64.3%) compared with C1 (55.2%) and H1 (61.4%). Also, significant differences (p-value: 0.015) were observed for the average effluent BPA removal (H2, C2, and P2), and the post hoc test showed that the *Heliconia*-CW had the highest average BPA removal efficiency (73.3%) compared withthe unplanted-CW (62.2%) and *Phragmites*-CW (70.2%), as shown in Table 2.

Both planted CWs had greater average effluent removal efficiencies than the unplanted control, and both planted CWs had greater effluent removal efficiencies than the unplanted-CW for all but one of the sampling events; thus, it is assumed that plant vegetation had a role in BPA removal. The specific removal mechanism by the plants was not investigated.

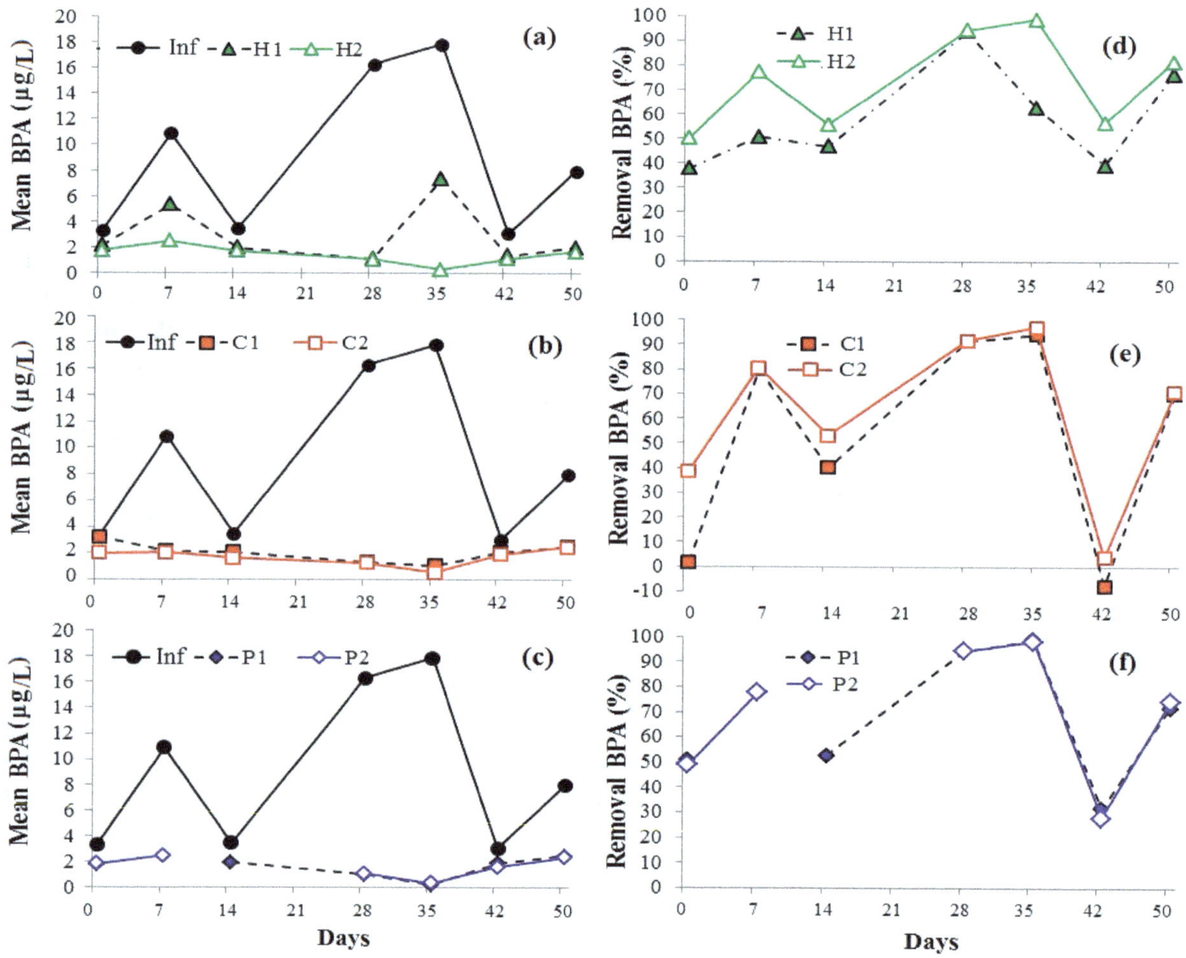

Figure 2. Longitudinal BPA concentrations in CWs of (**a**) *Heliconia* (H); (**b**) unplanted control (C), and (**c**) *Phragmites* (P), for influent (Inf), two-thirds internal (1), and effluent (2) sample points.Removal efficiency of BPA in (**d**) *Heliconia*; (**e**) unplanted control; and (**f**) *Phragmites*.

Table 2. Average effluent BPA removal for *Heliconia*-CW, unplanted-CW and *Phragmites*-CW.

Removal Efficiencies (%)	BPA		
	Heliconia	*Unplanted*	*Phragmites*
$\hat{y}$	73.3%	62.2%	70.2%
Maximum	98.6%	97.1%	98.3%
Minimum	50.0%	3.7%	27.9%
σ	19.6%	33.1%	27.1%
C.V.	0.27	0.53	0.39
n *	7	7	6

* The third *Phragmites* sample was broken in shipping. Notes: $\hat{y}$: mean value; σ: standard deviation; C.V.: coefficient of variation; and n number of samples.

3.2. Longitudinal Removal of NP in HSSF-CWs

The average total NP concentration in the influent was (1671 ± 838) μg·L^{-1}. A reduction of the NP concentration was observed at the internal sampling point (two-thirds of the length) of each wetland (H1, P1, and C1), with average concentrations of (724 ± 453), (1150 ± 499) and (1041 ± 446) μg·L^{-1}, respectively. The lowest final effluent concentrations were obtained for the *Heliconia*-CW (629 ± 318 μg·L^{-1}) and *Phragmites*-CW (736 ± 284 μg·L^{-1}), showing more NP removal in the final third of the CWs. The NP removal efficiencies were less than those of BPA. A Wilcoxon signed rank test was used to establish significance of differences between the internal sampling point and effluent concentrations in each CW. Although the plantedCWs had a higher average NP removal efficiency in the effluent than the internal sample point, these results are not statistically different.

The unplanted-CW displayed a different behavior, showing an increment of the NP concentration in the effluent (1103 ± 538 μg·L^{-1}) compared with its internal point. NP desorption was observed in the final third of the CW, despite NPs having high K_{ow} values (log K_{ow} 3.80 to 4.77). This negative removal efficiency in CW was also reported in other research [18] for seven different PPCPs, attributed to an initial retention and sorption, but subsequent release during passage of the wastewater through the medium of the CW.

The *Heliconia*-CW showed the highest average effluent NP removal efficiency of 62.8 ± 20.1%, while the *Phragmites*-CW and unplanted-CW had a removal efficiency of 25.3 ± 37.1% and 52.1 ± 23.2%, respectively (Table 3). These results confirm a statistically significant difference between planted and unplanted CWs with *p*-values of 0.042 for the internal sample point and 0.03 for the effluent. The posthoc test corroborated this conclusion, confirming that the *Heliconia*-CW showed a higher NP removal efficiency than the *Phragmites*-CW and the unplanted-CW.

Table 3. Average effluent NP removal for *Heliconia*-CW, unplanted-CW, and *Phragmites*-CW.

Removal Efficiencies (%)	NP		
	Heliconia	*Unplanted*	*Phragmites*
$\hat{y}$	62.8%	25.3%	52.1%
Maximum	90.0%	83.7%	80.2%
Minimum	28.0%	−12.3%	20.4%
σ	20.1%	37.1%	23.2%
C.V.	0.32	1.46	0.4
n *	7	7	6

* The third *Phragmites*-CW sample was broken in shipping. Notes: $\hat{y}$: mean value; σ: standard deviation; C.V.: coefficient of variation; and n number of samples.

3.3. EDC Removal Rate Against Mass Loading Rates

The total mass removed of each EDC compound was plotted against its total inlet mass loading rate (Figure 3). The removal rate of BPA in the *Heliconia*-CW, unplanted-CW and *Phragmites*-CW increased as the mass loading rate increased as well. BPA removal efficiency in all three CW types investigated was not sensitive to the mass loading rate and was almost completely removed in all CWs investigated (Figure 3a). However, NP removal decreased at a high mass loading rate. Moreover, NP was poorly removed in the unplanted-CW both at low and high mass loading rates (Figure 3b).

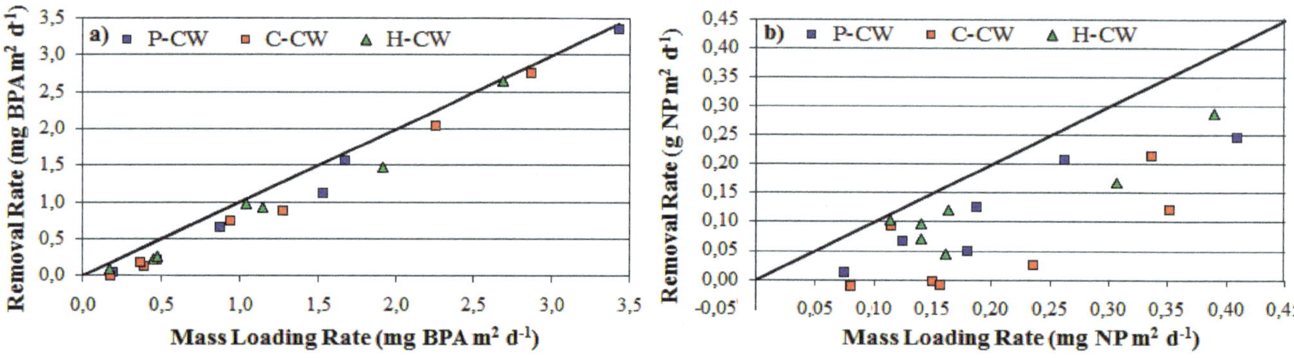

Figure 3. Endocrine disruptive compound(EDC) removal rate for H-CW, C-CW, and P-CW against their mass loading rates (**a**) BPA and (**b**) NP. Continuous line represents 100% removal.

3.4. Overall Performance of Each HSSF-CW in EDC Removal

Four scenarios (quadrants) were identified (Figure 4) according to the weekly effluent removal efficiencies of BPA and NP. Zone I is the worst scenario in which both compounds were removed with less than 50% efficiency. In Zone II, BPA removal was greater than 50%, but NP was less than 50%. Zone III shows a BPA removal efficiency less than 50% and NP greater than 50%. Finally, in Zone IV, both compounds were removed with efficiencies greater than 50%. The best CW removal efficiencies were observed for the *Heliconia*-CW, in which six of seven (85%) of the data points are in zone IV, compared with the unplanted-CW with more than five of seven (71%) of the data points in zone II. Overall, the BPA and NP removal efficiencies were in the following descending order: *Heliconia*-CW > *Phragmites*-CW > unplanted-CW. This performance is likely due to sorption in the HSSF-CW onto support media. Also, the rhizosphere of planted CWs likely generates benefits such as increasing DO concentrations and releasing organic exudates that serve as nutrient sources for the growth of microorganisms [25].

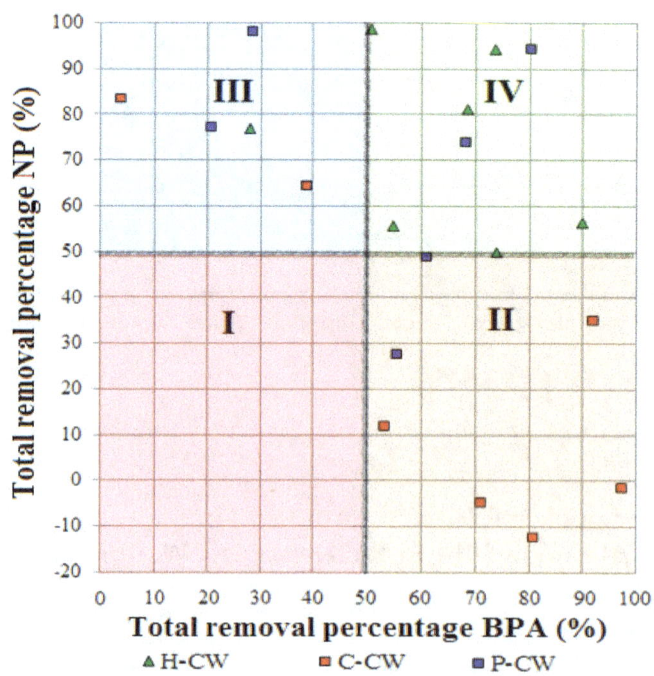

Figure 4. Quadrant chart for NP removal efficiency versus BPA removal efficiency for the *Heliconia*-CW (H), *Phragmites*-CW (P) and unplanted-CW (C).

This study showed that BPA and NP can beeffectively removed by planted HSSF CW under tropical conditions, with removal being more efficient withthe native and marketable plant *Heliconia* sp. in the CW.

4. Conclusions

The *Heliconia*-CW removed BPA (73%) and NP (62%) more efficiently than the *Phragmites*-CW (70% and 52%, respectively) and unplanted-CW (62% and 25%). The *Heliconia*-CW showed an improvement in BPA removal in the last third of the length of the wetland (*p*-value: 0.015). In contrast, the last third section of the *Phragmites*-CW and unplanted-CW did not contribute to additional BPA removal. Desorption of NP in the unplanted-CW was observed in the last third of the lengthof the CW, suggesting that a sorption–desorption equilibrium can be reached. This situation was not found for *Phragmites* sp. and *Heliconias* sp., corroborating that plants had a positive influence in the CW.

Acknowledgments: This study was carried out under the framework of UNESCO-IHE Partnership Research Fund (UpaRF), Natural System for Wastewater Treatment and Reuse: Technology Adaptations and Implementations in Developing Countries (NATSYS Project). The authors express special gratitude to ACUAVALLE (local water agency) and "Ricclisa: Programa para el Fortalecimiento de la Red Interinstitucional de Cambio Climático y Seguridad Alimentaria RC0853-2012 supported by Colciencias" for their contribution to this research.

Author Contributions: Andrés Toro-Vélez, Miguel Peña-Varón, Hector García-Hernández and Carlos Madera-Parra conceived and designed the experiments; Andrés Toro-Vélez performed the experiments; Andrés Toro-Vélez, analyzed the data; Wen Yee Lee and Shane Walker contributed reagents/materials/analysis tools; Andrés Toro-Vélez, Piet Lens and Shane Walker wrote the paper.

References

1. Zimmerman, M.J. *Occurrence of Organic Wastewater Contaminants, Pharmaceuticals, and Personal Care Products in Selected Water Supplies, 2005. Cape Cod, Massachusetts, June 2004*; Open-File Report 2005-1206; U.S. Geological Survey: Denver, CO, USA, 2005; p. 16.
2. Luo, Y.; Guo, W.; Ngo, H.H.; Nghiem, L.D.; Hai, F.I.; Zhang, J.; Wang, X.C. A review on the occurrence of micropollutants in the aquatic environment and their fate and removal during wastewater treatment. *Sci. Total Environ.* **2014**, *473*, 619–641. [CrossRef] [PubMed]
3. Terzić, S.; Senta, I.; Ahel, M.; Gros, M.; Petrović, M.; Barcelo, D.; Jabucar, D. Occurrence and fate of emerging wastewater contaminants in Western Balkan Region. *Sci. Total Environ.* **2008**, *399*, 66–77. [CrossRef] [PubMed]
4. Kasprzyk, B.; Dinsdale, R.M.; Guwy, A.J. The occurrence of pharmaceuticals, personal care products, endocrine disruptors and illicit drugs in surface water in South Wales, UK. *Water Res.* **2008**, *42*, 3498–3518. [CrossRef] [PubMed]
5. Suárez, S.; Carballa, M.; Omil, F.; Lema, J.M. How are pharmaceutical and personal care products (PPCPs) removed from urban wastewaters? *Rev. Environ. Sci. Bio/Technol.* **2008**, *7*, 125–138. [CrossRef]
6. Huang, Y.Q.; Wong, C.K.C.; Zheng, J.S.; Bouwman, H.; Barra, R.; Wahlström, B.; Wong, M.H. Bisphenol A (BPA) in China: A review of sources, environmental levels, and potential human health impacts. *Environ. Int.* **2012**, *42*, 91–99. [CrossRef] [PubMed]
7. Newbold, R.R.; Jefferson, W.N.; Padilla-Banks, E. Prenatal exposure to bisphenol at environmentally relevant doses adversely affects the murine female reproductive tract later in life. *Environ. Health Perspect.* **2009**, *117*, 879–885. [CrossRef] [PubMed]
8. United Nations Environment Programme and World Health Organization. *State of the Science of Endocrine Disrupting Chemicals–2012*; Bergman, A., Heindel, J.J., Jobling, S., Kidd, K.A., Zoeller, R.T., Eds.; WHO Press: Geneva, Switzerland, 2013.
9. Yan, P.; Qin, R.C.; Guo, J.S.; Yu, Q.; Li, Z.; Chen, Y.P.; Shen, Y.; Fang, F. Net-Zero-Energy Model for Sustainable Wastewater Treatment. *Environ. Sci. Technol.* **2017**, *51*, 1017–1023. [CrossRef] [PubMed]
10. Nahlik, A.M.; Mitsch, W.J. Tropical treatment wetlands dominated by free-floating macrophytes for water quality improvement in Costa Rica. *Ecol. Eng.* **2006**, *28*, 246–257. [CrossRef]

11. Hijosa, M.; Matamoros, V.; Sidrach, R.; Martín, J.; Bécares, E.; Bayona, J. Comprehensive Assessment of the Design Configuration of Constructed Wetlands for the Removal of Pharmaceuticals and Personal Care Products from Urban Wastewaters. *Water Res.* **2010**, *44*, 3669–3678. [CrossRef] [PubMed]
12. Abira, M.A.; van Bruggen, J.J.A.; Denny, P. A Potential of a tropical subsurface constructed wetland to remove phenol from pre-treated pulp and papermill wastewater. *Water Sci. Technol.* **2005**, *51*, 173–176. [PubMed]
13. Zhang, D.; Gersberg, R.M.; Ng, W.J.; Tan, S.K. Removal of pharmaceuticals and personal care products in aquatic plant-based systems: A review. *Environ. Pollut.* **2014**, *184*, 620–639. [CrossRef] [PubMed]
14. Fountoulakis, M.S.; Terzakis, S.; Kalogerakis, N.; Manios, T. Removal of polycyclic aromatic hydrocarbons and linear alkylbenzene sulfonates from domestic wastewater in pilot constructed wetlands and a gravel filter. *Ecol. Eng.* **2009**, *35*, 1702–1709. [CrossRef]
15. Matamoros, V.; Bayona, J.M. Elimination of pharmaceuticals and personal care products in subsurface flow constructed wetlands. *Environ. Sci. Technol.* **2006**, *40*, 5811–5816. [CrossRef] [PubMed]
16. Dordio, A.V.; Teimão, J.; Ramalho, I.; Carvalho, A.J.P.; Candeias, A.J.E. Selection of a support matrix for the removal of some phenoxyacetic compounds in constructed wetlands systems. *Sci. Total Environ.* **2007**, *380*, 237–246. [CrossRef] [PubMed]
17. Avila, C.; Reyes, C.; Bayona, J.M.; García, J. Emerging organic contaminant removal depending on primary treatment and operational strategy in horizontal subsurface flow constructed wetlands: Influence of redox. *Water Res.* **2013**, *47*, 315–325. [CrossRef] [PubMed]
18. Verlicchi, P.; Galletti, A.; Petrovic, M.; Barceló, D.; Al Aukidy, M.; Zambello, E. Removal of selected pharmaceuticals from domestic wastewater in an activated sludge system followed by a horizontal subsurface flow bed-analysis of their respective contributions. *Sci. Total Environ.* **2013**, *454–455*, 411–425. [CrossRef] [PubMed]
19. Imfeld, G.; Braeckevelt, M.; Kuschk, P.; Richnow, H. Monitoring and Assessing Processes of Organic Chemicals Removal in Constructed Wetlands. *Chemosphere* **2009**, *74*, 349–362. [CrossRef] [PubMed]
20. Clara, M.; Strenn, B.; Saracevic, E.; Kreuzinger, N. Adsorption of bisphenol-A, 17 beta-estradiole and 17 alpha-ethinylestradiole to sewage sludge. *Chemosphere* **2004**, *56*, 843–851. [CrossRef] [PubMed]
21. Zhang, B.Y.; Zheng, J.S.; Sharp, R.G. Phytoremediation in Engineered Wetlands: Mechanisms and Applications. *Procedia Environ. Sci.* **2010**, *2*, 1315–1325. [CrossRef]
22. Li, Y.; Zhu, G.; Ng, W.J.; Tan, S.K. A review on removing pharmaceutical contaminants from wastewater by constructed wetlands: Design, performance and mechanism. *Sci. Total Environ.* **2013**, *468–469*, 908–932. [CrossRef] [PubMed]
23. Toyama, T.; Yusuke, S.; Daisuke, I.; Kazunari, S.; Young, C.; Shintaro, K.; Michihiko, I. Biodegradation of Bisphenol A and Bisphenol F in the Rhizosphere Sediment of Phragmites Australis. *J. Biosci. Bioeng.* **2009**, *108*, 147–150. [CrossRef] [PubMed]
24. Dodgen, L.K.; Li, J.; Parker, D.; Gan, J.J. Uptake and accumulation of four PPCP/EDCs in two leafy vegetables. *Environ. Pollut.* **2013**, *182*, 150–156. [CrossRef] [PubMed]
25. Song, H.L.; Nakano, K.; Taniguchi, T.; Nomura, M.; Nishimura, O. Estrogen Removal from Treated Municipal Effluent in Small-Scale Constructed Wetland with Different Depth. *Bioresour. Technol.* **2009**, *100*, 2945–2951. [CrossRef] [PubMed]

7

Electrocoagulation: A Promising Method to Treat and Reuse Mineral Processing Wastewater with High COD

Gaogui Jing [1,2], Shuai Ren [1,2], Yuesheng Gao [3], Wei Sun [1,2] and Zhiyong Gao [1,2,*]

1 School of Minerals Processing and Bioengineering, Central South University, Changsha 410083, China; gaoguijing@csu.edu.cn (G.J.); renshuai@csu.edu.cn (S.R.); sunmenghu@126.com (W.S.)
2 Key Laboratory of Hunan Province for Clean and Efficient Utilization of Strategic Calcium-Containing Mineral Resources, Central South University, Changsha 410083, China
3 Department of Chemical Engineering, Michigan Technological University, Houghton, MI 49931, USA; ygao4@mtu.edu
* Correspondence: zhiyong.gao@csu.edu.cn

Abstract: Mineral processing wastewater contains large amounts of reagents which can lead to severe environmental problems, such as high chemical oxygen demand (COD). Inspired by the wastewater treatment in such industries as those of textiles, food, and petrochemistry, in the present work, electrocoagulation (EC) is applied for the first time to explore its feasibility in the treatment of wastewater with an initial COD of 424.29 mg/L from a Pb/Zn sulfide mineral flotation plant and its effect on water reuse. Typical parameters, such as anode materials, current density, initial pH, and additives, were characterized to evaluate the performance of the EC method. The results showed that, under optimal conditions, i.e., iron anode, pH 7.1, electrolysis time 70 min, 19.23 mA/cm^2 current density, and 4.1 g/L activated carbon, the initial COD can be reduced to 72.9 mg/L, corresponding to a removal rate of 82.8%. In addition, compared with the untreated wastewater, EC-treated wastewater was found to benefit the recovery of galena and sphalerite, with galena recovery increasing from 25.01% to 36.06% and sphalerite recovery increasing from 59.99% to 65.33%. This study confirmed that EC is a promising method for the treatment and reuse of high-COD-containing wastewater in the mining industry, and it possesses great potential for wide industrial applications.

Keywords: mineral processing; wastewater treatment; flotation; electrocoagulation (EC); chemical oxygen demand (COD)

---

1. Introduction

Wastewater from mineral processing remains a headache for the mining industry not only because of its large volume, but more importantly, its hazardous components. Among the complicated residual reagents, the organic reagents receive the most attention due to their high chemical oxygen demand (COD) levels. Direct discharge of wastewater causes serious environmental pollution, while its reuse as feed water without proper treatments causes great harm to the recovery and grade of valuable minerals [1]. To solve this problem, techniques for wastewater treatment have been intensively developed in the mining industry, such as natural settling, flocculating setting, chemical oxidation, adsorption, and biodegradation to remove COD [2]. However, these techniques possess such disadvantages as inefficiency in COD removal, extreme operating conditions, high operation cost, huge instrument investment, etc. It is worth noting that coagulation–flocculation, the most common technique in beneficiation plants because of its economy, usually has a relative low COD removal rate of only 35% [3].

Electrocoagulation (EC) has become widely used as a wastewater treatment technology in the past decade and has exhibited a satisfactory removal of COD in diverse fields, such as in textiles [4], the food industry [5], dairying [6], brackish water [7], and potable water [8]. Its advantages include easy operation, small area occupation, and a high degree of automation. In the process of EC, iron or aluminum is usually adopted as the anode material, and each one can lead to a series of similar redox reactions. Taking the iron electrode as an example, simplified reactions occurring in acidic and alkaline solutions can be described as follows [9]:

Under acidic conditions:

$$Fe_{(s)} \rightarrow Fe^{2+}_{(aq)} + 2e^- \quad (1)$$

$$2H_2O \rightarrow O_2 + 4H^+ + 4e^- \quad (2)$$

$$4Fe^{2+}_{(aq)} + 10H_2O_{(l)} + O_{2(g)} \rightarrow 4Fe(OH)_{3(s)} + 8H^+_{(aq)} \quad (3)$$

Under alkaline conditions:

$$Fe_{(s)} \rightarrow Fe^{2+}_{(aq)} + 2e^- \quad (4)$$

$$Fe^{2+} + 2OH^- \rightarrow Fe(OH)_2 \quad (5)$$

As shown in above equations, the iron anode is used to generate ions in the aqueous medium. The generated iron ions are then immediately hydrolyzed to form $Fe(OH)_3$, monomeric ions, and polymeric hydroxy complexes such as $Fe(H_2O)_6{}^{3+}$, $Fe(H_2O)_4(OH)^{2+}$, $Fe_2(H_2O)_6(OH)_4{}^{4+}$, $Fe_2(H_2O)_8(OH)_2{}^{4+}$, and $Fe(H_2O)_5(OH)^{2+}$, depending on the solution pH [10]. These hydroxides and polyhydroxides are excellent adsorbents for counter ions and organic pollutants. In some cases, they can even form complexes with some of the organics [11]. Pollutants in the wastewater can be treated by physical and chemical attachment to hydroxides or chemical reactions [12]. After the EC treatment, separation of precipitation from the liquid is traditionally achieved via a filter process. Then, the filtered and separated sludge is usually treated using the incineration and landfill treatments [13].

In this paper, EC treatment of sulfide mineral flotation wastewater was applied for the first time, and the effects of different parameters like anode materials, current density, initial pH, electrolysis time, and additives on COD removal rate were investigated. The treated water was then recycled as feed water in the sulfide ore flotation. For comparison, untreated wastewater and fresh water were used to assess the effects of EC-treated water on the grade and recovery of galena and sphalerite.

2. Materials and Methods

2.1. Samples, Reagents, and Analytical Methods

Samples of mineral processing wastewater and actual ores were all from a Pb-Zn beneficiation plant located in Southern China. The plant had three different types of wastewater, namely Pb-Zn flotation wastewater, pyrite concentrate wastewater, and pyrite tailing wastewater, accounting for 70%, 14%, and 16% of the total amount, respectively. In this study, three different types of wastewater samples were mixed at the above ratio to obtain the mixed wastewater samples. The pH and COD values of each wastewater are shown in Table 1.

Table 1. Chemical oxygen demand (COD) and pH values of mineral processing wastewater samples in a Pb-Zn beneficiation plant.

Sample Name	pH	COD (mg/L)
Pb-Zn flotation wastewater	11.90	434.7
Pyrite concentrate wastewater	9.11	245.3
Pyrite tailing wastewater	8.29	603.2
Mixed wastewater	11.83	424.9

The H_2O_2 (30% pure), H_2SO_4, HCl, and NaOH used in the EC experiments were all of analytical grades. The granular activated carbon (GAC) was made from a coconut shell.

The pH of the solution was measured by a pH meter (Leici, Shanghai, China). The COD values of the wastewater were analyzed by fast digestion–spectrophotometric method (HJ/T 399-2007).

The chemical compositions of the actual ore samples were relatively simple. SiO_2 and CaO were the main non-metallic ingredients [14]. The lead and zinc grades in the samples took up 4.17% and 7.18% of the total, respectively. The major metallic minerals were galena, sphalerite, and pyrite. Actual ore samples were crushed to below 3 mm using a stainless-steel hammer and then ground in a conical ball mill. The ground products were used as feed samples for flotation tests.

Diethyldithiocarbamate (95% purity) and sodium n-butylxanthate (92% purity) were purchased from Aladdin Industrial, Shanghai, China. The CaO (analytically pure) and $CuSO_4$ (analytically pure) used in this study were provided by Kermel Company, Tianjin, China. Terpenic oil (chemically pure) was obtained from Haoshen Chemical Reagent, Shanghai, China.

2.2. EC Experiments

The EC set-up used in this study is shown in Figure 1. The total volume of the electrolytic cell was 250 mL, and the experiments were carried out by treating 200 mL wastewater samples for each run at room temperature. Two pairs of plates (length = 6.5 cm; width = 5.6 cm; thickness = 0.2 cm) were placed on the electrolytic cell in parallel. The anode and cathode plates were connected to the positive and negative ports of the QW-MS3010D DC power supply (0–5 A, 0–80 V), with constant load current. NaOH and H_2SO_4 solutions were used to adjust the pH. At the end of the experiment, water was transferred to beakers for two hours of settling, and then the COD value of the supernatant was measured. Finally, the sludge produced by the EC was filtered, dried, and weighed.

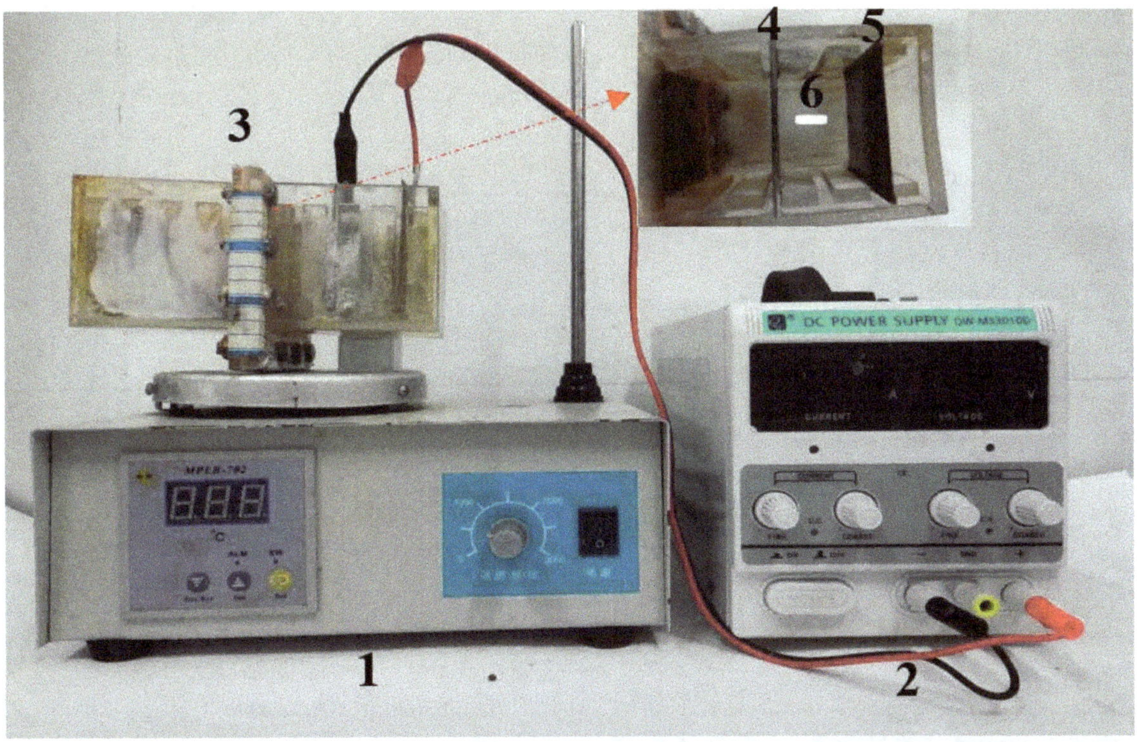

Figure 1. Electrocoagulation (EC) experimental set-up: (1) Magnetic stirrer; (2) DC power supply; (3) electrolytic cell; (4) anode plate; (5) cathode plate; (6) magnetic bar-stirrer.

The COD removal rate was employed to evaluate the electrocoagulation efficiency, which is calculated according to the following Equation (6).

$$R(\%) = \frac{COD_0 - COD_1}{COD_0} \times 100, \quad (6)$$

where R represents the removal rate of COD (%), COD_0 represents the initial COD value of the wastewater before electrocoagulation (mg/L), and COD_1 represents the COD value of the wastewater after electrocoagulation (mg/L).

2.3. Flotation Experiments

Fresh water, wastewater, and treated water were then used for flotation at room temperature. The whole process can be divided into two stages: Galena flotation followed by sphalerite flotation. For each flotation experiment, an 800 g ore sample was ground by a conical ball mill to obtain a product of 85 wt.% passing 74 μm. After grinding, the ore pulp was transferred to the galena flotation stage, including single-stage roughing, three-stage cleaning, and single-stage scavenging. The tailing of galena flotation was used later for sphalerite flotation, where single-stage roughing, two-stage cleaning, and single-stage scavenging were adopted. In both flotation stages, the concentrates of roughing were reground to 80 wt.% passing 44 μm in a Φ150 × 100 conical mill. According to the existing process of the beneficiation plant, sequential addition of regulator, collector, and frother to the flotation cell were carried out, as shown in Figure 2. After flotation, the products were dried, weighed, and assayed, and then the recovery was calculated on the basis of the mass and metal balance, as shown in Equation (7).

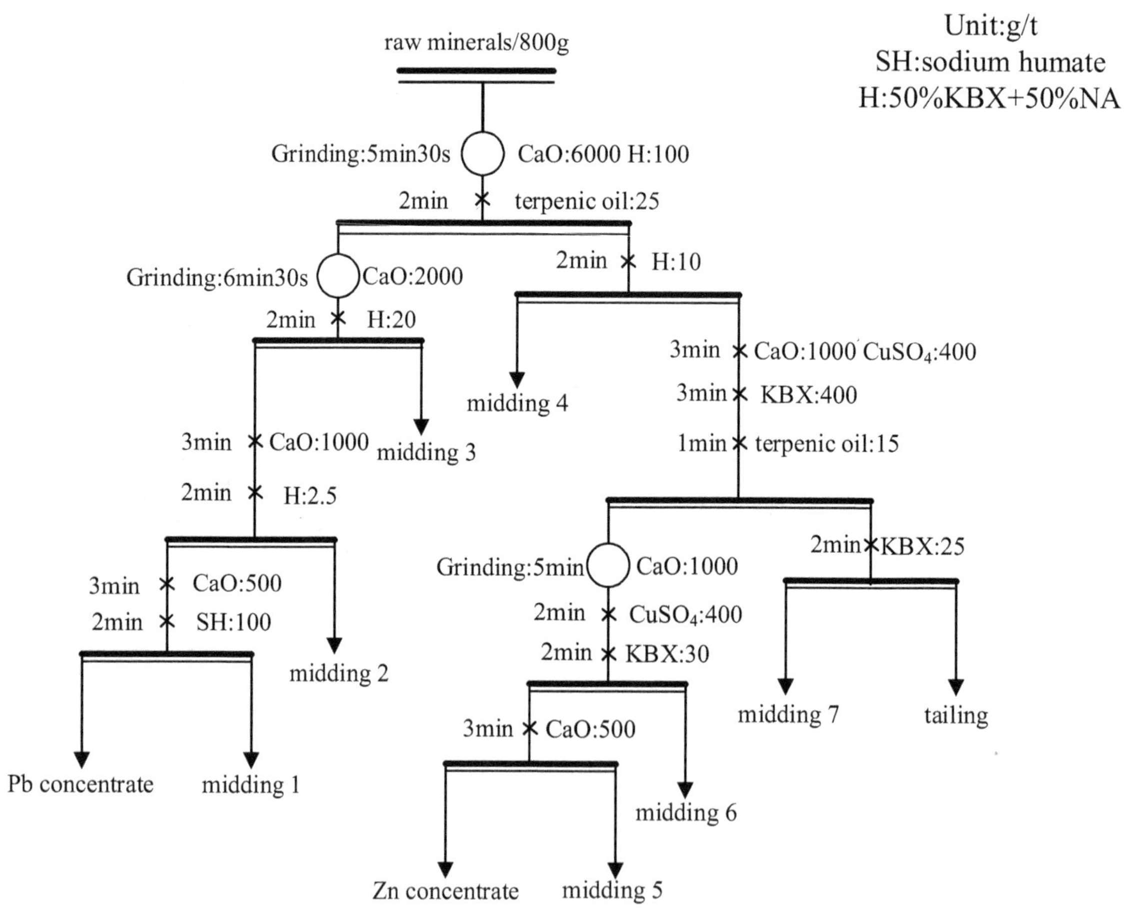

Figure 2. Flowsheet and corresponding experimental conditions of batch flotation tests using different types of water.

$$\varepsilon(\%) = \frac{\gamma_i \times \beta_i}{\sum_1^n (\gamma_i \times \beta_i)} \times 100, \quad (7)$$

where ε is the recovery of product, γ_i is the yield, and β_i is the grade.

3. Results and Discussion

3.1. *Effect of EC Parameters on COD Removal Rate of Mixed Wastewater*

Unlike anode materials, cathode materials do not have a significant effect on EC efficiency except in electro-Fenton processes [15]. We have learned that stainless steel is the most widely-used cathode material in industry, considering its economy and durability [16]. In addition, the distance between the anode and cathode is suggested to be 1~2 cm, according to previous publications and industrial experience [17]. After verification, 2 cm was the optimal distance to be adopted in our following experiments. As the anode material is a prime consideration for EC technique [18], iron and aluminum plates were used as anode materials in this study. The results of COD removal efficiency with the two anode materials are shown in Figure 3. When using an iron electrode, the COD value decreases from 424.9 to 158.6 mg/L, which is equivalent to a removal rate of 62.7%. However, when an aluminum anode is used, the COD value is reduced to 265.1 mg/L, corresponding to a removal rate of 37.7%. The results indicate that in this study, Fe performs better as the anode material of the EC process. Therefore, an Fe anode was adopted in the subsequent experiments.

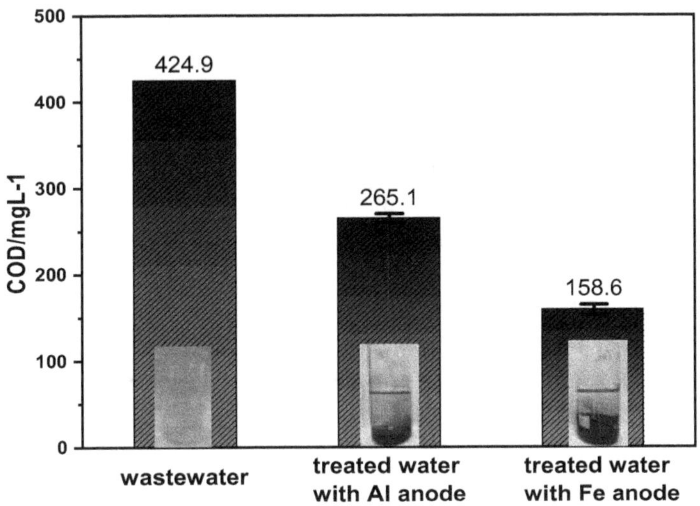

Figure 3. Effect of anode materials for EC treatment on mixed wastewater COD value (cathode material: Stainless steel, current density: 13.74 mA/cm^2, pH: 7.1, electrolysis time: 50 min).

Current density is another important parameter that determines the amount of coagulant produced in the reactor [19]. It has a significant effect on the EC efficiency and electricity consumption. The experimental results are summarized in Figure 4. Figure 4 shows that COD removal rate first increases and then declines with the increase of current density. COD removal rate increases from 49.41% to 77.46% as the current density rises to 24.73 mA/cm^2. Then, it decreases to 60.42% as the current density continues to increase to 30.22 mA/cm^2. The current density is related to the amount of dissolved iron ions and the formation of $Fe(OH)_3$, which determines the COD removal efficiency [20]. It can also affect the cell voltage through various overpotentials, which may be conducive to the removal efficiency of the COD. If the current density is relatively low, the lack of coagulant may restrict the removal efficiency. Therefore, the curve presents an upward trend at an early stage. However, when the current density continues to increase to 32.22 mA/cm^2, the temperature of treated wastewater increases significantly. High temperature causes the instability of coagulation, which can lead to a low EC efficiency. Furthermore, energy consumption grows gradually with the increase of current density [21]. Taking both the COD removal rate and electricity consumption into account, 19.23 mA/cm^2 was used as the current density for subsequent experiments.

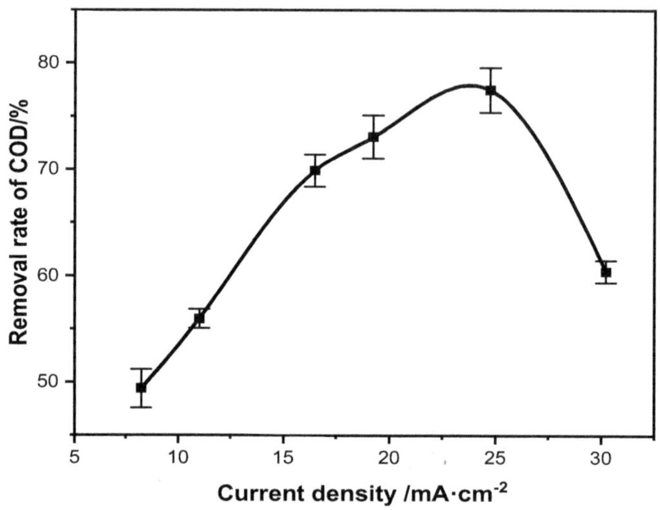

Figure 4. Effect of current density on mixed wastewater COD removal rate (anode material: Fe, cathode material: Stainless steel, pH: 7.1, electrolysis time: 50 min).

To evaluate the effect of the initial pH, the wastewater samples were adjusted to a preset pH value using sodium hydroxide or sulfuric acid. The effects of initial pH value on COD removal rate are shown in Figure 5. The results show that the COD removal rate in the EC process remains around 75.0% in a wide pH value range from 3 to 9. However, when the initial pH increases to 11, the removal efficiency presents a downward trend with a corresponding removal rate of 71.08%; this is probably because of the formation of soluble $Fe(OH)^{-4}$, which is not conducive to the floc formation. On the contrary, at a neutral or acidic pH, higher removal rates occur as most of the iron complexes are formed. Hence, subsequent experiments were carried out under neutral pH conditions. pH 7.1 was employed in the following experiments.

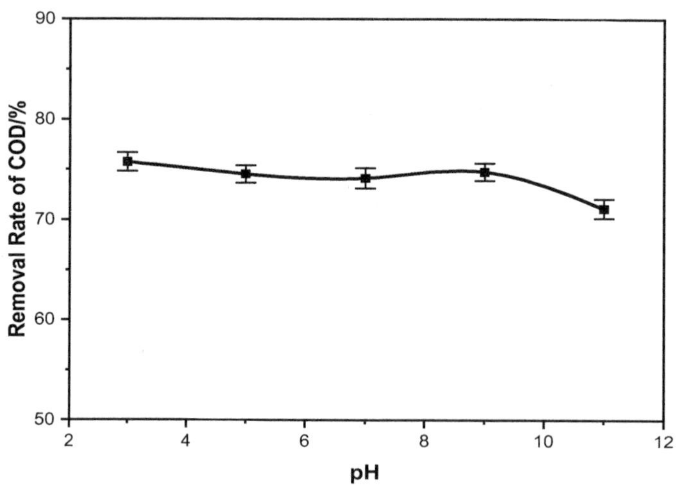

Figure 5. Effect of initial pH value on mixed wastewater COD removal rate (anode material: Fe, cathode material: Stainless steel, current density: 19.23 mA/cm^2, electrolysis time: 50 min).

Electrolysis time is another important parameter affecting COD removal efficiency [22]. The results of the effect of electrolysis time on the COD removal rate are provided in Figure 6. Figure 6 shows that the COD removal rate increases rapidly in the initial phase, which is due to the increase of coagulant;

yet, the growth trend of the COD removal rate gradually slows down with the extension of electrolysis time. When electrolysis time increases to 70 min, the COD removal rate reaches a plateau at 77.62% because the adsorption of organic pollutants reaches an equilibrium state. The electrolysis time of subsequent experiments was set to be 70 min.

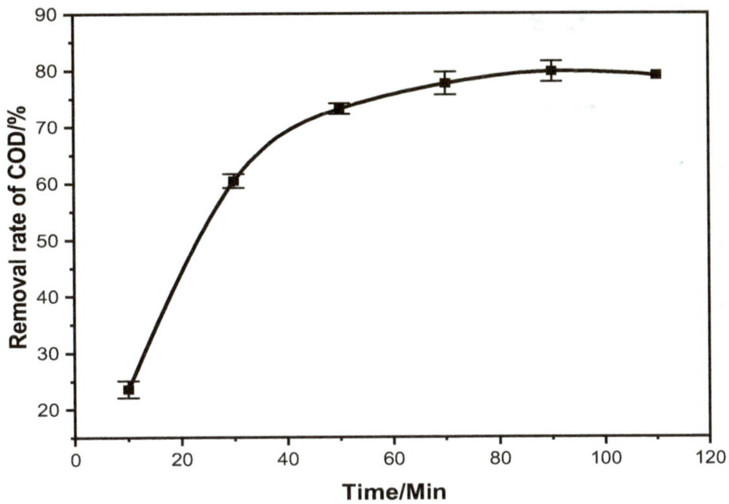

Figure 6. Effect of electrolysis time on mixed wastewater COD removal rate (anode material: Fe, cathode material: Stainless steel, current density: 19.23 mA/cm^2, pH: 7.1).

3.2. *Effect of Additives on COD Removal Rate of Mixed Wastewater*

Under the same current density, an increase of supporting electrolyte concentration can lead to a decrease of the interelectrode resistance and thus reduce energy consumption. Ghernaout reported that the addition of a supporting electrolyte can also affect the removal rate of COD [23]. Hence, the effect of supporting electrolyte concentration on EC efficiency was further studied in this study, and Na_2SO_4 was chosen as the supporting electrolyte because of its low cost. The Na_2SO_4 concentration of 1.0, 1.1, 1.2, and 1.3 g/L corresponded to the current densities of 8.24, 13.74, 19.23, and 24.73 mA/cm^2, respectively. The results of the effect of Na_2SO_4 addition on the COD removal rate are shown in Figure 7. The removal rates with the addition of Na_2SO_4 are much lower than those without Na_2SO_4. No positive effect on COD removal efficiency is observed by the addition of Na_2SO_4, which is consistent with previous reports [24]. The low removal efficiency can be attributed to the increase of the passivation layer caused by the addition of Na_2SO_4 [25].

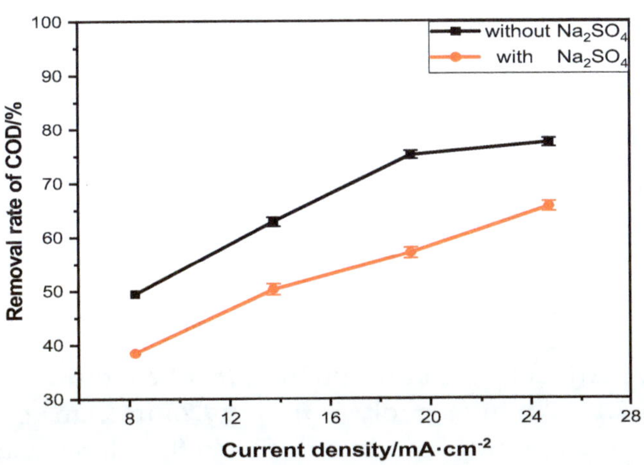

Figure 7. Effect of addition of Na_2SO_4 on mixed wastewater COD removal rate (anode material: Fe, cathode material: Stainless steel, pH: 7.1, electrolysis time: 70 min).

Activated carbon is widely used to adsorb organic compounds in wastewater [26]. In order to study the effect of activated carbon on the COD removal efficiency with EC, wastewater samples were treated by GAC adsorption, EC/GAC adsorption (EC plus GAC in one reactor), and EC + GAC adsorption (EC followed by GAC in different reactors) process, and their results are summarized in Figure 8.

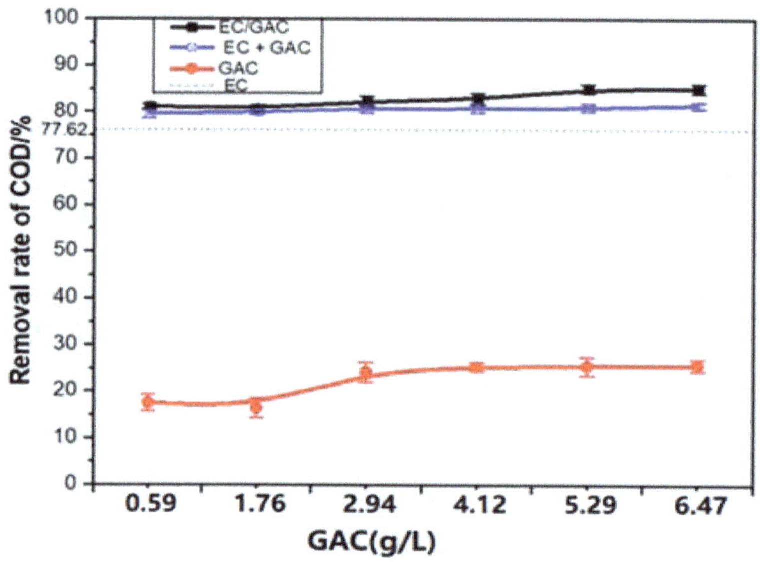

Figure 8. Effect of granular activated carbon (GAC) treatment on mixed wastewater COD removal rate (anode material: Fe, cathode material: Stainless steel, current density: 19.27 mA/cm^2, pH: 7.1, electrolysis time: 70 min).

The Figure 8 shows that the COD removal rate using GAC alone is low (only 25.7%) when the GAC dosage reaches 6.47 g/L, indicating that GAC adsorption has little effect on the residual reagents in mineral processing wastewater. When wastewater is treated by the EC process alone, the COD removal rate is 77.62%. Compared with EC or GAC alone, both the EC/GAC and EC + GAC processes can improve COD removal. However, it is worth noting that the COD removal rate of EC/GAC is higher than that of EC + GAC. At a dosage of 2.94 g/L GAC, the COD decreases from 424.9 to 74.9 and 83.3 mg/L, corresponding to removal rates of 82.37% for EC/GAC and 80.4% for EC/GAC, respectively. Therefore, the EC/GAC process was adopted in the subsequent experiments.

It can be concluded that the adsorption of GAC alone cannot be used as an effective treatment method for sulfide mineral flotation wastewater. In the EC/GAC system, GAC acts as a moving particle electrode to accelerate electron transfer rate and hence improve the processing efficiency [27,28]. The main reason is that it is essentially a three-dimensional electrochemical process which is not stable [29].

The sludge produced is around 5.8 kg/m^3 of wastewater under the optimal processing conditions. It can be considered as a valuable resource because of its high Fe content of about 45.5%. Further research on the sludge recycling utilization will soon be reported separately.

Furthermore, the effect of H_2O_2 concentration on EC efficiency was studied, and the results can be seen in the Supplementary Materials (Figure S1 and Figure S2).

3.3. *Effect of EC Treatment on COD Removal Rate of Different Types of Wastewater*

Through the above experiments, the optimum treatment conditions of mixed wastewater were obtained (iron anode, pH 7.1, 70 min electrolysis time, 19.23 mA/cm^2 current density, and 4.1 g/L activated carbon). Under these optimum conditions, Pb-Zn flotation wastewater, pyrite concentrate wastewater, pyrite tailing wastewater, and mixed wastewater were treated, respectively. The results are shown in Figure 9.

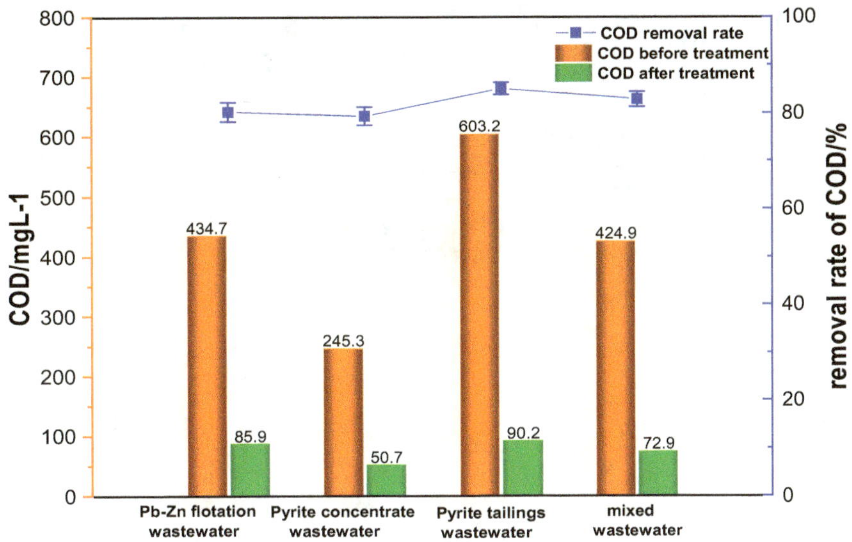

Figure 9. Effect of EC treatment on the COD values and COD removal rates of four different types of wastewaters.

The COD removal rates for the above four types of wastewaters are 80.24%, 79.33%, 85.05%, and 82.84%, respectively. The results indicate that the higher the COD value of wastewater is, the better efficiency the corresponding EC experiment obtains. The pyrite tailing wastewater with the highest COD value has the highest COD removal rate, followed by mixed wastewater, Pb-Zn flotation wastewater, and pyrite concentrate wastewater.

3.4. Effect of Water Type on the Grade and Recovery of Pb/Zn Sulfide Mineral Flotation

The above experimental results confirmed that EC is an effective technique for COD removal of the sulfide mineral flotation wastewater. To further investigate the possibility of reusing the EC-treated water as the feed water, flotation experiments were carried out, and their results are shown in Tables S1–S3, respectively. The comparison of the Pb/Zn grades and recovery rates using different types of water during flotation is presented in Figure 10.

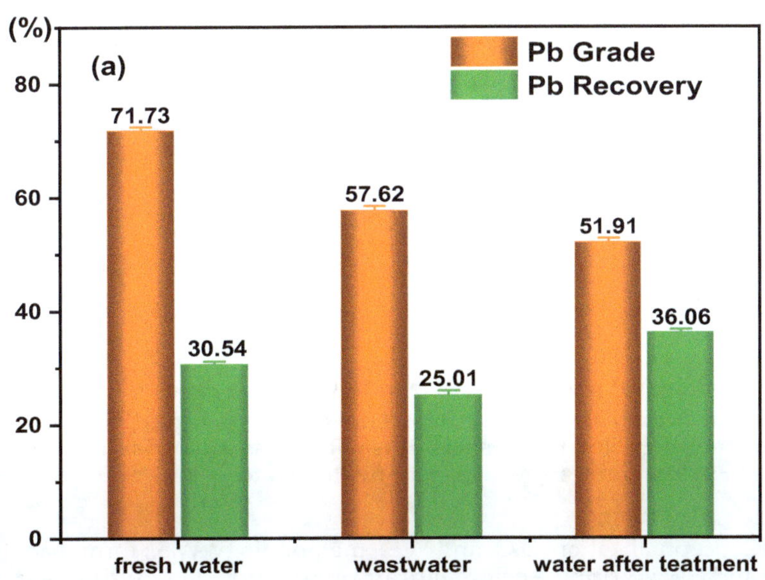

Figure 10. *Cont.*

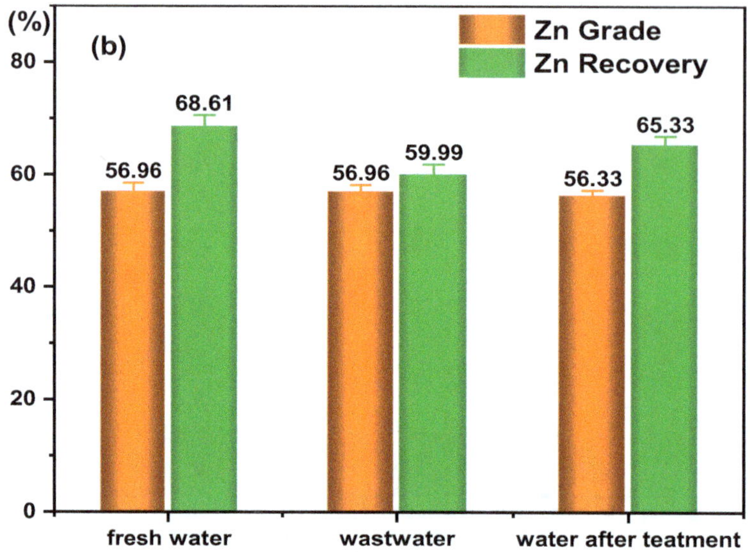

Figure 10. The grade and recovery of Pb (**a**) and Zn (**b**) sulfide flotation concentrates using different types of water.

The Pb recovery using different types of water obeys the following order: Treated water (36.06%) > fresh water (30.54%) > wastewater (25.01%), whereas the order of Pb grades is: Fresh water (71.73%) > wastewater (57.62%) > treated water (51.91%). The experimental results indicate that flotation using treated water can significantly improve the Pb recovery, while the Pb grade is reduced.

In terms of Zn concentrate, a similar Zn grade is obtained with different types of waters (around 56%). The Zn recovery obeys the following order: Fresh water (68.61%) > treated water (65.33%) > wastewater (59.99%). Compared with the untreated wastewater, EC-treated water is more conducive for the increase of Zn recovery without affecting the Zn grade.

4. Conclusions

This study provides the latest evidence that EC is a promising method for treating and recycling mineral processing wastewater. Different factors were investigated to obtain optimal experimental conditions, i.e., iron anode, pH 7.1, 19.23 mA/cm^2 current density, 70 min electrolysis time, and 4.1 g/L activated carbon. Under the optimal conditions, the COD of the mixed wastewater decreases from 424.9 to 72.9 mg/L. The addition of a supporting electrolyte like Na_2SO_4 can be harmful to the removal of COD. The COD removal rate of wastewater with the highest COD value is highest. Moreover, flotation results show that the recoveries of both Pb and Zn using EC-treated water are higher than those of directly using wastewater or fresh water. The results imply that the EC technique can provide a new direction for recycling mineral processing wastewater. This study provides novel insights into dealing with wastewater in the mining industry, and is meaningful in both environmental and economical ways.

Supplementary Materials:
Figure S1: Variation of COD removal rate with dosage of H_2O_2 (Anode material: Fe, Cathode material: Stainless steel, Current density: 19.23 mA/cm^2, pH: 3.0, Electrolysis time: 70 min), Figure S2: Effect of H_2O_2 on the boundary between mud and water, Table S1: Flotation tests results with fresh water, Table S2: Flotation tests results with mixed wastewater, Table S3: Flotation tests results with treated water.

Author Contributions: Conceptualization, Z.G. and W.S.; methodology, W.S.; formal analysis, Z.G.; investigation, G.J.; data curation, G.J. and S.R.; writing—original draft preparation, G.J.; writing—review and editing, Y.G.; supervision, Z.G.; project administration, W.S.; funding acquisition, Z.G. All authors have read and agreed to the published version of the manuscript.

Acknowledgments: The authors are grateful to the financial support from the National Natural Science Foundation of China (51634009, 51774328, and 51404300), the Young Elite Scientists Sponsorship Program by CAST (2017QNRC001), the Young Elite Scientists Sponsorship Program by Hunan province of China (2018RS3011), the Natural Science Foundation of Hunan Province of China (2018JJ2520), and the National 111 Project (B14034). All authors appreciate the School of Minerals Processing and Bioengineering, Central South University, Changsha, China for providing the facilities.

References

1. Farrokhpay, S.; Zanin, M. An investigation into the effect of water quality on froth stability. *Adv. Powder Technol.* **2012**, *23*, 493–497. [CrossRef]
2. Zhao, S.; Pan, J.Z. Overview on the Treatment Technology of Wastewater from Copper Mineral Processing (WCMP). *Adv. Mater. Res.* **2013**, *750*, 1369–1372.
3. Sitorus, F.; Cilliers, J.J.; Brito-Parada, P.R. Multi-criteria decision making for the choice problem in mining and mineral processing: Applications and trends. *Expert Syst. Appl.* **2019**, *121*, 393–417. [CrossRef]
4. Kobya, M.; Can, O.T.; Bayramoglu, M. Treatment of textile wastewaters by electrocoagulation using iron and aluminum electrodes. *J. Hazard. Mater.* **2003**, *99*, 163–178. [CrossRef]
5. Beck, E.C.; Giannini, A.P.; Ramirez, E.R. Electrocoagulation clarifies food wastewater. *Food Technol.* **1974**, *22*, 18–19.
6. Bazrafshan, E. Application of electrocoagulation process for dairy wastewater treatment. *J. Chem.* **2012**, *2013*, 1–8. [CrossRef]
7. Wang, S.; Wang, Q. Experimental studies on pretreatment process of brackish water using electrocoagulation (EC) method. *Desalination* **1987**, *66*, 353–364. [CrossRef]
8. Vik, E.A.; Carlson, D.A.; Eikum, A.S.; Gjessing, E.T. Electrocoagulation of potable water. *Water Res.* **1984**, *18*, 1355–1360. [CrossRef]
9. Eryuruk, K.; Un, U.T.; Ogutveren, U.B. Electrochemical treatment of wastewaters from poultry slaughtering and processing by using iron electrodes. *J. Clean. Prod.* **2019**, *121*, 1089–1095. [CrossRef]
10. Moreno-Casillas, H.A.; Cocke, D.L.; Gomes, J.A.G.; Morkovsky, P.; Parga, J.R.; Peterson, E. Electrocoagulation mechanism for COD removal. *Sep. Purif. Technol.* **2007**, *56*, 204–211. [CrossRef]
11. Mollah, M.Y.; Morkovsky, P.; Gomes, J.A.; Kesmez, M.; Parga, J.; Cocke, D.L. Fundamentals, present and future perspectives of electrocoagulation. *J. Hazard. Mater.* **2004**, *114*, 199–210. [CrossRef] [PubMed]
12. Johnson, P.N. Ferric chloride and alum as single and dual coagulants. *Acta Sci. Circumst.* **1983**, *75*, 232–239. [CrossRef]
13. Makarov, V.M.; Kalaeva, S.Z.; Zakharova, I.N.; Nevzorov, I.A.; Maltseva, M.S.; Shipilin, A.M.; Krzhizhanovskaya, M.G. Magnetic and X-ray studies of nanodispersed magnetite synthesized from chrome containing galvanic sludge. *J. Nano- Electron. Phys.* **2015**, *7*, 4–12.
14. Wei, X.; Cao, J.; Holub, R.F.; Hopke, P.K.; Zhao, S. TEM study of geogas-transported nanoparticles from the Fankou lead–zinc deposit, Guangdong Province, South China. *J. Geochem. Explor.* **2013**, *128*, 124–135. [CrossRef]
15. Yang, S.; Verdaguer-Casadevall, A.; Arnarson, L.; Silvioli, L.; Čolić, V.; Frydendal, R.; Rossmeisl, J.; Chorkendorff, I.; Stephens, I.E. Toward the decentralized electrochemical production of H_2O_2: a focus on the catalysis. *Acs Catal.* **2018**, *8*, 4064–4081. [CrossRef]
16. Arslan-Alaton, İ.; Kabdaşlı, I.; Vardar, B.; Tünay, O. Electrocoagulation of simulated reactive dyebath effluent with aluminum and stainless steel electrodes. *J. Hazard. Mater.* **2009**, *164*, 1586–1594. [CrossRef]
17. Ghalwa, A.; Nasser, M.; Farhat, N. Removal of abamectin pesticide by electrocoagulation process using stainless steel and iron electrodes. *J. Env. Anal. Chem.* **2015**, *2*, 134. [CrossRef]
18. Garg, K.K.; Prasad, B. Treatment of toxic pollutants of purified terephthalic acid waste water: A review. *Environ. Technol. Innov.* **2017**, *8*, 191–217. [CrossRef]
19. Lu, J.; Li, Y.; Yin, M.; Ma, X.; Lin, S. Removing heavy metal ions with continuous aluminum electrocoagulation: A study on back mixing and utilization rate of electro-generated Al ions. *Chem. Eng. J.* **2015**, *267*, 86–92. [CrossRef]
20. Doğan, D.; Türkdemir, H. Electrochemical oxidation of textile dye indigo. *J. Chem. Technol. Biotechnol.: Int. Res. Process, Environ. Clean Technol.* **2005**, *80*, 916–923.
21. Zongo, I.; Maiga, A.H.; Wéthé, J.; Valentin, G.; Leclerc, J.P.; Paternotte, G.; Lapicque, F. Electrocoagulation for

the treatment of textile wastewaters with Al or Fe electrodes: Compared variations of COD levels, turbidity and absorbance. *J. Hazard. Mater.* **2009**, *169*, 70–76. [CrossRef] [PubMed]
22. Holt, P.K.; Barton, G.W.; Mitchell, C.A. The future for electrocoagulation as a localised water treatment technology. *Chemosphere* **2005**, *59*, 355–567. [CrossRef] [PubMed]
23. Ghernaout, D.; Ghernaout, B. On the controversial effect of sodium sulphate as supporting electrolyte on electrocoagulation process: A review. *Desalination Water Treat.* **2011**, *27*, 243–254. [CrossRef]
24. Chen, X.; Chen, G.; Yue, P.L. Separation of pollutants from restaurant wastewater by electrocoagulation. *Sep. Sci. Technol.* **2007**, *42*, 819–833. [CrossRef]
25. Tezcan Un, U.; Topal, S.; Ates, F. Electrocoagulation of tissue paper wastewater and an evaluation of sludge for pyrolysis. *Desalination Water Treat.* **2016**, *57*, 28724–28733. [CrossRef]
26. Eldars, F.M.S.E.; Ibrahim, M.A.; Gabr, A.M.E. Reduction of COD in water-based paint wastewater using three types of activated carbon. *Desalination Water Treat.* **2014**, *52*, 2975–2986. [CrossRef]
27. Zhang, C.; Jiang, Y.; Li, Y.; Hu, Z.; Zhou, L.; Zhou, M. Three-dimensional electrochemical process for wastewater treatment: a general review. *Chem. Eng. J.* **2013**, *228*, 455–467. [CrossRef]
28. Jung, K.-W.; Hwang, M.-J.; Park, D.-S.; Ahn, K.-H. Performance evaluation and optimization of a fluidized three-dimensional electrode reactor combining pre-exposed granular activated carbon as a moving particle electrode for greywater treatment. *Sep. Purif. Technol.* **2015**, *156*, 414–423. [CrossRef]
29. Xiong, Y.; Strunk, P.J.; Xia, H.; Zhu, X.; Karlsson, H.T. Treatment of dye wastewater containing acid orange II using a cell with three-phase three-dimensional electrode. *Water Res.* **2001**, *35*, 4226–4230. [CrossRef]

8

Functionalization of a Hydrophilic Commercial Membrane using Inorganic-Organic Polymers Coatings for Membrane Distillation

Lies Eykens [1,2,*], Klaus Rose [3], Marjorie Dubreuil [1], Kristien De Sitter [1], Chris Dotremont [1], Luc Pinoy [4] and Bart Van der Bruggen [2,5]

1 VITO-Flemish Institute for Technological Research, Boeretang 200, 2400 Mol, Belgium; marjorie.dubreuil@vito.be (M.D.); kristien.desitter@vito.be (K.D.S.); chris.dotremont@vito.be (C.D.)
2 Department of Chemical Engineering, KU Leuven, Celestijnenlaan 200F, B-3001 Leuven, Belgium; bart.vanderbruggen@kuleuven.be
3 Fraunhofer Institute for Silicate Research ISC, Neunerplatz 2, 97082 Würzburg, Germany; Klaus.rose@isc.fraunhofer.de
4 Department of Chemical Engineering, Cluster Sustainable Chemical Process Technology, KU Leuven, Gebroeders Desmetstraat 1, B-9000 Ghent, Belgium; luc.pinoy@kuleuven.be
5 Faculty of Engineering and the Built Environment, Tshwane University of Technology, Private Bag X680, Pretoria 0001, South Africa
* Correspondence: lies.eykens@vito.be.

Academic Editor: Faisal Hai

Abstract: Membrane distillation is a thermal separation technique using a microporous hydrophobic membrane. One of the concerns with respect to the industrialization of the technique is the development of novel membranes. In this paper, a commercially available hydrophilic polyethersulfone membrane with a suitable structure for membrane distillation was modified using available hydrophobic coatings using ORMOCER® technology to obtain a hydrophobic membrane that can be applied in membrane distillation. The surface modification was performed using a selection of different components, concentrations, and application methods. The resulting membranes can have two hydrophobic surfaces or a hydrophobic and hydrophilic surface depending on the application method. An extensive characterization procedure confirmed the suitability of the coating technique and the obtained membranes for membrane distillation. The surface contact angle of water could be increased from 27° up to 110°, and fluxes comparable to membranes commonly used for membrane distillation were achieved under similar process conditions. A 100 h test demonstrated the stability of the coating and the importance of using sufficiently stable base membranes.

Keywords: hydrophobic coatings; direct contact membrane distillation (DCMD); polyethersulfone; ORMOCER®; wetting

---

1. Introduction

Membrane distillation (MD) is a thermal separation technique using a hydrophobic microporous membrane as a contactor between two liquid phases. The membrane allows vapors (e.g., water vapor) to permeate, whereas the liquid phase including the dissolved components (e.g., salts) is retained by the membrane. A temperature difference induces the driving force and allows vapors to permeate from the hot feed side to the cold permeate side. The technique was initially proposed as an alternative technology for reverse osmosis in seawater desalination. However, due to the benefits of very high retentions and less dependence on salinity, it is recently also proposed for applications beyond the scope of reverse osmosis. The applications can include but are not limited to desalination and brine treatment [1], waste water treatment [2,3], and resource recovery [4], where the dissolved components can be salts, proteins [5], acids [2,6], and minerals [4].

Currently, hydrophobic microfiltration membranes are used in membrane distillation, although these membranes are not optimized for the MD process [7,8]. The specific requirements for membrane distillation membranes are described in literature [9–11]. Most importantly, the membrane must consist of at least one layer that is not wetted by the liquid stream under the operational pressures used in the module. The minimum pressure required to wet a hydrophobic membrane is the liquid entry pressure (*LEP*), which depends on the membrane characteristics as well as on the feed composition and is defined by the following equation:

$$LEP = \frac{-2B\gamma_l \cos(\theta)}{r_{max}} \tag{1}$$

where γ_l is the surface tension (N·m^{-1}) of the liquid, θ the contact angle (°), r_{max} the maximum pore size (μm), and B is a geometric factor. To ensure proper operation under fluctuating pressures and temperatures, an *LEP* of 2.5 bar is required [12]. To achieve a sufficient *LEP*, membranes with maximum pore diameter between 0.1 and 1 μm with a water contact angle above 90° are recommended for membrane distillation [9,13,14]. Moreover, it is generally agreed that a high membrane porosity is one of the most important membrane parameters in membrane distillation for both flux and energy efficiency, regardless of the MD configuration [15–19]. Additionally, membranes with a thickness between 30 up to 60 μm are recommended; however, it was shown recently that this optimal value depends on salinity, and at high salinity, thicker membranes are preferred [20]. Currently, most commercial systems use membranes not specifically developed for membrane distillation (i.e., hydrophobic polyethylene (PE), polyvinylidene fluoride (PVDF) and polytetrafluoroethylene (PTFE) microfiltration membranes). However, in the literature, many efforts are described to improve the membrane performance. These efforts include the optimization of the phase inversion process, mainly using the hydrophobic polyvinylidene fluoride (PVDF) or poly (vinylidene fluoride-co-hexafluoropropylene) (PVDF-HFP) as polymer [21–24] and the use of surface modifying macromolecules [25–27] and electrospinning [28,29]. In addition to the optimization of the membrane structure, research is also oriented toward the enhancement of the surface properties for membrane distillation, including but not limited to plasma treatment [30,31], fluorination of a TiO_2 coating [32], or the use of fluoroalkylsilane coatings on Tunisian Clay membranes [33]. Currently, these coatings are only applied in lab scale experiments and are not yet commercially available. In this publication, the use of hydrophobic sol-gel coatings forming an organic-inorganic network on hydrophilic polyethersulfone (PES) membranes is presented. These types of coatings are already used on a commercial scale, showing excellent stability in other applications, including scratch- and abrasion-resistant coatings for plastics [34], functional coatings on glass [35], and gas-sensitive layers [36]. Because of its easy scalable production method, excellent stability, and ability to functionalize the surface properties, this coating material was selected to be applied on a commercially available hydrophilic membrane with the required structure for membrane distillation [14]. The inorganic network is formed by Si-O-Si bonds, whereas the organic network is formed by reactive and polymerizable organic functional groups. The choice of hydrophobic fluorosilanes results in a surface with a hydrophobic character, whereas the unique formation of an organic-inorganic network results in a scratch- and leach-resistant coating. For the first time, these readily available coatings are applied for tuning the hydrophobicity of a cheap hydrophilic membrane to enable application in membrane distillation.

2. Materials and Methods

2.1. Membranes

The commercial hydrophilic microfiltration membrane used as base material in this study is the MicroPES® 2F (3M, Wuppertal, Germany). PVDF GVHP (Merck Chemicals N.V., Overijse, Belgium) is a hydrophobic membrane commonly used in the membrane distillation literature as a reference for comparison of the performance of newly synthesized membrane distillation membranes [37–39]. PE (Solupor, Lydall, Manchester, CT, USA) is also added as a reference membrane, because it is used in commercially available membrane distillation modules.

2.2. Coatings

The commercially available and patented ORMOCER® technology was used to apply a hydrophobic coating on the MicroPES membrane. Three different combinations of silanes were explored. The three different fluorosilanes investigated in this study have a different structure, different chain length, and different hydrophobizing properties. The monomers exhibit a bifunctional character and are able to form a stable combined inorganic-organic network. The synthesis procedure is presented in Figure 1.

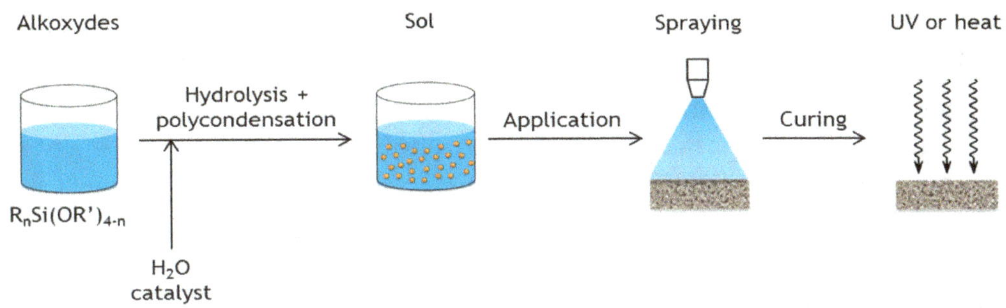

Figure 1. Coating procedure.

The process starts from an alcoholic solution of the $R'Si(OR)_3$ monomers, where R′ is a functional non-reactive hydrophobic group or a polymerizable group, e.g., acryl or vinyl. R represents simple aliphatic groups, e.g., methyl or ethyl. By the addition of water and catalyst, both hydrolysis and polycondensation reactions can occur, resulting in the formation of Si-O-Si covalent bonds forming the inorganic network (Figure 2). The hydrolysis reaction results in the cleavage of a chemical bond by the addition of water. During the polycondensation reactions, two molecules combine to form a larger molecule by splitting a small molecule. Two of the OH groups formed after hydrolysis on the silica components can form H_2O (polycondensation 1), or the unreacted OR group can react with an OH group to form ROH (polycondensation 2). After multiple polycondensation steps, an inorganic polymer network with a Si-O backbone is formed, resulting in a disperse solution, called the sol.

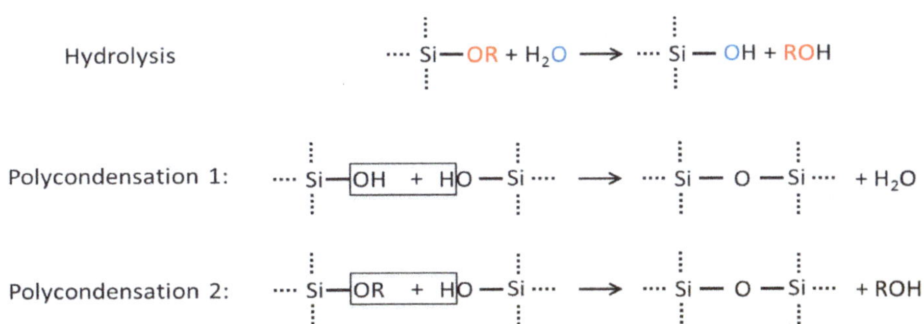

Figure 2. Mechanism of the hydrolysis and polycondensation reactions.

In a second step, this sol is applied on the membrane surface using a bar coater, a roll-to-roll system, or spray coating. Finally, the coating is cured using a thermal or photochemicalcuring step using ultraviolet (UV) light, in which the polymerizable groups present in the solution will form the organic network. The nature of the chemicals used (acryl, vinyl, etc.) determines the organic network type. More details on the system can be found in the literature [34,40].

Figure 3 shows the structure of the perfluorodecyl (PFD) silane. After the sol-gel processing this single component only forms the silica network in an alcoholic solution. No polymerizable group is present in this solution and hence, no organic network is formed in this system. The PFD system was only thermally dried at 80 °C for 30 min. i.e., thermal based evaporation of the solvents (water/alcohol), which led to a solid film.

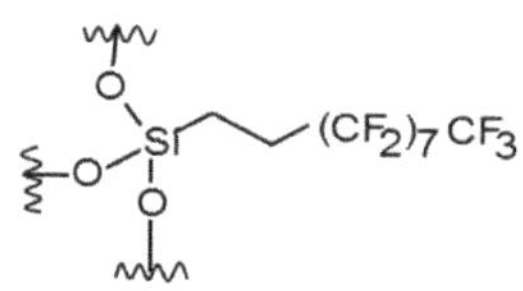

Figure 3. Single component system with perfluorodecyl (PFD), represents the inorganic network.

The second system is a four-component system where a mono-acrylic (Ak) component forms the reactive site for acrylic polymerization. The methyl (T) and dimethyl (D) silanes as well as the highly fluorinated silane (BTFO2N) are participating as hydrophobizing component in the organic network (Figure 4). A composition of Ak/T/D/BTFO2N of 20/44/34/2 wt. % was used.

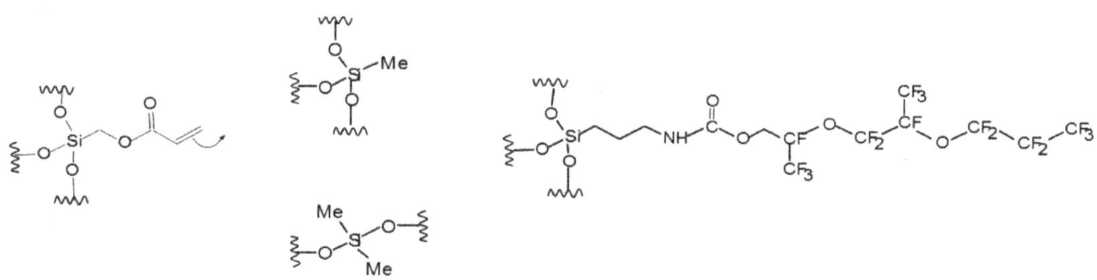

Figure 4. The second system (Ak/T/D/BTFO2N system), represents the inorganic network, represents the organic network formation.

The third is a three component system including a vinyl silane (V), a 3-mercaptopropyl silane (Mc) and a perfluoro-octyl silane (F13) with composition 49/49/2 wt. % V/Mc/F13 (Figure 5). The inorganic silica network is formed by the three components, while the organic network is formed by an addition reaction of the thiol-group to the vinyl group [41]. The hydrophobicity is provided by the perfluoro-octyl group. The UV-curing for the systems Ak/T/D/BTFO2N and V/Mc/F13 was performed using a mercury UV lamp, running with 1200 Watt power, a UV dose of 5000 mJ/m^2, and a UV curing duration 20 s. The UV-curing temperature is ca. 60–80 °C, which evolves from the UV-lamps.

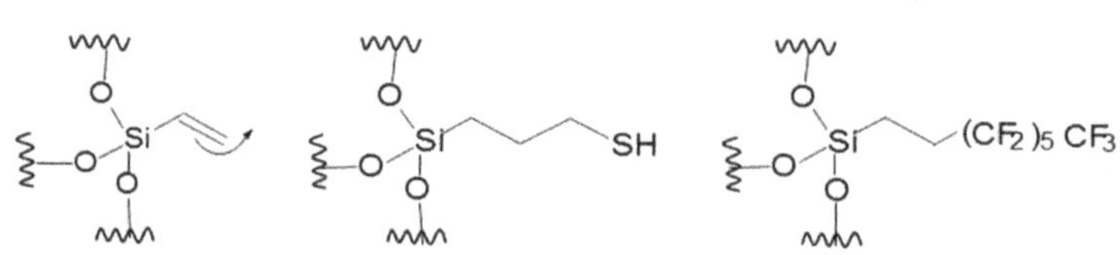

Figure 5. The third system (V/Mc/F13 system), represents the inorganic network, represents the organic network formation.

Table 1 shows the composition, the application method and the final network of the different coatings applied in this study. Coatings 1, 2, and 4 were applied using a roll-to-roll system, coatings 3 and 5 were applied using a bar coater system. Coatings 6 and 7 use the same components as for coating 5, but are applied using spray coating (6 as single side coating, 7 as double side coating). Coatings 1 and 2 have no reactive organic group and only differ in the mass fraction of the fluorinated alkylsiloxane. After thermal curing, these solutions only form an inorganic network. Coatings 3 to 7 are multiple component systems and contain an organic-inorganic network. For membrane 4, a higher mass fraction of the components was used.

Table 1. Composition of the coatings used in this study (perfluorodecyl: PFD; the second system: Ak/T/D/BTFO2N; the third system: V/Mc/F13).

Coating System	Coating Components	Mass Fraction in Coating Solution	Application Method	Network Formation
1	PFD solution	5 wt. %	Roll-to-roll system	No organic crosslinking
2	PFD solution	10 wt. %	Roll-to-roll system	No organic crosslinking
3	Ak/T/D/BTFO2N	5 wt. %	Bar coater	Acrylic polymerization
4	Ak/T/D/BTFO2N	30 wt. %	Roll-to-roll system	Acrylic polymerization
5	V/Mc/F13	5 wt. %	Bar coater	Vinyl + SH addition
6	V/Mc/F13	5 wt. %	Spray coater	Vinyl + SH addition
7	V/Mc/F13	5 wt. %	Spray coater	Vinyl + SH addition

2.3. Characterization Methods

The contact angle of the membranes was measured with the OCA 15EC Contact Angle System (DataPhysics Instruments GmbH, Filderstadt, Germany) using the static sessile drop method. The liquid entry pressure was determined as described by Khayet et al. [42]. The pressure was increased slowly by 0.1 bar each 30 s, until a flow was detected. A Porolux™ 1000 device (Porometer N.V., Eke, Belgium) using the wet/dry capillary flow porometry method measured the pore size distribution as described by Francis et al. [43]. Porefil with a liquid surface tension of 16 mN·m^{-1} was used as wetting liquid and the shape factor was assumed to be 1. The porosity of the membranes was calculated using the following equation suggested by Smolders and Franken [44]:

$$\epsilon = 1 - \frac{\rho_m}{\rho_{pol}} \quad (1)$$

with ρ_m and ρ_{pol} representing the density of the membrane and the polymer, respectively, in g·cm^{-3}. The density of the membrane was obtained by measuring the mass of a circular membrane cut with a circular mold with diameter of 5 cm. The density of the polymer was measured using gas pycnometry with a He-pycnometer (Micromeretics, Norcross, GA, USA) [20]. A cold field emission scanning electron microscope (SEM) type JSM6340F (JEOL, Tokyo, Japan) was used to study membrane cross-sections at an acceleration voltage of 5 keV. Cross-sections were obtained by a cross-section polisher type SM-09010 (JEOL, Tokyo, Japan) using an argon ion beam. All samples were coated with a thin Pt/Pd layer (~1.5 nm) using a Cressington HR208 high-resolution sputter-coater (Cressington Scientific Instruments, Watford, UK) to avoid charging by the e-beam. The images of the cross-sections were analyzed in ImageJ [20].

2.4. Membrane Distillation Setup

The membrane distillation performance was evaluated with a lab-scale DCMD setup (Figure 6). The flat-sheet module had a feed and permeate channel with dimensions of 6 cm width and 18 cm length. The channel height and spacer thickness was 2 mm. On the permeate side, purified water with electrical conductivity below 20 µS·cm^{-1} was used. The feed and distillate were circulated counter-currently on their respective sides of the membrane with a flow velocity of 0.13 m/s using peristaltic pumps (Watson-Marlow, 520DuN/R2, Zwijnaarde, Belgium). $T_{f,in}$ and $T_{p,out}$ was kept constant at 60 °C and 45 °C, respectively, for all experiments. The temperatures were kept constant using two heating baths (Huber, Ministat 230w-cc-NR, Offenburg, Germany) and monitored using four thermocouples (Thermo Electric Company, PT100 TF, Balen, Belgium). The flux was measured by evaluating the weight variations in the feed and distillate tank, using an analytical balance (Sartorius GmbH, ED8801-CW, Goettingen, Germany). The average of at least two experiments is reported. The electrical conductivity at the feed and permeate side were monitored by portable conductivity meters (WTW GmbH, pH/Cond 340i, Weilheim, Germany).

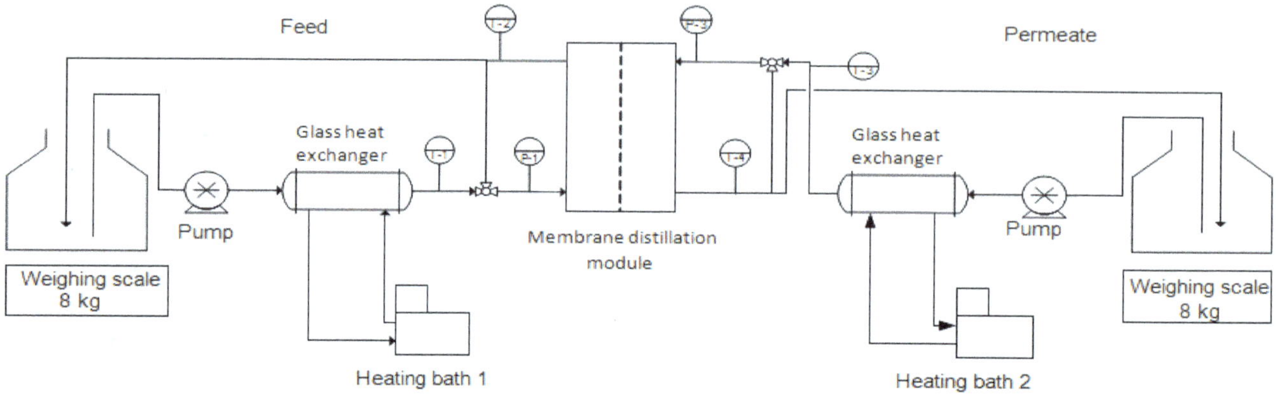

Figure 6. Schematic of the membrane distillation setup.

The energy efficiency (EE) of the process is defined in Equation (3). The total heat transfer through the membrane Q_m is considered to be equal to the heat transfer in the feed channel, as described by Khayet et al. [45].

$$EE\ (\%) = \frac{N \cdot \Delta H \cdot A}{F \cdot C_p \cdot (T_{in} - T_{out})} \quad (3)$$

N ($kg \cdot m^{-2} \cdot h^{-1}$) is the water flux and ΔH ($J \cdot kg^{-1}$) the enthalpy of evaporation. F is the mass flow rate in the channels expressed in $kg \cdot s^{-1}$, A (m^2) is the effective membrane surface area, C_p is the specific heat capacity of the solution ($J \cdot kg^{-1} \cdot {}^\circ C^{-1}$), T_{in} and T_{out} are the bulk temperatures at the channel inlet and outlet of the module expressed in °C, respectively. The calculations were carried out for the feed and permeate channel and the average and standard deviation are reported.

Long term stability tests were performed using 35 $g \cdot L^{-1}$ NaCl as feed concentration. The experimental conditions were chosen differently from the screening tests, because these experiments run overnight. The flux should be limited, to prevent spilling over of the permeate vessel. Therefore, lower temperatures have been chosen for these experiments: T_f and T_p were 45 °C and 40 °C respectively with a cross flow velocity of the feed of 0.1 $m \cdot s^{-1}$ and a salinity of 35 $g \cdot L^{-1}$. The goal of this experiment was mainly to investigate the stability of the coating under constant shear of the feed liquid, not the thermal stability. In general, the thermal stability of the hybrid coatings is 150 °C and higher, up to 300 °C. This thermal stability has been measured for another application elsewhere [4]. The temperature stability of the coating material is much higher compared to the temperatures generally used in MD up to 90 °C and the temperature stability of this coating is not an issue for these coatings.

3. Results

3.1. Characterization of the Membranes

The measured properties of the MicroPES 2F membrane used as base membrane for the coatings are shown in Table 2, together with the properties of the commercial hydrophobic membranes commonly used in membrane distillation.

SEM images reveal a difference in the surface porosity on both sides of the PES-membrane (Table 3). While the pore size is larger on the surface side 2, the pore density and porosity are the highest on surface side 1. The cross-section shows an hourglass-shaped pore structure [46]. The densest zone of the membrane is located in the region of 10 to 50 μm distance from surface 1. All coatings are applied on the surface 1 because this side has the lowest pore size, which is preferred to increase the wetting resistance.

Table 2. Characteristics of the hydrophilic support membrane (polyethersulfone (PES)) and the membranes commonly used for membrane distillation (polyethylene (PE) and polyvinylidene fluoride (PVDF)).

Membrane	Θ (°)	d_{max} (μm)	d_{av} (μm)	Δ (μm)	E (%)	LEP (bar)
PES (Membrana, MicroPES)	26 ± 4	0.61 ± 0.04	0.56 ± 0.03	115 ± 1	75.2 ± 0.3	0
PE (Lydall, Solupor)	120 ± 1	0.32 ± 0.02	0.43 ± 0.02	99 ± 1	75.6 ± 0.6	3.9 ± 0.1
PVDF (Millipore, GVHP)	120 ± 3	0.44 ± 0.01	0.60 ± 0.01	119 ± 1	65.7 ± 0.9	2.3 ± 0.1

Table 3. Surface and cross-section images of the commercial polyethersulfone (PES) membrane and properties obtained using image analysis.

SEM	Surface 1	Surface 2	Cross-Section
Images	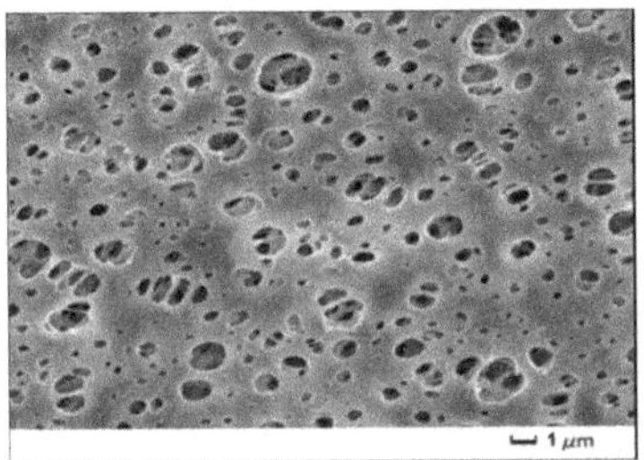	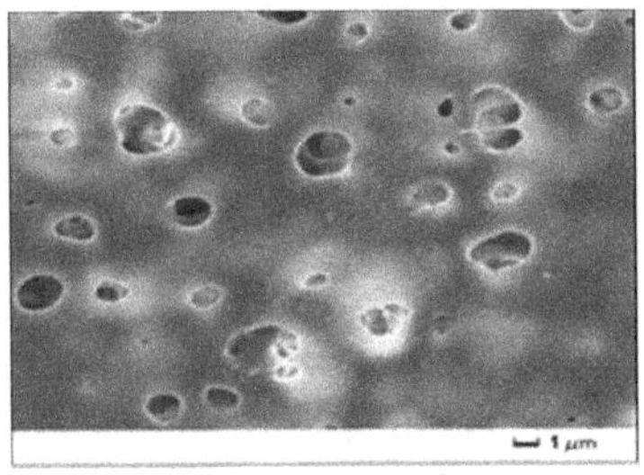	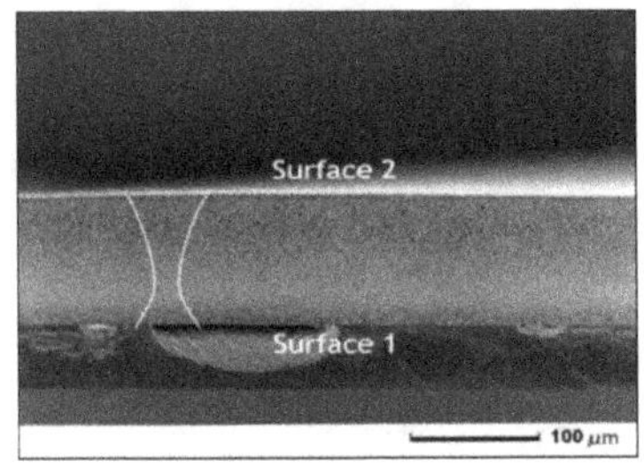
Properties	$\varepsilon_s = 22 \pm 4\%$ $d_{min,s} = 0.03 \pm 0.1$ μm $d_{av,s} = 0.47 \pm 0.2$ μm $d_{max,s} = 1.90 \pm 0.5$ μm Pore density = 4.1 ± 0.7	$\varepsilon_s = 8.8 \pm 0.5\%$ $d_{min,s} = 0.24 \pm 0.1$ μm $d_{av,s} = 0.91 \pm 0.7$ μm $d_{max,s} = 2.2 \pm 0.3$ μm Pore density = 1.2 ± 0.2	$\delta = 113 \pm 1$ μm

The SEM images in Table 4 show a difference in membrane structure for the stretched PE membrane, which is more porous compared to the PVDF membrane. Both membranes have a symmetric cross-section. The PVDF membrane has a more open surface structure compared to the PES-membrane, while it has lower bulk porosity.

Table 4. SEM (Cold field emission scanning electron microscope) images of PE (polyethylene) and PVDF (polyvinylidene fluoride) commercial hydrophobic membrane.

PE1 Stretched	PVDF Phase Inversion
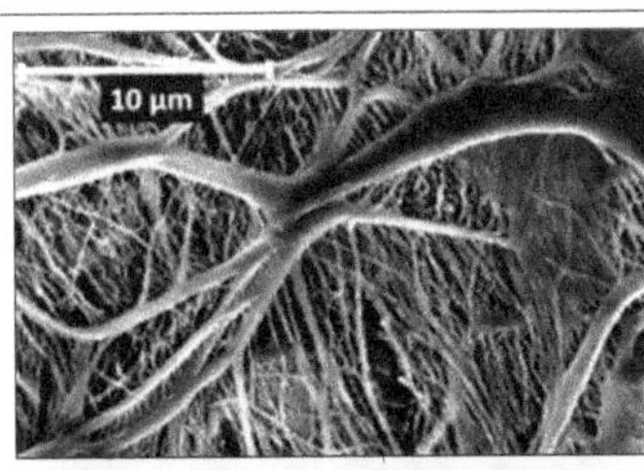	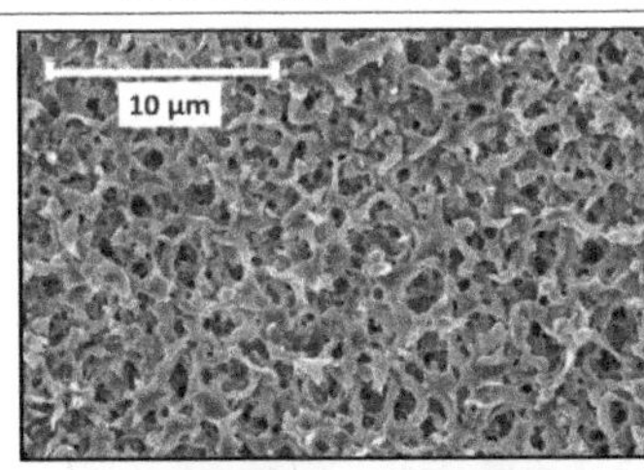
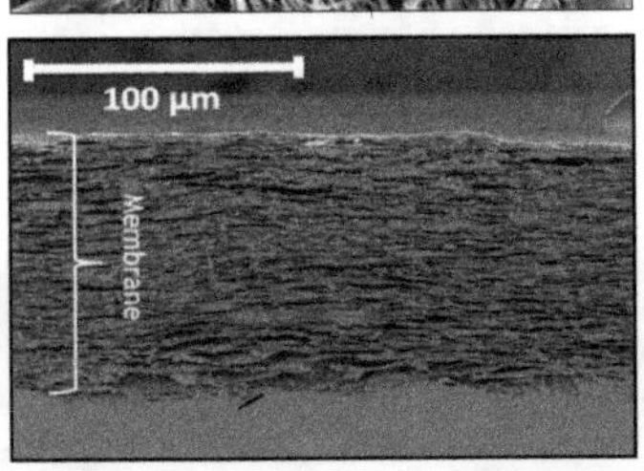	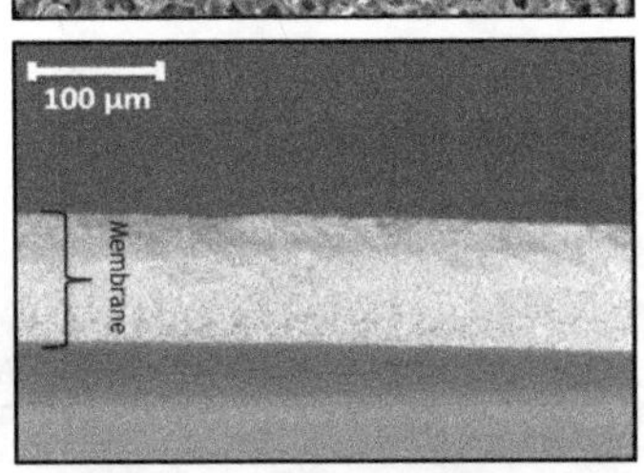

3.2. Wetting Resistance of the Coating

Table 5 shows the water contact angle of the membranes coated in this study. The uncoated membrane has a hydrophilic water contact angle of 27°. After the application of the coatings, surface 1 is hydrophobic and has a water contact angle of at least 100°, confirming the hydrophobic character of the coatings. Without organic crosslinking agent (PFD solution only), the highest water contact angle of 117° for 5 wt. % solution (membrane 1) and 118° for a 10 wt. % solution (membrane 2) are achieved. This difference in water contact angle is statistically insignificant and therefore, the contact angle is considered to be independent of the concentration in the range from 5 and 10 wt. % for this system. When adding components with polymerizable functionalities to form the polymeric network, the fluorinated fraction decreases (membranes 3–5). This is visualized in a slightly reduced hydrophobicity on the coated surface compared to membrane 1 and 2, with a water contact angle ranging from 100° to 110°. For the Ak/T/D/BTFO2N system (3 and 4), the water contact angle is increased from 100° to 110° using a 5 and 30 wt. % solution respectively. The V/Mc/F13 system (5) achieves a water contact angle of 109° with a concentration of only 5 wt. % and including a polymeric network. The water contact angle of untreated surface 2 equals 28° after 0.5 s, whereas after the coating procedure the water contact angle after 0.5 s varies from 83° to 100° for membrane 1–4. For these membranes, the water contact angle continuously decreases over time. For membrane 1–3 the droplet disperses in the membrane only after 2 min. For membrane 4, the droplet sinks into the membrane within 2 min of contact time. For membrane 5, the uncoated side even shows a stable contact angle of 110° similar to the contact angle on the coated side. These observations show that the surface hydrophobicity on the uncoated side is also affected by the coating process, indicating that a part of the coating is able to pass through the entire membrane cross-section and is also applied (partially) on the uncoated side of the membrane. However, despite the increase in water contact angle, the membranes 1–4 are still wetted in membrane distillation and are considered as hydrophobic/hydrophilic membranes. Membrane 5 is considered as an entirely hydrophobic membrane. The measured liquid entry pressure for the different membranes varies from 1.6 to 3.5 bar, with a large variation for multiple measurements of the same membrane. This variation in liquid entry pressure is attributed to the inhomogeneity of the coatings, meaning that the coating is not applied equally on the entire membrane surface on some parts of the surface are not sufficiently hydrophobic. This becomes visible after submerging the membranes in water. While most regions are not wetted, some areas of the membrane surface are wetted, showing that the bar coater and the roll-to-roll system are not the preferred application methods when applying the coatings on porous membranes. The contact angles reported in Table 5 only consider the area of the membrane that was not wetted after submersion of the membrane in water. However, the inhomogeneous application of the coating with barcoater or the roll-to-roll system was also visible in the large spreading of the contact angle measurements of 20° when measuring at random spots on the membrane.

Table 5. Contact angle.

Membrane	Coating Components	Mass Fraction in Coating Solution	θ (°)	
			Surface 1 (Coated Side)	Surface 2 (Uncoated Side)
Uncoated	-	-	27 ± 6 [1]	28 ± 4 [1]
1	PFD solution	5 wt. %	117 ± 1	$90 \pm 5 \rightarrow 62 \pm 5$ [2]
2	PFD solution	10 wt. %	118 ± 1	$96 \pm 1 \rightarrow 73 \pm 6$ [2]
3	Ak/T/D/BTFO2N	5 wt. %	100 ± 3	$100 \pm 1 \rightarrow 85 \pm 5$ [2]
4	Ak/T/D/BTFO2N	30 wt. %	110 ± 1	83 ± 2 [1]
5	V/Mc/F13	5 wt. %	109 ± 1	110 ± 1

Legend: [1] Droplet wets the membrane within 2 min; [2] Contact angle after 0.5 s → Contact angle after 2 min.

To improve the coating homogeneity, coating system 5 (V/Mc/F13, 5 wt. %) was selected for spray coating application. The single component systems (1 and 2) are not sufficiently stable (see Section 3.4), whereas the Ak/T/D/BTFO2N system (3 and 4) requires a higher amount of silica components to achieve the same contact angle (Table 5). This spray coating process ensures an improved homogeneity compared to the bar coater system. The coating was applied in two ways: (a) only on surface 1, resulting in a membrane with a hydrophobic and a hydrophilic side and (b) on both sides of the membrane, producing a membrane with two hydrophobic sides (Table 6). Membrane 6, coated on surface 1 only shows a slightly lower liquid entry pressure compared to membrane 7, coated on both sides.

Table 6. Contact angle and LEP of the coatings applied by the spray system.

Membrane	Coated Side	θ (°)		*LEP* (bar)
		Surface 1	Surface 2	
6	Surface side 1 only	97 ± 1	41 ± 3 *	1.8 ± 0.2
7	Both sides	102 ± 1	107 ± 1	2.2 ± 0.1

* Droplet wets the membrane within 2 min.

3.3. Structure of the Coating

Apart from the contact angle measurements, another important issue is the possibility that the coating will block the pores. This reduces the porosity and pore size, and in the ultimate case, a dense membrane is obtained, obstructing the mass transport. Pore blockage can easily be seen by porometry because it reduces both the gas flow through the membrane and the pore size compared to the untreated membrane. For membranes 1–3 and 5–7, the pore sizes and the gas flows of the uncoated and coated membranes are equal. As an example, the pore size distribution obtained using porometry is presented for an untreated membrane and membrane 3 and 4 are given in Figure 7. For membrane 4, coated with a 30 wt. % solution, no pores are detected using porometry, indicating that the pores are completely blocked by the coating in this case. These measurements reveal that a 30% solution is not suitable for membrane modification since it significantly affects the porosity.

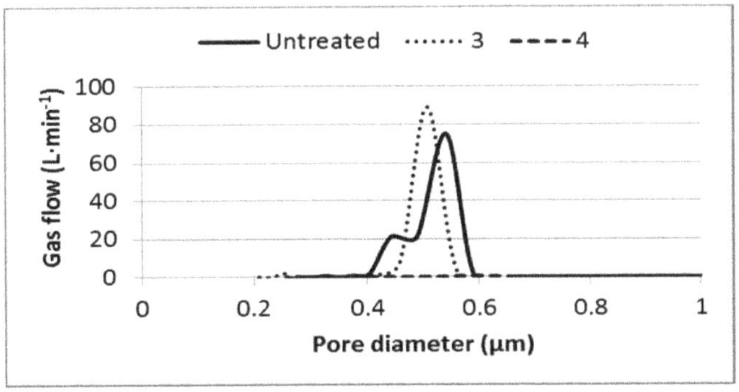

Figure 7. Pore size distribution untreated PES-membrane, membrane 3, and membrane 4 using porometry.

The position of the coating is investigated using energy-dispersive X-ray spectroscopy (EDX). As an example, Figure 8 shows the EDX spectrum of membrane 3. The peaks of the oxygen and sulfur atoms in the spectrum correlate with the presence of the PES membrane material or pores at the measured position. In the first 20 µm at surface 1, a first increase of the silicon and fluorine atoms is observed, which correlates with the position of the coating. A second increase is observed between 30 and 50 µm, which is also the densest zone of the membrane (Table 3) and therefore, more surface is available to deposit the coating. In the part between 50 and 120 µm, the silicon and fluorine content is lower. The permeation of the coating might be inhibited by the denser structure of the membrane. Further in the membrane cross-section, the silicon and fluorine content decreases, but are not equal to zero. The increase of the oxygen and silicon at 120 to 140 µm is caused by the silicon glue used to fix the sample. Based on the EDX on the chemical composition, the structure is hydrophobic until a depth of at least 50 µm. Unfortunately, it is impossible to indicate the exact hydrophobicity or hydrophilicity in terms of a water contact angle at a certain point of the membrane cross-section based on the elemental composition obtained with EDX. Therefore, the exact hydrophobic thickness cannot be derived from these EDX figures.

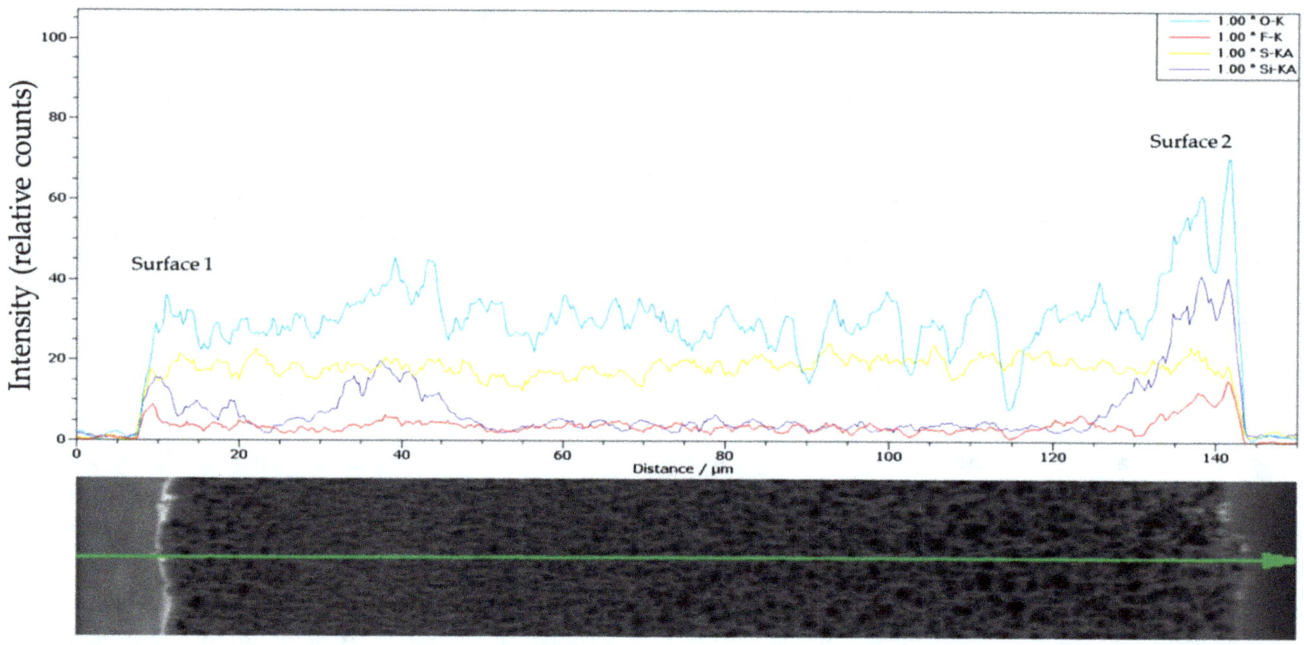

Figure 8. EDX (energy-dispersive X-ray spectroscopy) spectra of the cross section for membrane 3.

In summary, these results indicate that no difference in membrane structure in terms of thickness, pore size, and porosity was found for the coatings applied using less than 10 wt. % silanes in the coating solution. Only coating 4, with very high load of silanes, the MD-flux and N2 flux during porometry decreased to zero, showed a strong difference, which is a strong indication for pore blocking.

3.4. Membrane Distillation Performance

The flux and energy efficiency of the membranes produced in this study are compared to the PVDF GVHP membrane from Millipore commonly used in the literature and to a PE membrane currently used in pilot scale membrane distillation modules (Table 2).

All coated membranes are coated on the same base membrane structure, except for membrane 4, where pore blocking occurs. The other membranes have equal porosity, pore size, and total thickness. However, the position of the coating and hence the thickness of the hydrophobic layer can be different. Since this thickness can affect the flux and energy efficiency of the membrane for desalination applications [20], a difference in MD performance is expected. Flux and energy efficiencies of the different membranes are summarized in Table 7.

Membranes 1 and 2 are single component systems without organic network formation and show immediate breakthrough of the salts. The membrane wetting is also visually observed after demounting the module. This indicates that the inorganic network formed by these coatings is not sufficient to provide stable coatings for the process conditions used in membrane distillation.

Membranes 3 and 4 use the same components, but with a difference in mass fraction of 5 wt. % and 30 wt. % respectively (Table 1). Membrane 3 shows a much higher flux of 16.2 kg·m^{-2}·h^{-1} compared to membrane 4 with a flux of only 0.4 kg·m^{-2}·h^{-1}. This low flux of membrane 4 is caused by the pore blocking shown in Figure 7, which hinders the transport of water vapor through the membrane. This shows that a mass fraction of 5 wt. % silanes in the coating solutions is balancing between a sufficient hydrophobicity and avoiding obstruction of the pores, which occurred at concentrations of 30 wt. %.

The hydrophobic/hydrophilic membranes 3 and 6 show higher fluxes of about 16 kg·m^{-2}·h^{-1} compared to membranes 5 and 7 with a flux of about 14 kg·m^{-2}·h^{-1}, which have two hydrophobic sides (Table 7). The effect of partial pore wetting on flux and heat transfer is comprehensively described in reference [47], where it was shown that an increase of the depth of pore wetting results in an increase of the flux. This difference in flux is explained by a different in hydrophobic non-wetted membrane thickness, which imposes the mass transport resistance for vapor transport. As discussed in the literature, the optimal hydrophobic thickness ranges from 10 to 60 μm using 35 g·L^{-1} NaCl [20,48–50]. While the membranes with two hydrophobic surfaces (5 and 7) are probably fully hydrophobic or at least they do not contain a wetted part, the cross-section of the hydrophobic/hydrophilic membrane structures (3 and 6) is partially wetted by the permeate liquid on surface side 2. Therefore, the hydrophobic layer thickness is much closer to the optimal values of 10–60 μm for the hydrophobic/hydrophilic membranes. The fluxes achieved are higher compared to the commercial PVDF membrane, whereas the commercial PE membrane shows a flux slightly higher compared to the membranes with two hydrophobic sides, but lower compared to the membranes with a hydrophobic/hydrophilic structure.

The energy efficiency varies from 43% to 55% and is lower compared to the commercial membranes and appears mainly to depend on the membrane base structure and porosity (Section 3.1). PE has the highest porosity and surface porosity and shows the highest energy efficiency as well. The energy efficiency of the PVDF membrane in negatively affected by the lower bulk porosity [14]. The coated PES membrane has relatively high bulk porosity, but as shown in Table 3, the membrane is not symmetric and has a more dense structure at the surface and in the first 100 μm. This causes less mass transport and more heat transport through the membrane in the first 60 μm, reducing the energy efficiency. Membranes applied with the same application systems (membrane 3 and 5 and membrane 6 and 7) have equal energy efficiency, regardless of the fact that the resulting membrane is a hydrophobic/hydrophilic membrane or a membrane with 2 hydrophobic surfaces. This can be explained by the independence of the energy efficiency as function of membrane thickness, which is shown by different authors [20,48]. This independency occurs because both heat transfer due to flux and heat transfer due to conduction are approximately inversely proportional to the membrane thickness. A high salt retention above 99.9% was measured for all membranes.

Table 7. Flux, energy efficiency and salt retention. Process conditions: T_f = 60 °C, T_p = 45 °C, v = 0.13 m·s^{-1}, NaCl concentration = 35 g·L^{-1}.

Membrane	Structure	Flux (kg·m^{-2}·h^{-1})	Energy Efficiency (%)	Salt Retention (%)
1	hydrophobic/hydrophilic		Wetted	
2	hydrophobic/hydrophilic			
3	hydrophobic/hydrophilic	16.2 ± 0.5	55 ± 2	99.98 ± 0.01
4	hydrophobic/hydrophilic	0.4 ± 0.3	-	99.99 ± 0.01
5	2 hydrophobic surfaces	14.5 ± 0.5	50 ± 2	99.99 ± 0.01
6	hydrophobic/hydrophilic	16.1 ± 0.1	44 ± 3	99.92 ± 0.06
7	2 hydrophobic surfaces	13.9 ± 0.1	43 ± 5	99.99 ± 0.01
PVDF	hydrophobic	12.0 ± 0.1	52 ± 2	99.99 ± 0.01
PE	hydrophobic	15.3 ± 0.4	67 ± 4	99.99 ± 0.01

3.5. Medium Term Performance

The coating material applied on membranes 5–7 is advantageous for membrane distillation based on its superior wetting resistance and homogeneity. Membranes 5 and 6 were tested for a longer period. The flux and salt retention for membrane 6 are shown as an example in Figure 9, but similar results are obtained for membrane 5. The steady decrease in flux is caused by the increasing salt concentration, while jump in flux is explained by the addition of water to the feed solution after a certain amount of time to maintain the concentration at 35 g·L^{-1}. During a period of 80 h, salt retention was always above 99.9% and flux remained constant. To confirm the stability of the coating on the active membrane surface, the LEP of the used membrane was measured. The initial LEP was 1.8, while the LEP of the membrane after 80 h of operation equals 1.4 bar. This reduction is caused by the reduced surface tension after long term exposure to salts, which is also reported in the literature [51]. However, this reduction is not severe enough to indicate that the coating is washed off during operation. In that case, 0.1 bar would already result in the penetration of liquid through the membrane.

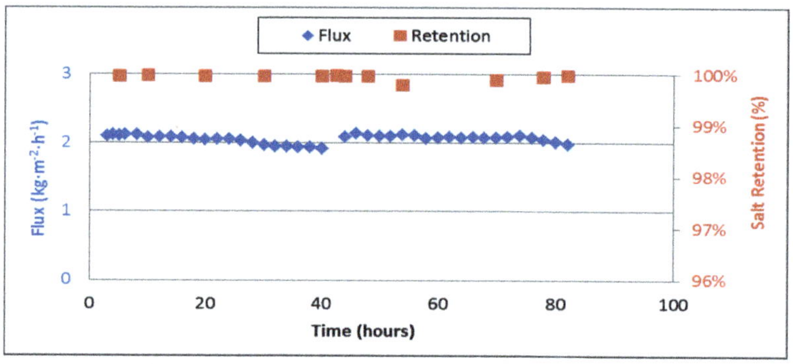

Figure 9. Long term experiment of membrane 6, T_f = 45 °C, T_p = 40 °C, v = 0.13 m·s^{-1}, NaCl concentration = 35 g·L^{-1}.

4. Conclusions

Sol-gel coatings have proven their stability and excellent performance in many other applications and were successfully applied for the first time on a hydrophilic PES membrane with a suitable pore size, porosity, and thickness for application in membrane distillation. The sol-gel coatings provide sufficient hydrophobicity and resistance against membrane wetting, and the surface contact angle can be increased from 27° up to 110°. Based on this study, the V/Mc/F13 system was recommended because of its higher hydrophobicity at 5 wt. % loading of siloxanes. This allows for keeping the pores open for vapor transport. Moreover, it is possible to produce a membrane with two hydrophobic sides using the spraying technique on both sides of the membrane, whereas a membrane with a hydrophobic/hydrophilic structure is obtained when spraying the coating on only one side of the membrane. The membranes with a hydrophobic/hydrophilic structure are recommended for seawater desalination because the hydrophobic layer thickness is closer to the optimal thickness for flux. While the coated membranes achieve comparable fluxes, the energy efficiency is relatively low compared to the commercial membranes in the same conditions. The energy efficiency was found to be independent of the coating procedure, but is dependent on the base membrane structure. Therefore, further optimization of the base membrane structure is required to further improve the membrane performance of these types of membranes in membrane distillation.

Acknowledgments: L. Eykens thankfully acknowledge a Ph.D. scholarship provided by VITO.

Author Contributions: L.E. performed the membrane characterization and M.D. testing and wrote the paper. K.R. prepared the inorganic-organic coatings on the PES-substrate and corrected the manuscript. M.D., K.D.S., C.D., L.P. and B.V.D.B. guided the experiments, analysis and the writing process.

References

1. Zhang, P.; Knötig, P.; Gray, S.; Duke, M. Scale reduction and cleaning techniques during direct contact membrane distillation of seawater reverse osmosis brine. *Desalination* **2015**, *374*, 20–30. [CrossRef]
2. Kesieme, U.K. Mine Waste Water Treatment and Acid Recovery Using Membrane Distillation and Solvent Extraction. Ph.D. Thesis, Victoria University, Footscray, Australia, 2015.
3. Calabrò, V.; Drioli, E.; Matera, F. Membrane distillation in the textile wastewater treatment. *Desalination* **1991**, *83*, 209–224. [CrossRef]
4. Hickenbottom, K.L.; Cath, T.Y. Sustainable operation of membrane distillation for enhancement of mineral recovery from hypersaline solutions. *J. Membr. Sci.* **2014**, *454*, 426–435. [CrossRef]
5. Christensen, K.; Andresen, R.; Tandskov, I.; Norddahl, B.; du Preez, J.H. Using direct contact membrane distillation for whey protein concentration. *Desalination* **2006**, *200*, 523–525. [CrossRef]
6. Tomaszewska, M.; Gryta, M.; Morawski, A.W. Study on the concentration of acids by membrane distillation. *J. Membr. Sci.* **1995**, *102*, 113–122. [CrossRef]
7. Teoh, M.M.; Chung, T.S. Membrane distillation with hydrophobic macrovoid-free PVDF-PTFE hollow fiber membranes. *Sep. Purif. Technol.* **2009**, *66*, 229–236. [CrossRef]
8. Bonyadi, S.; Chung, T.S. Flux enhancement in membrane distillation by fabrication of dual layer hydrophilic-hydrophobic hollow fiber membranes. *J. Membr. Sci.* **2007**, *306*, 134–146. [CrossRef]
9. El-Bourawi, M.S.; Ding, Z.; Ma, R.; Khayet, M. A framework for better understanding membrane distillation separation process. *J. Membr. Sci.* **2006**, *285*, 4–29. [CrossRef]
10. Alkhudhiri, A.; Darwish, N.; Hilal, N. Membrane distillation: A comprehensive review. *Desalination* **2012**, *287*, 2–18. [CrossRef]
11. Lawson, K.W.; Lloyd, D.R. Membrane distillation. *J. Membr. Sci.* **1997**, *124*, 1–25. [CrossRef]
12. Schneider, K.; Hölz, W.; Wollbeck, R.; Ripperger, S. Membranes and modules for transmembrane distillation. *J. Membr. Sci.* **1988**, *39*, 25–42. [CrossRef]
13. Khayet, M.; Matsuura, T. *Membrane Distillation: Principles and Applications*; Elsevier B.V.: Amsterdam, The Netherland, 2011.
14. Eykens, L.; De Sitter, K.; Dotremont, C.; Pinoy, L.; van der Bruggen, B. How to Optimize the Membrane Properties for Membrane Distillation: A Review. *Ind. Eng. Chem. Res.* **2016**, *55*, 9333–9343. [CrossRef]
15. Alklaibi, A.M.; Lior, N. Transport analysis of air-gap membrane distillation. *J. Membr. Sci.* **2005**, *255*, 239–253. [CrossRef]
16. Abu Al-Rub, F.A.; Banat, F.; Beni-Melhim, K. Parametric Sensitivity Analysis of Direct Contact Membrane Distillation. *Sep. Sci. Technol.* **2002**, *37*, 3245–3271. [CrossRef]
17. Jönsson, A.S.; Wimmerstedt, R.; Harrysson, A.C. Membrane distillation—A theoretical study of evaporation through microporous membranes. *Desalination* **1985**, *56*, 237–249. [CrossRef]
18. Ali, M.I.; Summers, E.K.; Arafat, H.A.; Lienhard, J.H.V. Effects of membrane properties on water production cost in small scale membrane distillation systems. *Desalination* **2012**, *306*, 60–71. [CrossRef]
19. Bahmanyar, A.; Asghari, M.; Khoobi, N. Numerical simulation and theoretical study on simultaneously effects of operating parameters in direct contact membrane distillation. *Chem. Eng. Process. Process Intensif.* **2012**, *61*, 42–50. [CrossRef]
20. Eykens, L.; Hitsov, I.; de Sitter, K.; Dotremont, C.; Pinoy, L.; Nopens, I.; van der Bruggen, B. Influence of membrane thickness and process conditions on direct contact membrane distillation at different salinities. *J. Membr. Sci.* **2016**, *498*, 353–364. [CrossRef]
21. Thomas, R.; Guillen-Burrieza, E.; Arafat, H.A. Pore structure control of PVDF membranes using a 2-stage coagulation bath phase inversion process for application in membrane distillation (MD). *J. Membr. Sci.* **2014**, *452*, 470–480. [CrossRef]
22. Song, Z.W.; Jiang, L.Y. Optimization of morphology and perf ormance of PVDF hollow fiber for direct contact membrane distillation using experimental design. *Chem. Eng. Sci.* **2013**, *101*, 130–143. [CrossRef]
23. Tang, Y.; Li, N.; Liu, A.; Ding, S.; Yi, C.; Liu, H. Effect of spinning conditions on the structure and performance of hydrophobic PVDF hollow fiber membranes for membrane distillation. *Desalination* **2012**, *287*, 326–339. [CrossRef]

24. García-Payo, M.C.; Essalhi, M.; Khayet, M. Effects of PVDF-HFP concentration on membrane distillation performance and structural morphology of hollow fiber membranes. *J. Membr. Sci.* **2010**, *347*, 209–219. [CrossRef]
25. Kim, Y.; Rana, D.; Matsuura, T.; Chung, W.J. Influence of surface modifying macromolecules on the surface properties of poly(ether sulfone) ultra-filtration membranes. *J. Membr. Sci.* **2009**, *338*, 84–91. [CrossRef]
26. Essalhi, M.; Khayet, M. Surface segregation of fluorinated modifying macromolecule for hydrophobic/hydrophilic membrane preparation and application in air gap and direct contact membrane distillation. *J. Membr. Sci.* **2012**, *417–418*, 163–173. [CrossRef]
27. Khayet, M.; Matsuura, T. Application of surface modifying macromolecules for the preparation of membranes for membrane distillation. *Desalination* **2003**, *158*, 51–56. [CrossRef]
28. Tijing, L.D.; Choi, J.S.; Lee, S.; Kim, S.H.; Shon, H.K. Recent progress of membrane distillation using electrospun nanofibrous membrane. *J. Membr. Sci.* **2014**, *453*, 435–462. [CrossRef]
29. Essalhi, M.; Khayet, M. Self-sustained webs of polyvinylidene fluoride electrospun nanofibers at different electrospinning times: 1. Desalination by direct contact membrane distillation. *J. Membr. Sci.* **2013**, *433*, 167–179. [CrossRef]
30. Tian, M.; Yin, Y.; Yang, C.; Zhao, B.; Song, J.; Liu, J.; Li, X.M.; He, T. CF_4 plasma modified highly interconnective porous polysulfone membranes for direct contact membrane distillation (DCMD). *Desalination* **2015**, *369*, 105–114. [CrossRef]
31. Wu, Y.; Kong, Y.; Lin, X.; Liu, W.; Xu, J. Surface-modified hydrophilic membranes in membrane distillation. *J. Membr. Sci.* **1992**, *72*, 189–196. [CrossRef]
32. Razmjou, A.; Arifin, E.; Dong, G.; Mansouri, J.; Chen, V. Superhydrophobic modification of TiO_2 nanocomposite PVDF membranes for applications in membrane distillation. *J. Membr. Sci.* **2012**, *415–416*, 850–863. [CrossRef]
33. Khemakhem, S.; Amar, R.B. Modification of Tunisian clay membrane surface by silane grafting: Application for desalination with Air Gap Membrane Distillation process. *Colloids Surf. A Physicochem. Eng. Asp.* **2011**, *387*, 79–85. [CrossRef]
34. Haas, K.-H.; Amberg-Schwab, S.; Rose, K. Functionalized coating materials based on inorganic-organic polymers. *Thin Solid Films* **1999**, *351*, 198–203. [CrossRef]
35. Kron, J.; Amberg-schwab, S.; Schottner, G. Functional coatings on glass using ORMOCER®-systems. Code: BP20. *J. Sol-Gel Sci. Technol.* **1994**, *2*, 189–192. [CrossRef]
36. Matějec, V.; Rose, K.; Hayer, M.; Pospíšllová, M.; Chomát, M. Development of organically modified polysiloxanes for coating optical fibers and their sensitivity to gases and solvents. *Sens. Actuators B Chem.* **1997**, *39*, 438–442. [CrossRef]
37. Liao, Y.; Wang, R.; Tian, M.; Qiu, C.; Fane, A.G. Fabrication of polyvinylidene fluoride (PVDF) nanofiber membranes by electro-spinning for direct contact membrane distillation. *J. Membr. Sci.* **2013**, *425–426*, 30–39. [CrossRef]
38. Khayet, M.; Khulbe, K.C.; Matsuura, T. Characterization of membranes for membrane distillation by atomic force microscopy and estimation of their water vapor transfer coefficients in vacuum membrane distillation process. *J. Membr. Sci.* **2004**, *238*, 199–211. [CrossRef]
39. Hou, D.; Dai, G.; Wang, J.; Fan, H.; Zhang, L.; Luan, Z. Preparation and characterization of PVDF/nonwoven fabric flat-sheet composite membranes for desalination through direct contact membrane distillation. *Sep. Purif. Technol.* **2012**, *101*, 1–10. [CrossRef]
40. Haas, K.H.; Amberg-Schwab, S.; Rose, K.; Schottner, G. Functionalized coatings based on inorganic-organic polymers (ORMOCER S) and their combination with vapor deposited inorganic thin films. *Surf. Coat. Technol.* **1999**, *111*, 72–79. [CrossRef]
41. Schottner, G.; Rose, K.; Posset, U. Scratch and Abrasion Resistant Coatings on Plastic Lenses—State of the Art, Current Developments and Perspectives. *J. Sol-Gel Sci. Technol.* **2003**, *27*, 71–79. [CrossRef]
42. Khayet, M.; Matsuura, T. Preparation and Characterization of Polyvinylidene Fluoride Membranes for Membrane Distillation. *Ind. Eng. Chem. Res.* **2001**, *40*, 5710–5718. [CrossRef]
43. Francis, L.; Ghaffour, N.; Alsaadi, A.S.; Nunes, S.P.; Amy, G.L. Performance evaluation of the DCMD desalination process under bench scale and large scale module operating conditions. *J. Membr. Sci.* **2014**, *455*, 103–112. [CrossRef]

44. Smolders, K.; Franken, A.C.M. Terminology for Membrane Distillation. *Desalination* **1989**, *72*, 249–262. [CrossRef]
45. Khayet, M. Solar desalination by membrane distillation: Dispersion in energy consumption analysis and water production costs (a review). *Desalination* **2013**, *308*, 89–101. [CrossRef]
46. Khare, V.P.; Greenberg, A.R.; Krantz, W.B. Vapor-induced phase separation—Effect of the humid air exposure step on membrane morphology: Part I. Insights from mathematical modeling. *J. Membr. Sci.* **2005**, *258*, 140–156. [CrossRef]
47. Gilron, J.; Ladizansky, Y.; Korin, E. Silica Fouling in Direct Contact Membrane Distillation. *Ind. Eng. Chem. Res.* **2013**, *52*, 10521–10529. [CrossRef]
48. Martínez, L.; Rodríguez-Maroto, J.M. Membrane thickness reduction effects on direct contact membrane distillation performance. *J. Membr. Sci.* **2008**, *312*, 143–156. [CrossRef]
49. Laganà, F.; Barbieri, G.; Drioli, E. Direct contact membrane distillation: Modelling and concentration experiments. *J. Membr. Sci.* **2000**, *166*, 1–11. [CrossRef]
50. Wu, H.Y.; Wang, R.; Field, R.W. Direct contact membrane distillation: An experimental and analytical investigation of the effect of membrane thickness upon transmembrane flux. *J. Membr. Sci.* **2014**, *470*, 2257–2265. [CrossRef]
51. Boubakri, A.; Bouguecha, S.A.-T.; Dhaouadi, I.; Hafiane, A. Effect of operating parameters on boron removal from seawater using membrane distillation process. *Desalination* **2015**, *373*, 86–93. [CrossRef]

9

Optimization of Green Extraction and Purification of PHA Produced by Mixed Microbial Cultures from Sludge

Guilherme A. de Souza Reis [1,*], Michiel H. A. Michels [1], Gabriela L. Fajardo [1,2], Ischa Lamot [1] and Jappe H. de Best [1]

1 Centre of Expertise Biobased Economy, Avans University of Applied Sciences, Lovensdijkstraat 63, 4818 AJ Breda, The Netherlands; mha.michels@avans.nl (M.H.A.M.); gabrielalfajardo@gmail.com (G.L.F.); i.lamot@avans.nl (I.L.); jh.debest@avans.nl (J.H.d.B.)

2 Department of Chemical Engineering, Universidade Federal de Minas Gerais, Belo Horizonte 31270-901, Brazil

* Correspondence: guilhermeasreis@gmail.com.

Abstract: Sludge from municipal wastewater treatment systems can be used as a source of mixed microbial cultures for the production of polyhydroxyalkanoates (PHA). Stored intracellularly, the PHA is accumulated by some species of bacteria as energy stockpile and can be extracted from the cells by reflux extraction. Dimethyl carbonate was tested as a solvent for the PHA extraction at different extraction times and biomass to solvent ratios, and 1-butanol was tested for purifying the obtained PHA at different purification times and PHA to solvent ratios. Overall, only a very small difference was observed in the different extraction scenarios. An average extraction amount of 30.7 ± 1.6 g of PHA per 100 g of biomass was achieved. After purification with 1-butanol, a visual difference was observed in the PHA between the tested scenarios, although the actual purity of the resulting samples did not present a significant difference. The overall purity increased from 91.2 ± 0.1% to 98.0 ± 0.1%.

Keywords: polyhydroxyalkanoates; PHA; PHBV; mixed microbial culture; green extraction; dimethyl carbonate; purification; 1-butanol; wastewater valorization

1. Introduction

The constant search for environmentally-friendly alternatives for fossil-based materials has been backed up lately by an increase of research in the field of bioplastics such as polyhydroxyalkanoates (PHAs). PHAs are biodegradable polymers synthesized by a variety of bacteria in intracellular granules that serve as energy storage [1,2]. Industrial PHA production has been made feasible by using selected strains of pure microbial cultures to ferment refined substrates [3]. However, a much more sustainable, and perhaps cheaper, option can be found in the use of industrial residue streams as a source of bacterial feed [4–6]. VFAs are organic acids with an aliphatic chain of less than five carbons which can be present in or derived from a large variety of residue streams. VFAs have been shown to be an interesting and very feasible feedstock in the PHA production process by both pure and mixed microbial cultures [7].

Applying a mixed microbial culture (MMC) in the process could furthermore reduce the costs of PHA production, because sterilization of the substrate and reactors is not needed. It has been observed that activated sludge of municipal wastewater treatment plants can be used as a source of MMC with a good PHA-accumulating potential [8,9]. Certain fermentation strategies can be used to explore this accumulating potential and generate a PHA-rich biomass [10–12]. A very useful method is a dynamic fed-batch fermentation with alternating repeated periods of feast and famine, which can also be combined with an aerobic or anaerobic environment [13–15]. Through application of a pulsed VFA-feeding regime, it is possible to reduce the effects of too extreme pH variations caused by the addition of VFAs to the medium, as well as to favor the maintenance of the PHA-accumulating bacterial population over other non-accumulating species during the famine periods. This method also stimulates the PHA-accumulating bacteria to stockpile the biopolymer intracellularly throughout multiple feast and famine cycles, which highlights this feed-on-demand process amongst other approaches even on an industrial scale [8].

When studying the use of different feedstocks for PHA production, it is also important to understand the relationship between the feed composition and the monomeric proportions of the resulting polymer. When VFAs are the bacterial feed, acetic and propionic acid are the main precursors in a mechanism for the production of the monomers 3-hydroxybutyrate (3HB) and 3-hydroxyvalerate (3HV) in PHA [16–19]. Acetic acid can be converted into 3HB via acetyl-CoA, while acetic acid together with the odd numbered propionic acid are used to form 3HV via the conversion to acetyl-CoA and propionyl-CoA [17,20]. The combined production of 3HB and 3HV as monomers leads to the synthesis of copolymer poly(3-hydroxybutyrate-co-3-hydroxyvalerate) (PHBV) (Figure 1).

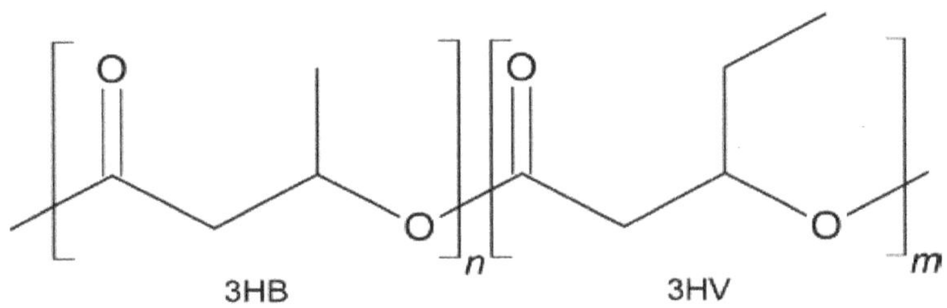

Figure 1. Representation of molecular structure of poly (3-hydroxybutyrate-co-3-hydroxyvalerate) (PHBV) (Drawn on ACD/ChemSketch Freeware, version 2019.2.1, Advanced Chemistry Development, Inc., Toronto, ON, Canada).

Extraction is the next step in the production process of PHA, which is done by solubilizing the intracellular PHA followed by separation of the extracted residual biomass and isolation of PHA from the solvent. Reflux and Soxhlet extractions, with and without biomass pretreatments, have been described in literature for many different solvents [21–24]. Non-halogenated solvents have been the focus of many researches for their reduced toxicity, although the chlorinated ones, such as chloroform and dichloromethane, are still considered reference solvents because of their high efficiency [22,25,26]. The non-halogenated solvent dimethyl carbonate performs much better than a range of solvents such as diethyl carbonate, propylene carbonate and ethyl acetate and it achieved satisfactory yields of PHA recovery when compared to dichloromethane [27,28]. Furthermore, dimethyl carbonate (DMC) is considered to be a green solvent for its low toxicity [29] when compared to chloroform and dichloromethane [30]. Therefore, DMC was chosen as the solvent for PHA extraction in the present work. 1-Butanol has also been shown to be efficient as a solvent for PHA extraction [9], with the advantage of leading to a simple separation process through gelation of the polymer when cooling down the mixture [31]. Due to this easiness in separating the solvent from the solid PHA, 1-butanol was chosen as a purification agent in this study.

Total PHA content in biomass and purity evaluation of the obtained polymer can both be done with various simple analytical techniques. Thermal gravimetric analysis (TGA) [32] can be applied for a quick overview of these parameters when focusing on the degradation temperature of the produced PHA. Gas chromatography combined with mass spectrometry (GC-MS) is a more accurate technique that allows the investigation of the monomer concentration, composition and purity of the product [33]. Some studies made a parallel in between these techniques, showing that, even though different PHA content values were obtained by each method, there is a direct correlation between them [34]. For the GC-MS analysis, a pre-treatment step has to be added for the PHA to be able to be analyzed.

In the present work, the aim was to optimize a green extraction and purification of PHA from a mixed microbial culture, using dimethyl carbonate and 1-butanol, respectively. The biomass to solvent ratio or PHA to solvent ratio and the extraction or purification time were the parameters to be optimized.

2. Material and Methods

2.1. Fermentation Process for PHA Production

The fermentation process was adapted from the work of Valentino (2015) and the patent of Werker (2013) [8,12]. The accumulation of PHA was done in two identical 2.5 L bioreactors (Infors™ MINI-2.5-BACT, Bottmingen, Switzerland) set at 25 °C and 200 rpm and aerated with 1.5 L min^{-1} of air. Thickened secondary sludge from the wastewater treatment plant of water board Brabantse Delta in Bath, the Netherlands, was used as a source of PHA-accumulating bacteria. A synthetic feed was made with a solution of acetic and propionic acid in a molar proportion of 3:1, respectively, and a total chemical oxygen demand (COD) of 20 g L^{-1}. This feed composition guarantees that the PHA produced is the co-polymer PHBV in an expected monomer distribution of 50% of each [17]. To start the process, 500 mL of sludge with volatile suspended solids (VSS) concentration of about 8 g L^{-1} was mixed with 500 mL of tap water in the reactor. The mixture was then left for 2 h without any feed under the fermentation conditions in order to stabilize the biomass. A single feed pulse of 10 mL was then given, followed by the first starvation period of 1 hour. By the end of this period, the dissolved oxygen (DO) level was then considered to be the maximum achieved. The feed-on-demand was then automatically controlled by a DO value corresponding to 80% of the maximum as the set point as the condition for the next pulse feed. The feed pulses corresponded to a volume of 10 mL given over 1 min every time the mentioned condition was met. The whole process was set to a total of 22 h. The process was automatically stopped by dosing hydrochloric acid until a pH of at least 2 was reached in order to cease the bacterial metabolism. At that point, stirring and air inlet were shut down.

2.2. Homogenization of Biomass

The mixture inside the bioreactors was left for a few minutes to settle and the PHA-rich biomass was centrifuged (5810 series, Eppendorf™, Nijmegen, The Netherlands) and washed twice with tap water. The biomass was then freeze-dried (CHRIST Alpha 1-4 LD Plus) overnight. The dried biomass resulted from 20 runs (10 in each bioreactor) was mixed and made homogeneous using a mortar and pestle.

2.3. Extraction Process

The PHA was extracted from the biomass via reflux by adding 25 mL of dimethyl carbonate (DMC) as a solvent to different amounts of biomass in a round-bottom flask connected to a cooling column. Different biomass to solvent ratios, 0.01 g mL^{-1}, 0.025 g mL^{-1}, 0.05 g mL^{-1}, and 0.1 g mL^{-1}, were tested in duplicate by adjusting the amount of biomass. These biomass to solvent ratios will be further referred as 1%, 2.5%, 5%, and 10%, respectively. The round-bottom flask was immersed in a pan filled with glycerin, previously heated up to the boiling point of the solvent (90 °C for the DMC). Different times of extraction were tested (0.5 h, 1 h, 1.5 h, and 2 h). After the extraction, the pan was removed to allow the flask to naturally cool down to room temperature. A vacuum filtration with Whatman™ paper filters was then used to separate the biomass from the solution. A rotary evaporator (Hei-VAP Value, Heidolph™, Schwabach, Germany) was used to recover the solvent and to separate the PHA in the form of a film attached to the wall of the flask. The obtained PHA was left to dry overnight and then weighed. The experimental data obtained under the different extraction conditions was compared using first an F-test for variances and then the adequate t-test for the p values. Chloroform and dichloromethane were used as reference extraction solvents [26–28], for which a biomass to solvent ratio of 1% and an extraction time of 1 h were used.

2.4. Purification Process

The extracted PHA was purified with 1-butanol (≥99%, Sigma-Aldrich™, Zwijndrecht, The Netherlands) via reflux. 20 mL of the solvent was added to different amounts of the polymer in a round-bottom flask which was connected to a cooling column. The different PHA to solvent ratios tested in duplicate were 0.01 g mL^{-1}, 0.02 g mL^{-1}, and 0.04 g mL^{-1} (1, 2 and 4%). The flask was immersed in a pan with glycerin and heated up to the boiling point of 1-butanol (117.7 °C). The purification times tested were 0.5 h, 1 h, and 2 h. The round-bottom flask was then taken from the pan, closed with a cap and allowed to cool down at room temperature overnight. The solvent was then separated from the jelly-like polymer by physically pressing it out [31] with a piece of cloth. The PHA was allowed to dry at room temperature overnight. The 1-butanol was separated from the solubilized contaminants and recycled with a rotary evaporator (Hei-VAP Value, Heidolph™, Schwabach, Germany). The experimental data obtained under the different purification conditions was compared using first an F-test for variances and then the adequate t-test for the p values.

2.5. Analytical Methods

2.5.1. Thermal Gravimetric Analysis (TGA)

Samples of produced biomass were analyzed in duplicate for PHA content and every sample of extracted and purified PHA was analyzed for purity with TGA (TGA 500Q, TA instruments, Etten-Leur, The Netherlands). Because of the feed composition, all PHA produced was considered to be the co-polymer PHBV. A mass of around 10 mg was placed into platinum pans and a ramp mode of 5 °C per minute until 600 °C was set under nitrogen atmosphere, with flow rates of 40 mL min^{-1} on the balance, 60 mL min^{-1} on the sample (adapted from Hahn and Chang, 1995) [34]. The mass of PHA in each sample was determined as the mass loss in the temperature range between 250 °C and 270 °C. A commercial sample of PHBV was used to establish this range.

2.5.2. Gas Chromatography/Mass Spectrometry (GC-MS)

Sample Preparation

Following the procedure in Lo et al. (2009) [33], an acidic methanolysis reaction was used to hydrolyze the polymeric chain and to convert the monomers into their methylated ester form. A PHA sample of each experiment was weighed for a mass between 2 and 5 mg and brought into a 5 mL reaction vial. 1 mL of chloroform (99%, Sigma-Aldrich™, Zwijndrecht, The Netherlands), 0.95 mL of methanol (Technical grade, BOOM™, Meppel, The Netherlands) and 0.05 mL of sulfuric acid (95%–98%, Sigma-Aldrich™, Zwijndrecht, The Netherlands) were added into the reaction vial which was closed tightly and shaken. The reaction vial was placed into a heating block (Techne Dri-block™, Staffordshire, England, UK) at 100 °C for 6 h and shaken every 1 h. The vial was then left to cool down to room temperature. The solution was transferred from the vial to a centrifuge tube and 1 mL of 1 M NaCl (Extra pure, BOOM™, Meppel, The Netherlands) solution was added. The tube was shaken and centrifuged (5810 series, Eppendorf™, Nijmegen, The Netherlands) for 3 minutes at 4000 rpm. The aqueous layer was discarded and another 1 mL of the NaCl solution was added. The tube was shaken and centrifuged again. The organic phase was then taken from the tube with a needle attached to a 1 mL syringe. The solution volume was measured with the syringe graduation, filtered (0.45 μm × 13 mm PTFE filter) and transferred to a vial.

GC-MS Settings

A calibration curve was made using standards of methyl-(R)-3-hydroxybutyrate and methyl-(R)-3-hydroxyvalerate (>98%, Sigma-Aldrich™, Zwijndrecht, The Netherlands) using chloroform as a solvent to achieve different dilutions. The GC-MS used was a 7820A GC System/5977E MSD Agilent Technologies™ with a HP-5MS capillary GC column (30 m × 0.25 mm, 0.25 μm), flow of

2 mL min^{-1} helium, sample injection of 1 μL, temperature of 250 °C at the injector and detector with a heating rate of 10 °C min^{-1}. The software NIST MS Search Program was used to identify the samples components through their mass spectrum.

3. Results and Discussion

3.1. PHA Accumulation

Although there were slight variations throughout the runs, the fermentation process was consistent, with an initial volatile suspended solids concentration of about 4 g L^{-1}. A range of 23 to 25 pulse feeds were given during each of the 20 runs. Each fermentation run resulted in around 5.5 g of dry biomass. It was observed that the biomass easily settled to the bottom of the vessel.

The TGA analysis of the dry biomass revealed a content of around 40% of PHA in mass, represented by a degradation peak at the temperature range of 265 to 277 °C, which was proved to be the right degradation temperature by comparing the analysis with a commercial sample of PHBV. Hahn and Chang (1995) [34] discovered a correlation between the PHA content measured through TGA and the PHA content measure through GC analysis, where the result from GC analysis are considered to be more accurate. This correlation is expressed as a linear model:

$$y = 1.16x - 15.27 \quad (1)$$

where x is the PHA content by TGA and y the real content. Using this correlation, the total PHA content is around 32%. This result is slightly lower, but still close, to the ones mentioned in the PHARIO report [9], for which the same MMC source was used and the PHA accumulation resulted in values around 39 g of PHA per 100 g of VSS. A difference, however, that might explain the higher production result in that work is the nitrogen and phosphorous supplement in the feed composition and the extra feeds during acclimation process of the biomass.

3.2. Extraction

The results of PHA extraction in all studied conditions were around 31 g of PHA per 100 g of biomass, with slightly higher extraction values at lower biomass to solvent ratios and longer extraction times (Table 1). This result indicates a very high polymer recovery, which contrasts with what is discussed in Samorì (2015) [27], where only about half of all the PHA inside the MMC biomass could be extracted with dimethyl carbonate without any cell pretreatment. It is important to mention, however, that the MMC used in that work for PHA accumulation has a different source and it was submitted to an extensive process of bacterial selection over time, which might affect the general composition of the biomass and, perhaps, the efficiency of the DMC as a solvent for PHA extraction.

Table 1. Extraction results of PHA with dimethyl carbonate. Freeze dried biomass after polyhydroxyalkanoates (PHA) accumulation was extracted in 25 mL dimethyl carbonate at its boiling point at different biomass to solvent ratios for different extraction times. The biomass to solvent ratio is expressed as a percentage of grams of biomass per 100 ml of solvent. Values represent averages ± standard deviation of duplicates in g of PHA per 100 g of biomass.

Biomass to Solvent Ratio (%)	0.25 h	0.5 h	1 h	1.5 h	2 h
1%	31.4 ± 1.0^{abc}	32.2 ± 0.1^{b}	31.7 ± 0.2^{bd}	32.9 ± 0.2^{b}	
2.5%	30.6 ± 0.4^{cd}	30.4 ± 0.3^{cd}	31.5 ± 0.1^{ad}	32.3 ± 0.4^{bd}	
5%		30.5 ± 0.1^{c}	30.0 ± 1.0^{abc}	30.8 ± 0.7^{abc}	30.4 ± 1.3^{abc}
10%			29.3 ± 1.1^{abc}	27.8 ± 0.4^{a}	28.2 ± 0.5^{ac}

[abcd] Average values not sharing a common superscript were significantly different ($p < 0.05$).

The scenarios with a 10% biomass to solvent ratio presented some practical issues because of the relatively high amount of biomass that settled in the bottom part of the extraction flask in direct contact with the heating source and with low or no contact to the solvent.

The PHA extraction process with DMC was compared with chloroform [23] and dichloromethane [28] as the reference solvents (Table 2).

Table 2. Comparative extraction results for the different solvents. The reflux extraction of PHA with dimethyl carbonate at 1% biomass to solvent ratio for 1 h was compared with the reflux extraction with chloroform and with dichloromethane at the same solvent ratio and time. Values represent averages ± standard deviation of duplicates of g of PHA per 100 g of biomass.

Solvent	g PHA/100 g Biomass
Dimethyl carbonate	31.7 ± 0.2
Chloroform	37.5 ± 0.2
Dichloromethane	39.0 ± 0.2

Although the amount of extracted PHA seems higher with chloroform or dichloromethane compared with dimethyl carbonate as a solvent, when manually stretched, the polymer films produced with chloroform and dichloromethane were both very brittle and not much elastic. They would immediately break apart when pulled. The brittleness was caused by a higher percentage of impurities in the PHA extracted with chloroform and dichloromethane, as it is further discussed in the results obtained with TGA. The PHA plastic films produced in the process with dimethyl carbonate, on the other hand, had a much higher plastic deformation capability, similar to a common strong plastic bag. However, regardless of the solvent used, the resulting solid PHA had a green/brown color after the solvent recovery in all the produced samples (Figure 2).

Figure 2. A sample of PHA film produced directly after the solvent recovery. The dark green color was common to all of the samples produced.

A TGA of PHA samples obtained directly by extraction with dimethyl carbonate reveals a purity of 91.2 ± 0.1% versus a purity of 82.5 ± 3.3% for the extraction with chloroform and of 86.4% ± 3.7% with dichloromethane. These results could explain the higher yields obtained for the extraction process with chloroform and dichloromethane meaning that these reference solvents are solubilizing not only the PHA, but also higher amounts of other compounds present in the biomass, which results in lower overall purity in these samples.

3.3. Purification

Purification of the extracted PHA with 1-butanol revealed that the whitest product was obtained with a PHA to solvent ratio of 1% and after 0.5 h of purification time (Figure 3).

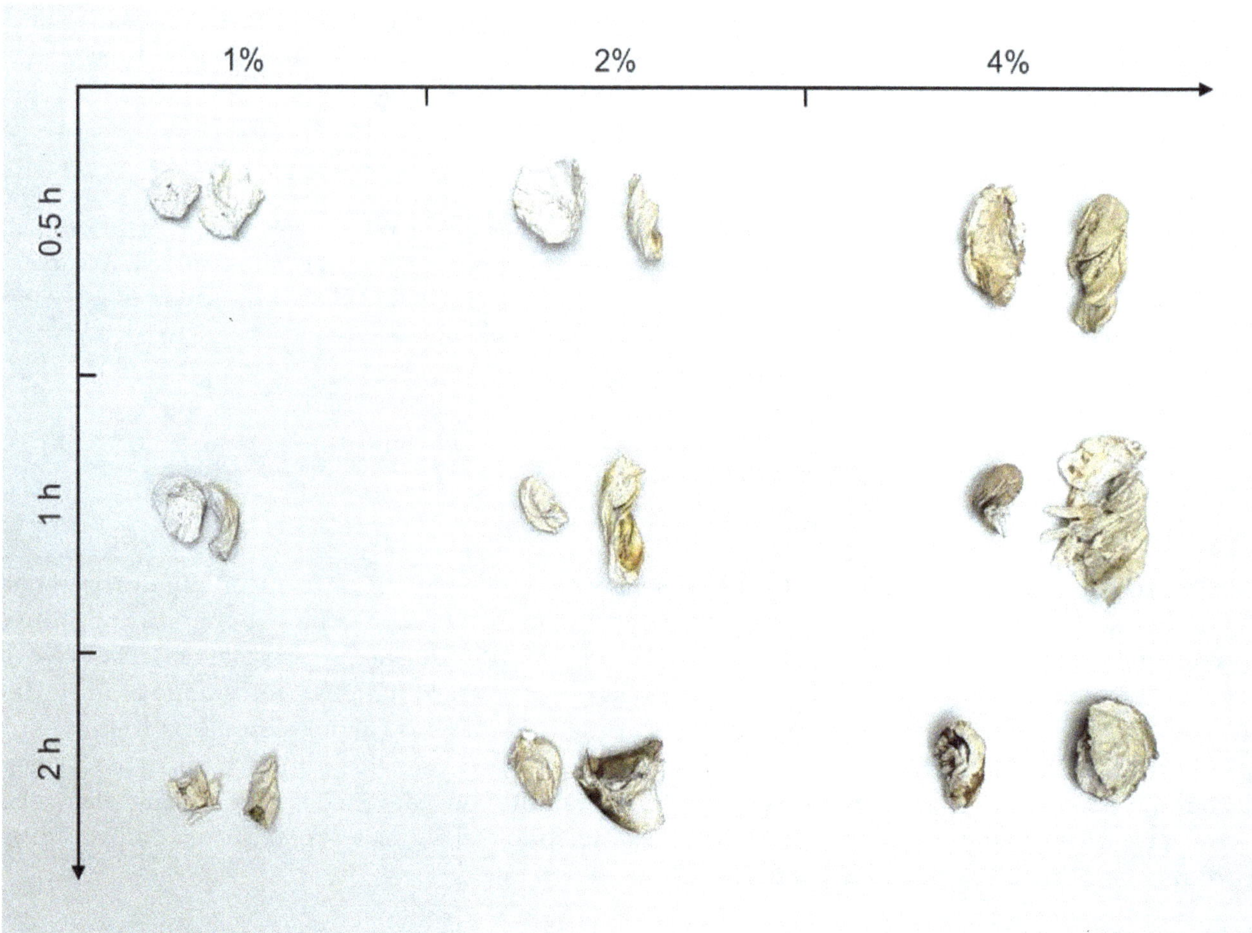

Figure 3. Comparison between the purified PHA at different experimental points.

It was expected that a lower PHA to solvent ratio led to PHA with less impurities. However, a longer purification time led to a darker-colored product, although not much difference was registered in the actual purity of the samples (Table 3).

Table 3. Purity results by TGA of the samples submitted to the purification process by 1-butanol. The PHA to solvent ratio is written as in grams of PHA per 100 ml of solvent Values represent averages ± standard deviation of triplicates of percentage of purity of purified PHA.

PHA to Solvent Ratio (%)	0.5 h	1 h	2 h
1%	98.3 ± 0.1^{acd}	98.5 ± 0.1^{a}	98.2 ± 0.2^{acd}
2%	98.4 ± 0.1^{bc}	98.9 ± 0.1^{ac}	97.9 ± 0.4^{acd}
4%	98.3 ± 0.3^{acd}	98.2 ± 0.1^{bd}	98.4 ± 0.3^{acd}

^{abcd} Average values not sharing a common superscript were significantly different ($p < 0.05$).

The evaluation of PHA by TGA revealed an increase in purity from 91.2 ± 0.1% to 98.0 ± 0.1% after purification with 1-butanol (Figure 4). The peak degradation temperature of the PHA was identified to be 253.4 ± 7.3 °C which is comparable to literature about different monomer compositions of the PHBV copolymer [35–37].

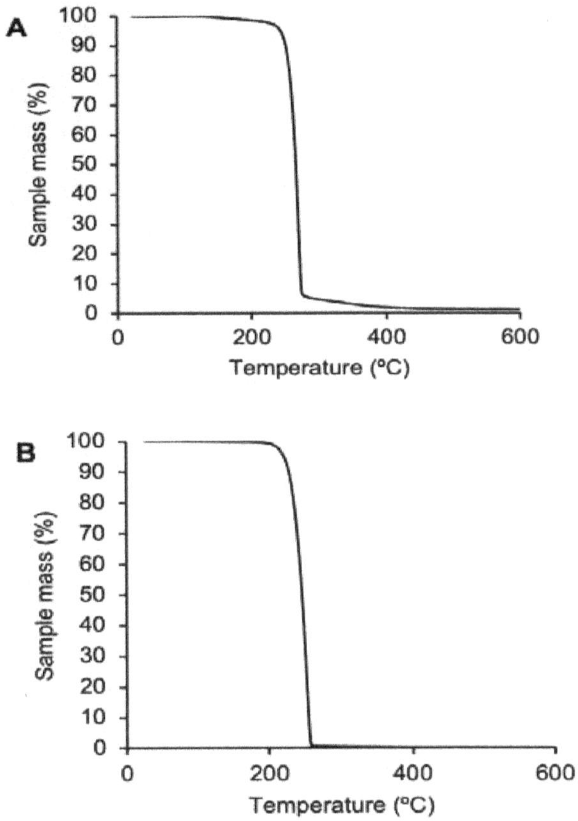

Figure 4. Thermal gravimetric analysis (TGA) of extracted PHA before (**A**) and after (**B**) purification with 1-butanol.

3.4. Analysis by Gas Chromatography-Mass Spectrometry (GC-MS)

Samples of the extracted PHA before and after purification were analyzed with GC-MS for its monomeric composition and identification of impurities (Figure 5). Besides 3-hydroxybutyrate and 3-hydroxyvalerate, the monomer 3-hydroxy-2-methylvalerate was also present in minor quantities in the samples. This monomer has been reported already as a common component of polymers synthesized by enriched cultures of glycogen-accumulating organisms (GAO) [35,38,39].

The GC-MS analysis revealed a monomer composition of 35.6 ± 2.5% 3-hydroxybutyrate and 64.4 ± 2.5% 3-hydroxyvalerate. Given the feed composition, a monomer distribution of the produced PHA of 50% 3-hydroxybutyrate and 50% 3-hydroxyvalerate was expected [17]. However, less energy is needed to metabolize propionic acid than acetic acid [40], which explains the higher percentage of 3-hydroxyvalerate in the PHA.

The non-purified PHA (Figure 5A) contained a bigger variety of impurities than the purified PHA (Figure 5B). Not much can be said about the absolute concentration of impurities before and after the purification process, as no calibration curves were made for the non-PHA related compounds. However, a reduction of 71.4%, 71.6%, and 63.7% in the areas of the impurities III, IV and V, respectively, was calculated, indicating a significant reduction in the overall concentration of such impurities.

Although the quantity of impurities was reduced after the 1-butanol treatments, hexadecanoic acid, octadecanoic acid and dehydroabietic acid were still found in all purified samples. Hexadecanoic and octadecanoic acids have been reported as storage compounds produced by mixed bacterial cultures [41]. The source of the dehydroabietic acid is unknown.

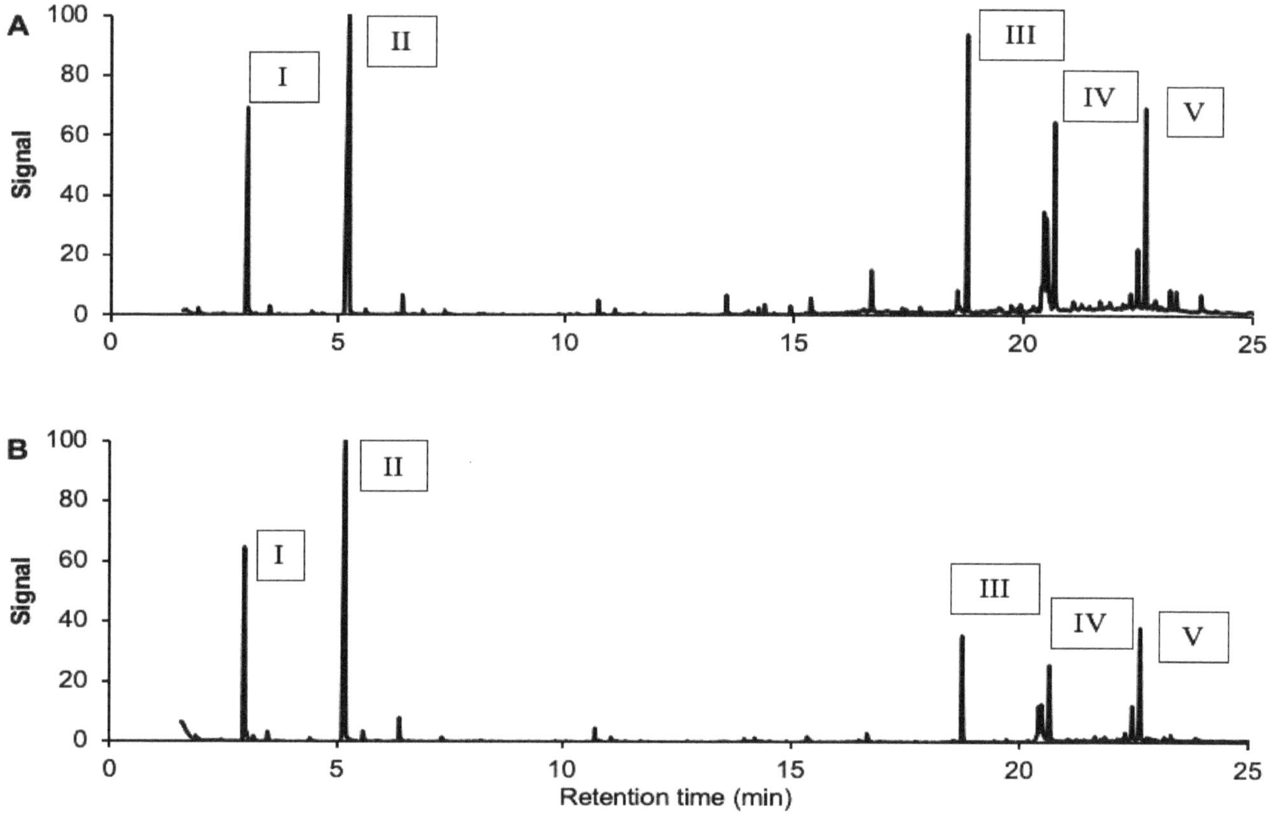

Figure 5. Chromatograms of PHA before (**A**) and after (**B**) purification. The identified compounds are: I—3-hydroxybutyric acid; II—3-hydroxyvaleric acid; III—Hexadecanoic acid; IV—Octadecanoic acid; V—Dehydroabietic acid.

3.5. Applicability

In this work, a mixed microbial culture was used for a PHA accumulation procedure followed by an extraction and purification with different solvents for obtaining a high purity final product. However, the extraction and purification are a two-step process that can be very costly when it comes to an industrial setting. For certain PHA applications where high purity is not a major factor, a single extraction with DMC could be enough for the commercial feasibility of the process. For applications where high purity PHA is required, the purification step can be added, although higher costs should be expected.

4. Conclusions

The extraction of PHA from mixed microbial cultures can be successfully done with dimethyl carbonate via reflux extraction. Overall, a very small variance of PHA yield was observed for different extraction times or biomass to solvent ratios. A ratio of 0.05 g ml^{-1} is considered to be ideal, as the use of higher amounts of biomass lead to practical difficulties. Although higher extraction yields can be obtained with chloroform or dichloromethane as solvents, that also leads to a decrease in the purity of PHA and a less sustainable extraction.

A purification of the extracted PHA with 1-butanol resulted in an increase in purity from 91.2 ± 0.1% to 98.0 ± 0.1%. Although the total purity is approximately the same for different purification times and PHA to solvent ratios, a 0.01 g mL^{-1} ratio for 0.5 h of purification time led to a whiter PHA.

Dimethyl carbonate is a great alternative to conventional hazardous solvents in the extraction process of PHA and 1-butanol can be used to increase the purity of PHA if necessary, leading to a more commercially attractive product, although this could lead to higher production costs. In terms of a circular economy, the whole process opens new environmentally friendly possibilities for the bioplastic industry.

Author Contributions: Conceptualization, G.A.d.S.R., M.H.A.M., G.L.F. and I.L.; Data curation, G.A.d.S.R. and G.L.F.; Formal analysis, G.A.d.S.R., M.H.A.M. and G.L.F.; Funding acquisition, J.H.d.B.; Investigation, G.A.d.S.R. and G.L.F.; Methodology, G.A.d.S.R., M.H.A.M. and G.L.F.; Project administration, J.H.d.B.; Supervision, M.H.A.M. and I.L.; Writing—original draft, G.A.d.S.R. and G.L.F.; Writing—review and editing, G.A.d.S.R., M.H.A.M., I.L. and J.H.d.B. All authors have read and agreed to the published version of the manuscript.

Acknowledgments: The authors would like to thank the technicians at Avans University of Applied Sciences for their technical assistance in this research and the Living Lab Biobased Brazil for connecting the authors. We are also grateful for the editors and the reviewers for their constructive comments which helped to improve the quality of this publication.

Abbreviations

3HB	3-hydroxybutyrate
3HV	3-hydroxyvalerate
COD	Chemical oxygen demand
DMC	Dimethyl carbonate
DO	Dissolved oxygen
GC-MS	Gas chromatography—mass spectrometry
MMC	Mixed Microbial Culture
PHAs	Polyhydroxyalkanoates
PHBV	Poly(3-hydroxybutyrate-co-3-hydroxyvalerate)
TGA	Thermal gravimetric analysis
VFAs	Volatile fatty acids
VSS	Volatile suspended solids

References

1. Laycock, B.; Halley, P.; Pratt, S.; Werker, A.; Lant, P. The chemomechanical properties of microbial polyhydroxyalkanoates. *Prog. Polym. Sci.* **2014**, *39*, 397–442. [CrossRef]
2. Shrivastav, A.; Kim, H.; Kim, Y. Advances in the Applications of Polyhydroxyalkanoate Nanoparticles for Novel Drug Delivery System. *Biomed Res. Int.* **2013**, *2013*, 1–12. [CrossRef] [PubMed]
3. Chen, G.-Q. Plastics from Bacteria: Natural Functions and Applications. *Microbiol. Monogr.* **2009**, *14*, 121–132.
4. Elain, A.; Le Grand, A.; Corre, Y.-M.; Le Fellic, M.; Hachet, N.; Le Tilly, V.; Loulergue, P.; Audic, J.-L.; Bruzaud, S. Valorisation of local agro-industrial processing waters as growth media for polyhydroxyalkanoates (PHA) production. *Ind. Crops Prod.* **2016**, *80*, 1–5. [CrossRef]
5. Koller, M.; Marsalek, L.; de Sousa Dias, M.M.; Braunegg, G. Producing microbial polyhydroxyalkanoate (PHA) biopolyesters in a sustainable manner. *New Biotechnol.* **2017**, *37*, 24–38. [CrossRef]
6. Tamis, J.; Luzkov, K.; Jiang, Y.; van Loosdrecht, M.C.M.; Kleerebezem, R. Enrichment of *Plasticicumulans acidivorans* at pilot-scale for PHA production on industrial wastewater. *J. Biotechnol.* **2014**, *192*, 161–169. [CrossRef]
7. Nikodinovic-Runic, J.; Guzik, M.; Kenny, S.T.; Babu, R.; Werker, A.; Connor, K.E. Carbon-Rich Wastes as Feedstocks for Biodegradable Polymer (Polyhydroxyalkanoate) Production Using Bacteria. *Adv. Appl. Microbiol.* **2013**, *84*, 139–200. [PubMed]
8. Werker, A.G.; Johansson, P.S.T.; Magnusson, P.O.G.; Maurer, F.H.J.; Jannasch, P. Method for Recovery of Stabilized Polyhydroxyalkanoates from Biomass that Has been Used to Treat Organic Waste. US2013/0203954A1, 2013. Available online: https://patents.google.com/patent/US20130203954A1/fr (accessed on 2 March 2020).
9. Bengtsson, S.; Werker, A.; Visser, C.; Korving, L. *PHARIO—Stepping Stone for a Sustainable Value Chain for PHA Bioplastic Using Municipal Activated Sludge*; Stowa: Amersfoort, The Netherlands, 2017; p. 15.

10. Oliveira, C.S.S.; Silva, C.E.; Carvalho, G.; Reis, M.A.M. Strategies for efficiently selecting PHA producing mixed microbial cultures using complex feedstocks: Feast and famine regime and uncoupled carbon and nitrogen availabilities. *New Biotechnol.* **2017**, *37*, 69–79. [CrossRef] [PubMed]
11. Carvalho, G.; Oehmen, A.; Albuquerque, M.G.E.; Reis, M.A.M. The relationship between mixed microbial culture composition and PHA production performance from fermented molasses. *New Biotechnol.* **2014**, *31*, 257–263. [CrossRef]
12. Valentino, F.; Karabegovic, L.; Majone, M.; Morgan-Sagatsume, F.; Werker, A. Polyhydroxyalkanoate (PHA) storage within a mixed-culture biomass with simultaneous growth as a function of accumulation substrate nitrogen and phosphorus levels. *Water Res.* **2015**, *77*, 49–63. [CrossRef]
13. Lemos, P.C.; Serafim, L.S.; Reis, M.A.M. Synthesis of polyhydroxyalkanoates from different short-chain fatty acids by mixed cultures submitted to aerobic dynamic feeding. *J. Biotechnol.* **2006**, *122*, 226–238. [CrossRef] [PubMed]
14. Bengtsson, S. The utilization of glycogen accumulating organisms for mixed culture production of polyhydroxyalkanoates. *Biotechnol. Bioeng.* **2009**, *104*, 698–708. [CrossRef] [PubMed]
15. Dionisi, D.; Levantesi, C.; Renzi, V.; Tandoi, V.; Majone, M. PHA storage from several substrates by different morphological types in an anoxic/aerobic SBR. *Water Sci. Technol.* **2002**, *46*, 337–344. [CrossRef]
16. Kessler, B.; Witholt, B. Factors involved in the regulatory network of polyhydroxyalkanoate metabolism. *J. Biotechnol.* **2001**, *86*, 97–104. [CrossRef]
17. Wang, J.; Yue, Z.-B.; Sheng, G.-P.; Yu, H.-Q. Kinetic analysis on the production of polyhydroxyalkanoates from volatile fatty acids by *Cupriavidus necator* with a consideration of substrate inhibition, cell growth, maintenance, and product formation. *Biochem. Eng. J.* **2010**, *49*, 422–428. [CrossRef]
18. Pardelha, F.; Albuquerque, M.G.E.; Reis, M.A.M.; Dias, J.M.L.; Oliveira, R. Flux balance analysis of mixed microbial cultures: Application to the production of polyhydroxyalkanoates from complex mixtures of volatile fatty acids. *J. Biotechnol.* **2012**, *162*, 336–345. [CrossRef] [PubMed]
19. Albuquerque, M.G.E.; Martino, V.; Pollet, E.; Avérous, L.; Reis, M.A.M. Mixed culture polyhydroxyalkanoate (PHA) production from volatile fatty acid (VFA)-rich streams: Effect of substrate composition and feeding regime on PHA productivity, composition and properties. *J. Biotechnol.* **2011**, *151*, 66–76. [CrossRef]
20. Hao, J.; Wang, H.; Wang, X. Selecting optimal feast-to-famine ratio for a new polyhydroxyalkanoate (PHA) production system fed by valerate-dominant sludge hydrolysate. *Appl. Microbiol. Biotechnol.* **2018**, *102*, 3133–3143. [CrossRef]
21. Chen, G.Q.; Zhang, G.; Park, S.J.; Lee, S.Y. Industrial scale production of poly(3-hydroxybutyrate-co-3-hydroxyhexanoate). *Appl. Microbiol. Biotechnol.* **2001**, *57*, 50–55.
22. Manangan, T.; Shawaphun, S. Quantitative extraction and determination of polyhydroxyalkanoate accumulated in *Alcaligenes latus* dry cells. *Scienceasia* **2010**, *36*, 199–203. [CrossRef]
23. Riedel, S.L.; Brigham, C.J.; Budde, C.F.; Bader, J.; Rha, C.K.; Stahl, U.; Sinskey, A.J. Recovery of poly(3-hydroxybutyrate-co-3-hydroxyhexanoate) from *Ralstonia eutrophacultures* with non-halogenated solvents. *Biotechnol. Bioeng.* **2012**, *110*, 461–470. [CrossRef] [PubMed]
24. Jiang, X.; Ramsay, J.A.; Ramsay, B.A. Acetone extraction of mcl-PHA from *Pseudomonas putida* KT2440. *J. Microbiol. Methods* **2006**, *67*, 212–219. [CrossRef] [PubMed]
25. Fernández-Dacosta, C.; Posada, J.A.; Kleerebezem, R.; Cuellar, M.C.; Ramirez, A. Microbial community-based polyhydroxyalkanoates (PHAs) production from wastewater: Techno-economic analysis and ex-ante environmental assessment. *Bioresour. Technol.* **2015**, *185*, 368–377. [CrossRef] [PubMed]
26. Fei, T.; Cazeneuve, S.; Wen, Z.; Wu, L.; Wang, T. Effective Recovery of Poly-b-Hydroxybutyrate (PHB) Biopolymer from *Cupriavidus necator* Using a Novel and Environmentally Friendly Solvent System. *Biotechnol. Prog.* **2016**, *32*, 678–685. [CrossRef]
27. Samorì, C.; Abbondanzi, F.; Galletti, P.; Giorgini, L.; Mazzocchetti, L.; Torri, C.; Tagliavini, E. Extraction of polyhydroxyalkanoates from mixed microbial cultures: Impact on polymer quality and recovery. *Bioresour. Technol.* **2015**, *189*, 195–202. [CrossRef]
28. Samorì, C.; Basaglia, M.; Casella, S.; Favaro, L.; Galletti, P.; Giorgini, L.; Marchi, D.; Mazzocchetti, L.; Torri, C.; Tagliavini, E. Dimethyl carbonate and Switchable Anionic Surfactants: Two effective tools for the extraction of polyhydroxyalkanoates from microbial biomass. *Green Chem.* **2015**, *17*, 1047–1056. [CrossRef]

29. National Center for Biotechnology Information. PubChem Database. Dimethyl Carbonate, CID:12021. Available online: https://pubchem.ncbi.nlm.nih.gov/compound/Dimethyl-carbonate (accessed on 2 March 2020).
30. Hansch, C.; Leo, A.; Hoekman, D. *Exploring QSAR—Hydrophobic, Electronic and Steric Constants*; American Chemical Society: Washington, DC, USA, 1995; pp. 3–14.
31. Werker, A.G.; Johansson, P.S.T.; Magnusson, P.O.G. Process for the Extraction of Polyhydroxyalkanoates from Biomass. US2015/0368393A1, 2015. Available online: https://patents.google.com/patent/US20150368393/en17 (accessed on 2 March 2020).
32. Carrier, M.; Loppinet-Serani, A.; Denux, D.; Lasnier, J.-M.; Ham-Pichavant, F.; Cansell, F.; Aymonier, C. Thermogravimetric analysis as a new method to determine the lignocellulosic composition of biomass. *Biomass Bioenergy* **2011**, *35*, 298–307. [CrossRef]
33. Lo, C.-W.; Wu, H.-S.; Wei, Y.-H. Optimizing acidic methanolysis of poly(3-hydroxyalkanoates) in gas chromatography analysis. *Asia Pac. J. Chem. Eng.* **2009**, *4*, 487–494. [CrossRef]
34. Hahn, S.K.; Chang, Y.K. A thermogravimetric analysis for poly(3-hydroxybutyrate) quantification. *Biotechnol. Tech.* **1995**, *9*, 873–878. [CrossRef]
35. Bengtsson, S.; Pisco, A.R.; Johansson, P.; Lemos, P.C.; Reis, M.A.M. Molecular weight and thermal properties of polyhydroxyalkanoates produced from fermented sugar molasses by open mixed cultures. *J. Biotechnol.* **2010**, *147*, 172–179. [CrossRef]
36. Carrasco, F.; Dionisi, D.; Beccari, M.; Majone, M.; Martinelli, A. Thermal stability of polyhydroxyalkanotaes. *J. Appl. Polym. Sci.* **2006**, *100*, 2111–2121. [CrossRef]
37. Keenan, T.M.; Tanenbaum, S.W.; Stipanovic, A.J.; Nakas, J.P. Production and characterization of poly-beta-hydroxyalkanoate copolymers from *Burkholderia cepacia* utilizing xylose and levulinic acid. *Biotechnol. Prog.* **2004**, *20*, 1697–1704. [CrossRef] [PubMed]
38. Jiang, Y.; Chen, Y. The effects of the ratio of propionate to acetate on the transformation and composition of polyhydroxyalkanoates with enriched cultures of glycogen-accumulating organisms. *Environ. Technol.* **2009**, *30*, 241–249. [CrossRef] [PubMed]
39. Dai, Y.; Lambert, L.; Yuan, Z.; Keller, J. Characterisation of polyhydroxyalkanoate copolymers with controllable four-monomer composition. *J. Biotechnol.* **2008**, *134*, 137–145. [CrossRef] [PubMed]
40. Montano-Herrera, L.; Laycock, B.; Werker, A.; Pratt, S. The Evolution of Polymer Composition during PHA Accumulation: The Significance of Reducing Equivalents. *Bioengineering* **2017**, *4*, 20. [CrossRef]
41. Castro, A.R.; Silva, P.T.S.; Castro, P.J.G.; Alves, E.; Domingues, M.R.M.; Pereira, M.A. Tuning culturing conditions towards the production of neutral lipids from lubricant-based wastewater in open mixed bacterial communities. *Water Res.* **2018**, *144*, 532–542. [CrossRef]

10

Removal of *Escherichia coli* by Intermittent Operation of Saturated Sand Columns Supplemented with Hydrochar Derived from Sewage Sludge

Jae Wook Chung [1,*], Oghosa Charles Edewi [1], Jan Willem Foppen [1], Gabriel Gerner [2], Rolf Krebs [2] and Piet Nicolaas Luc Lens [1]

1 UNESCO-IHE Institute for Water Education, P.O. BOX 3015, 2601 DA Delft, The Netherlands; charlesedewi@yahoo.com (O.C.E.); j.foppen@unesco-ihe.org (J.W.F.), p.lens@unesco-ihe.org (P.N.L.L.)

2 Institute of Natural Resource Sciences, Zurich University of Applied Sciences, Grüental, 8820 Wädenswil, Switzerland; gabriel_gerner@hotmail.com (G.G.); krbs@zhaw.ch (R.K.)

* Correspondence: shoutjx@gmail.com.

Featured Application: Bacterial removal in water treatment using a sand column supplemented with adsorbents derived from hydrothermally treated sewage sludge.

Abstract: Hydrothermal carbonization (HTC) technology can convert various types of waste biomass into a carbon-rich product referred to as hydrochar. In order to verify the potential of hydrochar produced from stabilized sewage sludge to be an adsorbent for bacterial pathogen removal in water treatment, the *Escherichia coli*'s removal efficiency was determined by using 10 cm sand columns loaded with 1.5% (w/w) hydrochar. Furthermore, the removal of *E. coli* based on intermittent operation in larger columns of 50 cm was measured for 30 days. Since the removal of *E. coli* was not sufficient when the sand columns were supplemented with raw hydrochar, an additional cold-alkali activation of the hydrochar using potassium hydroxide was applied. This enabled more than 90% of *E. coli* removal in both the 10 cm and 50 cm column experiments. The enhancement of the *E. coli* removal efficiency could be attributed to the more hydrophobic surface of the KOH pre-treated hydrochar. The idle time during the intermittent flushing experiments in the sand-only columns without the hydrochar supplement had a significant effect on the *E. coli* removal ($p < 0.05$), resulting in a removal efficiency of 55.2%. This research suggested the possible utilization of hydrochar produced from sewage sludge as an adsorbent in water treatment for the removal of bacterial contaminants.

Keywords: *Escherichia coli*; bacterial removal; sewage sludge; chloride tracer; hydrothermal carbonization; hydrochar

---

1. Introduction

Hydrothermal carbonization (HTC) is considered an emerging technology for effective waste conversion and/or treatment. In the HTC, also known as "wet pyrolysis", process, feed stock (organic stock immersed in water) is heated in a pressure-resistant reactor. Autogenous pressure inside the reactor allows subcritical water temperatures (180–350 °C), and the feed stock is converted into a mixture of process water containing water-soluble organics and a carbonaceous solid called hydrochar [1]. Several distinctive features of HTC, such as minimal environmental impact, simplicity, cost effectiveness, low greenhouse gas emission, and energy efficiency, make the technology attractive [2].

In 1913, Bergius performed the first HTC experiment to facilitate the natural coalification process of organic feedstock under laboratory conditions [3]. Since then, extensive research has investigated

HTC for the conversion of a wide range of feedstock, from pure substances (e.g., cellulose and glucose) to more complex materials, such as fruit shells and paper [4–7]. The advantage of HTC over other pyrolysis techniques is that a feedstock with high moisture content can be directly converted into a solid carbonaceous material with high yield. The additional cost for feedstock drying in other conventional dry pyrolysis methods can be saved, and this enables various continuously generated waste materials, such as animal and human faeces, municipal sewage sludge, and activated sludge from wastewater treatment plants, to be used as potential feedstock [1].

The traditional disposal methods of sewage sludge have limitations, due to potential environmental risks that could result from various pollutants, such as pathogenic micro-organisms, organic pollutants, and heavy metals [8]. Recently, HTC has been suggested as a cost-effective and eco-friendly solution for this sewage sludge management challenge [9–11]. Also, the carbonaceous solid output (hydrochar) from the HTC process, with a competitive pollutant adsorption capacity, could replace commercial activated carbon in water treatment [12].

The research on the use of hydrochar derived from sewage as a low-cost adsorbent in water treatment is still in its infancy. It was reported that the hydrochars derived from industrial sludge and anaerobically digested sludge removed Pb(II) (q_m = 11 mg/g) in laboratory conditions [13]. Also, sewage sludge-derived hydrochar showed a comparable Pb(II) removal efficiency (q_m = 15 mg/g) [14]. For the removal of pathogenic contaminants, hydrochar derived from sewage sludge was able to achieve a 2 to >3 log removal of human pathogenic rotavirus and adenovirus [15]. However, the removal of faecal bacteria has not yet been reported. Therefore, the main objective of this research is to evaluate the hydrochar derived from sewage sludge as an additive adsorbent in sand filtration setups for *Escherichia coli* removal during water treatment.

2. Materials and Methods

2.1. *Escherichia coli* Suspension

As a surrogate of enteric bacterial pathogens, a wild-type *E. coli* strain UCFL-94 obtained from a grazing field was provided from previous research [16]. Previous investigations reported a relatively low sticking efficiency of the strain. UCFL-94 was inoculated in 50 mL nutrient broth (OXOID, Basingstoke, Great Britain), and the culture medium was stored at 37 °C for 24 h under agitation at 150 rpm using an orbital shaker. Fresh *E. coli* stocks were prepared every week during the experimental period. The influent for the *E. coli* removal experiments was prepared by dilution of the *E. coli* stock into artificial groundwater (AGW). AGW was prepared by dissolving 526 mg/L $CaCl_2{\cdot}2H_2O$, 184 mg/L $MgSO_4{\cdot}7H_2O$, 8.5 mg/L KH_2PO_4, 21.75 mg/L K_2HPO_4, and 17.7 mg/L Na_2HPO_4 in demineralized (DI) water [17]. The results from pH and electrical conductivity measurements on the AGW ranged between 6.6–6.8 and 1012–1030 μS/cm, respectively.

The influent was prepared at room temperature (23 ± 2 °C), and stabilized for >24 h before its use in the experiments. Since there was no significant change in the *E. coli* concentration of the influent during 4 days of observation (data not shown), the natural die-off of the *E. coli* was not considered. The *E. coli* concentration of the influent was controlled to be ~10^6 CFU/mL or ~10^3 CFU/mL for small and large column experiments, respectively. In this research, the *E. coli* concentration was measured by using the conventional heterotrophic plate counting method [18] employing Chromocult agar plates (ChromoCult® Coliform Agar, Merck, Darmstadt, Germany). After the incubation period of 24 h at 37 °C, the colonies on the agar plates were counted using a Colony Counter (IUL, Barcelona, Spain).

2.2. *Hydrochar*

The Zurich University of Applied Sciences (ZHAW, Wädenswil, Switzerland) provided the hydrochar stock used in this research. Briefly, the hydrothermal conversion of stabilized sewage sludge from a wastewater treatment plant was carried out with sulphuric and acetic acid supplements for 5 h at a median temperature of ~210 °C. The autogenous pressure inside the reactor ranged between

21 and 24 bar. The output of the hydrothermal reaction had the form of thick slurry. It was chilled to 20 °C and dehydrated using a membrane filter press. The resulting filter cake was oven-dried at 105 °C overnight, and manually powdered using a mortar and pestle. Finally, the hydrochar powders were rinsed by a few rounds of suspension in DI water and successive centrifugation at 2700 g for 3 min (236 HK, Hermle, Wehingen, Germany).

In order to increase the adsorptive capacity, the raw hydrochar was activated by KOH solution [17,19,20]. Washed hydrochar powders were suspended in a 1 M KOH solution at a concentration of 5 g hydrochar (dry weight)/L and stirred for 1 h at room temperature. The hydrochar-KOH suspension was washed as described earlier until a neutral pH was observed [17]. The KOH concentration for the activation was selected based on the results from preliminary experiments (data not shown) as described previously [17]. The concentration of each hydrochar suspension was determined by measuring the dry weight (at 105 °C). All hydrochar stocks were stored at 4 °C until needed.

2.3. *Material Characterization*

2.3.1. Zeta Potential

The zeta potential of hydrochar and *E. coli* in a certain pH range (4–10 for hydrochar and 5.5–8.5 for *E.* coli) was measured using a Zetasizer Nano ZS (Malvern, Worcestershire, UK) equipped with an auto-titration unit MPT-2. Hydrochar and *E. coli* samples were conditioned in AGW by several rounds of washing prior to the experiments. All of the test samples were diluted to a certain extent in order to have an adequate attenuator selection (6–8) of the instrument.

2.3.2. Elemental Composition

Both the raw and activated hydrochar samples were analyzed by X-ray fluorescence (XRF) spectroscopy in order to assess their elemental compositions using a SPECTRO-XEPOS XRF spectrometer (SPECTRO, Kleve, Germany). The instrument was equipped with a 10 mm^2 Si-Drift Detector with Peltier cooling and a spectral resolution (FWHM) at Mn Ka $\leq$155 eV for determination in the element range of Na–U (SPECTRO, 2014). The hydrochar samples were powdered (grain size <100 μm) using a milling instrument, MM 400 (Retsch, Haan, Germany), for 5 min at a frequency of 25 s^{-1}. Then, sample pellets with a 32 mm diameter were prepared by mixing of 4 g hydrochar powder with 0.9 g Licowax C micro powder PM (Clariant, Muttenz, Switzerland) and successive pressing under 15 tonnes pressure. The analysis was performed using the TurboQuant-screening method. Each side of the pellet was subjected to one exposure to X-ray radiation, and the mean results of both sides (two analyses) were recorded.

Also, the carbon (C), hydrogen (H), nitrogen (N), and oxygen (O) contents were measured based on the dry combustion method using a TruSpec analyzer (LECO, St. Joseph, MI, USA). The samples for the C, H, N, and O analysis were thoroughly dried at 105 °C and powdered using the mixer mill for 5 min at a frequency of 25 s^{-1}. Then, 100 mg of each sample was incinerated at 950 °C and recorded by a TruSpec CHN Macro Analyzer. Furthermore, the O content was assessed by combusting 3 mg of sample material at 1300 °C using the additional high-temperature pyrolysis furnace TruSpec Micro Oxygen Module. All of the samples were analyzed in duplicate, and the mean results were recorded.

2.3.3. Surface Functional Groups

The raw and activated hydrochar samples were analyzed by Photoacoustic Fourier transform Infrared Spectroscopy (FTIR-PAS) in order to investigate their surface functional groups. A Tensor 37 FTIR spectrometer (Bruker Optics, Fällanden, Switzerland) equipped with a photoacoustic optical cantilever microphone PA301 detector (Gasera, Turku, Finland) was used for recording infrared spectra in the range of 4000–400 cm^{-1}. In order to enhance the signal-to-noise ratio, an average determination of 32 single spectra was performed after the analyses. In addition, the interference resulting from CO_2

was minimized by subtracting the CO_2 spectrum from the sample spectrum. Prior to the measurements, samples were desiccated by drying at 105 °C for 1 h and successive storing in an exsiccator at room temperature. For each analysis, the PA301 sample cell was refreshed with helium gas of 99.999% purity.

2.3.4. Specific Surface Area and Pore Size Distribution

In order to investigate the surface area and pore size distribution of the hydrochars, gas sorption analyses were carried out using an Autosorb-iQ (Quantochrome, Boynton Beach, FL, USA). Prior to the analysis, the residual water and organic contents of the hydrochar samples were removed by placing them in dynamic vacuum conditions at 120 °C. Gas sorption experiments were performed at 77 K (−196 °C) employing N_2 gas as an adsorbing gas. The surface area was derived based on the multi-point Brunauer–Emmett–Teller (BET) method. Also, density functional theory (DFT) using the quench solid DFT (QSDFT) method was used for analyzing micro and mesoporous pore size distribution [21]. Macropores were not subjected to the analysis.

2.3.5. Surface Morphology

The surface morphology of the hydrochar variants was investigated by performing a Scanning Electron Microscopy (SEM) analysis using a Quanta 250 FEG (FEI, Hillsboro, OR, USA). The hydrochar sample was placed on an adhesive side of a carbon tape fixed on an aluminium stub. Successive flushing with nitrogen gas of 99.999% purity removed detached particles.

2.3.6. Hydrophobicity

The hydrophobic property of both hydrochar samples was investigated by measuring the static contact angle using a DSA100 (KRÜSS, Hamburg, Germany). In order to obtain a flat surface of a hydrochar sample, identical pelletizing procedures employed in XRF analysis were used (see Section 2.3.2), except for the addition of Licowax C [22]. The average results of triplicated measurements were recorded.

2.4. *Column Experiments*

2.4.1. Experimental Setup

A 99.1% pure-quartz sand (Kristall quartz-sand, Gebrüder Dorfner, Hirschau, Germany) was used as a packing material for the columns. A sieve analysis on the sand stock reported an effective grain size (D_{10}) of 0.45 mm, a uniformity coefficient of 2.0, and a maximum grain size of 2.0 mm [23]. The size of the hydrochar particles was smaller than 0.425 mm. In order to remove undesirable impurities, the sand stock was immersed in 5% HCl overnight and rinsed with DI water until the pH of the washing water became stabilized. Finally, the residual water was removed by draining and successive drying at 105 °C.

A certain amount of hydrochar suspension was added to the washed sand grains to achieve a 1.5% (dry w/w) hydrochar concentration, and the mixture was then thoroughly mixed. The hydrochar–sand mixture was then loaded into two types (small or large) of columns: Omnifit borosilicate glass columns (250 mm length × 25 mm inner diameter; Diba industries, Cambridge, UK) for supporting short (10 cm) column beds, and acrylic poly vinyl chloride (PVC) pipes with an inner diameter of 5.6 cm for supporting long (50 cm) column beds. During the packing process, the columns were manually tapped and agitated in order to prevent channelling in the column bed. Also, the packing material loaded in the column was compacted carefully using either a glass rod or a steel rod to avoid possible air entrapment. Each type of column was assembled using appropriate connectors and fitting materials. Throughout the experiments, the packing materials were tightly fixed in the columns by stoppers at both ends, without under-drain media and exposure to the atmosphere. Then, the columns were washed with DI water overnight using a Masterflex pump (model 77201-60, Vernon Hills, IL, USA). During the washing process, the DI water was fed into columns at an upward flow rate of 1 mL

(0.2 cm)/min for small columns, or 33.3 mL (1.35 cm)/min for large columns. Chloride tracer tests, performed on both the small and large columns (data not shown), reported a void ratio of ~40%.

The columns were vertically positioned using tripod stands or a steel frame rack. Prior to the main *E. coli* flushing experiment, chloride tracer tests were performed to estimate the pore volume (PV) and to verify a stable aqueous flow in the column bed. The columns were flushed with the tracer solution (0.02 M NaCl) and successively with the DI water. The tracer concentration in the effluent was measured using ion chromatography (ICS-1000, Dionex, Sunnyvale, CA, USA).

2.4.2. Small Column Experiments

The *E. coli* removal performance of the small columns (10 cm long filtration bed) packed with either sand, the sand–raw hydrochar mixture, or the sand–activated hydrochar mixture was investigated by column flushing experiments. For each type of column medium, a pair of columns loaded with fresh packing materials was used to generate two breakthrough curves (BTC). Each flushing consisted of a loading (feeding of 50 mL AGW seeded with *E. coli*) and successive deloading (feeding of 50 mL *E. coli*-free AGW) phase. The influent was fed into the column at an upward flow rate of 1 mL/min (0.2 cm/min). The *E. coli* concentration in the effluent was measured every 5 min. After the column flushing experiments, the vertical distribution of the hydrochar concentration was measured as described previously [17].

2.4.3. Large Column Experiments

Three pairs of large columns (50 cm long filtration bed) were prepared for testing the sand as well as the sand–raw and sand–activated hydrochar mixtures. The columns were subjected to intermittent daily flushings with *E. coli*-seeded AGW for 30 days. For each week, five consecutive daily flushings were performed, and 2 days of pause were given. This flushing regime was designed to give four sets of 24 h idle time (in the first 5 days of each week) and one set of 72 h idle time per week. For each flushing experiment, 1 L influent (~2 PV, calculated from the chloride tracer test, data not shown) was flushed into each column, and the effluent was sampled for every 100 mL.

During the flushing, the first PV of the influent expelled the residual pore water stored in the column media from the previous flushing experiment, and the second PV was stored in the column media until the next flushing. It is considered that the first five effluent samples reflect the effect of idle time in *E. coli* attenuation. The following five effluent samples represented the direct *E. coli* removal that occurred when passing through the column bed. One-way ANOVA and Tukey's test ($p < 0.05$) were used to verify the statistical differences of the results from the three types of setup. The statistical analyses were carried out using IBM SPSS Statistics for Windows Version 22.0 released 2013 (Armonk, NY, USA).

3. Results

3.1. Material Characterization

3.1.1. Zeta Potential

The zeta potential assessment performed on the test *E. coli* strain UCFL-94, raw, and activated hydrochar used in the column experiments showed negative values in the pH range of 6.6–6.8 (Figure 1). The zeta potential of raw hydrochar was slightly increased by KOH activation from −15 mV to −13 mV. Also, a negative zeta potential of −11 mV was measured for the *E. coli* UCFL-94 suspension.

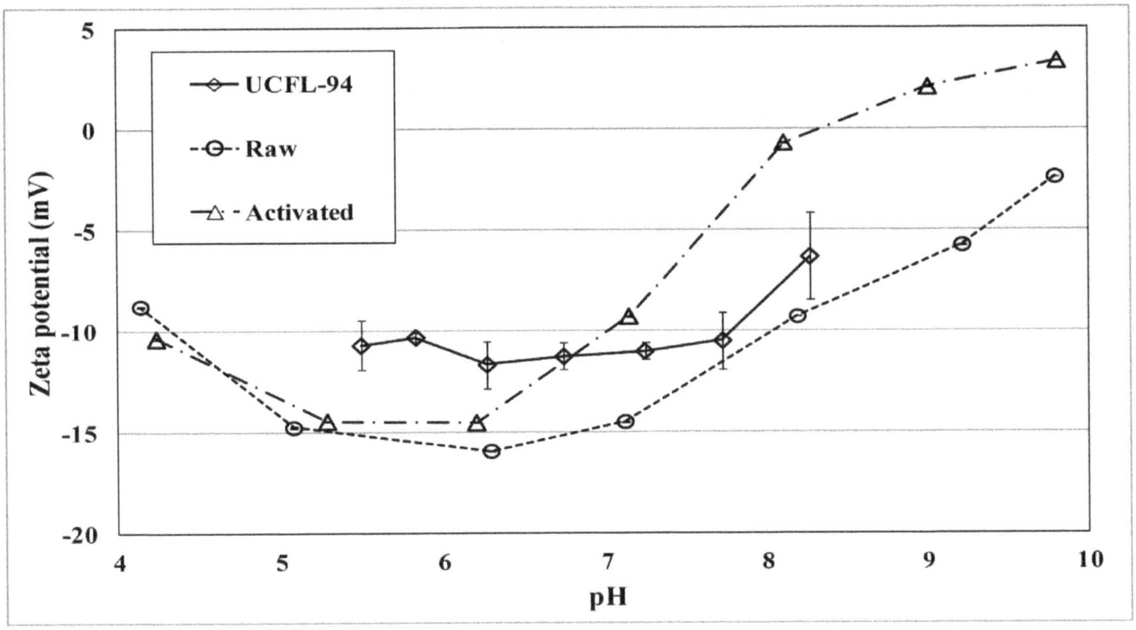

Figure 1. Zeta potential of the hydrochar and *Escherichia coli* strain over a function of pH. The lines represent the mean zeta potential value and error bars indicate the standard deviation.

3.1.2. Elemental Composition

Table 1 shows the elemental composition of both hydrochar samples obtained from the analyses using XRF and a TruSpec analyzer. The KOH activation increased the potassium (K) content by 0.4%, which supports the association of K on the surface of raw hydrochar. It was speculated that the increase in iron (Fe), Magnesium (Mg), and calcium (Ca) content was derived from impurities of the KOH reagent used for the activation, and/or due to an increase in relative concentration due to the loss of other elements. The decreases in aluminium (Al) and phosphorous (P) contents resulted from the washing out during the activation process. The alterations in the constitution of the carbon (C), hydrogen (H), Nitrogen (N), and oxygen (O) contents were considered insignificant.

Table 1. Elemental composition (%) of hydrochar variants used in this study.

	C	H	N	O	Ca	Mg	Al	K	Fe	P
Raw	28.6	3.6	2.0	22.3	5.1	0.9	2.6	0.5	5.0	4.5
Activated	29.4	3.7	1.9	21.1	5.8	1.0	2.0	0.9	5.5	2.6

Obtained from TruSpec analyzer (C, H, N, and O) or XRF (Ca, Mg, Al, K, Fe, and P).

3.1.3. Surface Functional Groups

The qualitative composition of surface functional groups was not significantly altered by the KOH activation. The raw and activated hydrochar samples showed similar peaks in the FTIR-PAS analyses (Figure 2). The spectra in the region between 1700 and 1300 cm^{-1} and the bands at 3800 cm^{-1} were assigned to the residual water content in the hydrochar samples. The bands in the range of 2930–2850 cm^{-1} corresponded to aliphatic C-H stretching vibrations [24–26]. The bands at 1600 and 1446 cm^{-1} referred to aromatic C=C stretching vibrations [26,27] and aliphatic CH_2 scissoring vibrations [28,29], respectively. A modest vibration shift was observed from the bands in the range of 1446–1400 cm^{-1} of the activated hydrochar samples. This referred to the deprotonation of the OH at the surface of hydrochar. Te bands in the 1110–1010 cm^{-1} region were derived from O stretching vibrations [25,27,28]. The bands at 780 cm^{-1} was derived from out-of-plane bending vibrations of aromatic C-H bonds [24,26].

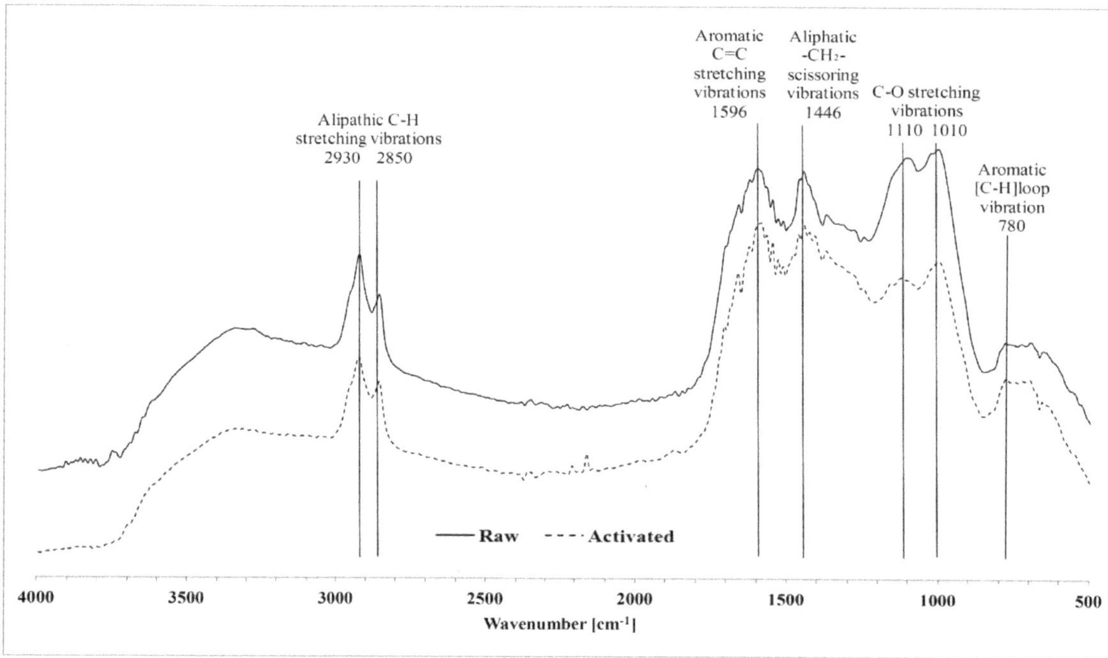

Figure 2. FTIR-PAS spectra of raw and activated hydrochar.

3.1.4. Specific Surface Area and Pore Size Distribution

The BET analysis performed on the raw and activated hydrochar showed a specific surface area of, respectively, 25.3 and 18.5 m^2/g. Figure 3 shows the results of the QSDFT pore size (differential pore volume) distribution, which demonstrates the pore volume composition attributed to the specific pore widths. The pore volume derived from the micro and mesopore range was negligible. A clear decrease in the surface area derived from the KOH activation was observable in the pore fraction, with a size >15 nm.

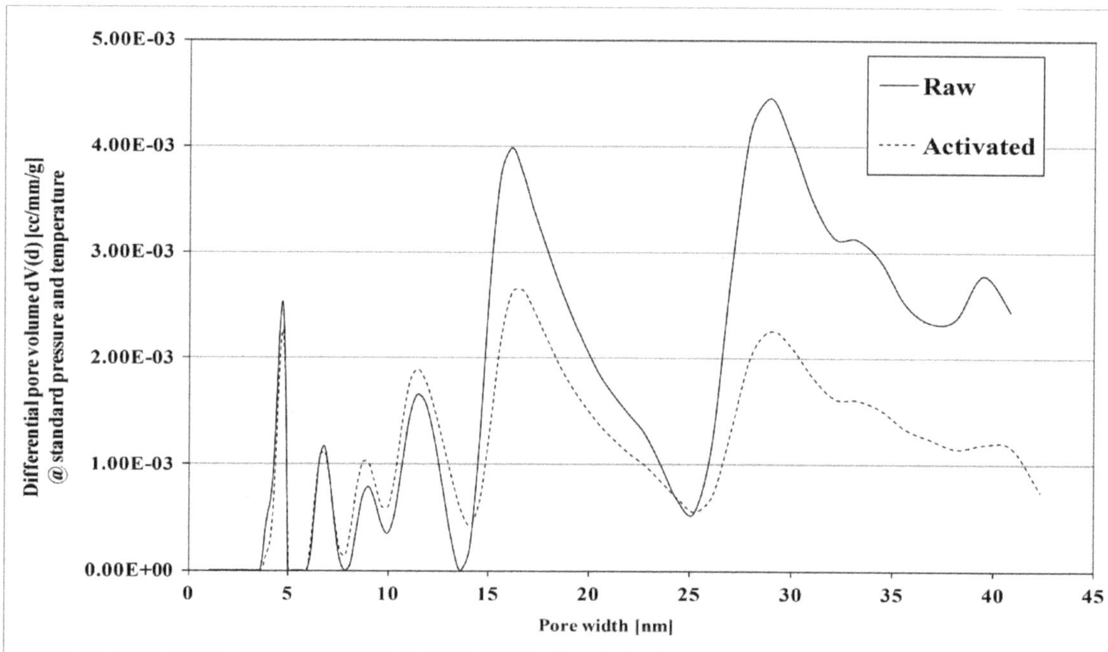

Figure 3. Pore size distribution curves obtained from quench solid density functional theory (QSDFT) analysis.

3.1.5. Surface Morphology

The SEM analyses for the raw and activated hydrochar samples were performed in order to investigate the morphological change derived from the KOH activation. The images from the SEM analyses indicated insignificant physical alterations (Figure 4). The surfaces of both types of hydrochar had a relatively rough structure, consisting of macropores (>50 nm) which could provide attachment sites for *E. coli* cells with a size of ~2 μm (measured by Zeta-sizer nano, Malvern, data not shown).

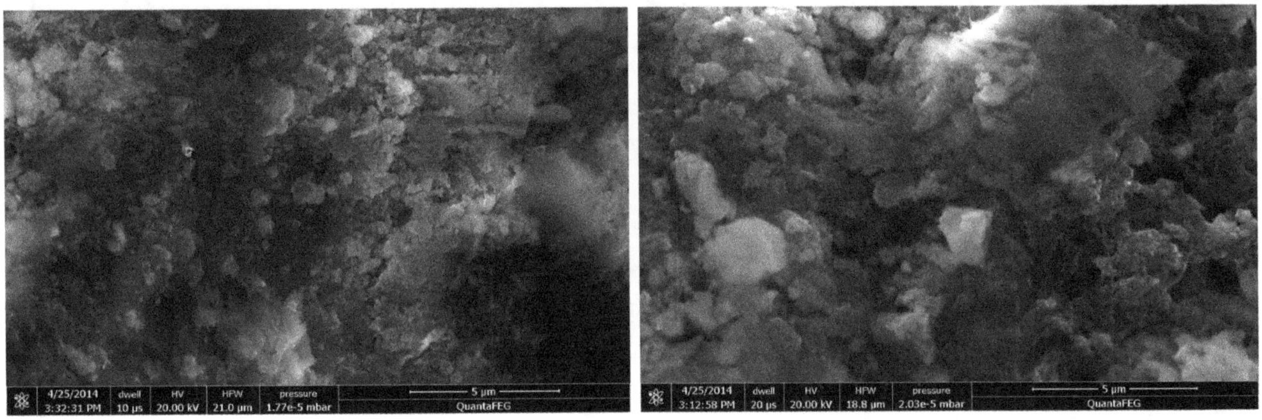

Figure 4. SEM images of raw (**left**) and activated (**right**) hydrochar.

3.1.6. Hydrophobicity

The results from the contact angle measurements on the raw and activated hydrochar samples indicated that the surfaces of both materials had hydrophobic characteristics. The hydrophobicity of the raw hydrochar was increased by KOH activation. The contact angle values of the raw and activated hydrochar were 126.5° (±2.9) (average of triplicate ± standard deviation) and 135.4° (±4.7), respectively.

3.2. E. coli Flushing Test

3.2.1. Breakthrough Analysis in Small Column Experiments

Figure 5 presents the BTCs obtained from the small column flushing experiments. All BTCs showed a clear pattern, consisting of a rising limb, a plateau phase, and a declining limb. The supplementation of raw hydrochar in the sand media resulted in early *E. coli* breakthrough. In contrast to the rising limb of the sand-only column observed after 15 min, the one from the raw hydrochar-amended column was observed already after 10 min. This was similar in the declining limbs: the one in the sand-only column started 5 min later than the column supplemented with the raw hydrochar. It was apparent that the decrease in the pore space of the sand media induced by the filling of pores by raw hydrochar amendment facilitated the *E. coli* transportation. While the effect of raw hydrochar addition in the sand media for the *E. coli* removal was insignificant, the amendments with the activated hydrochar showed an important increase in the removal efficiency. The C/C_0 ratios in the plateau phase of both the sand-only and raw hydrochar-amended columns were similar at ~0.9. However, the C/C_0 ratio of the sand column with the activated hydrochar supplement was only around 0.1. The average *E. coli* removal efficiencies of the sand and the raw and activated hydrochar-supplemented columns were 9.2%, 9.6%, and 90.1%, respectively. The measurements on the vertical hydrochar distribution in the column showed no significant migration of hydrochar particles during the flushing.

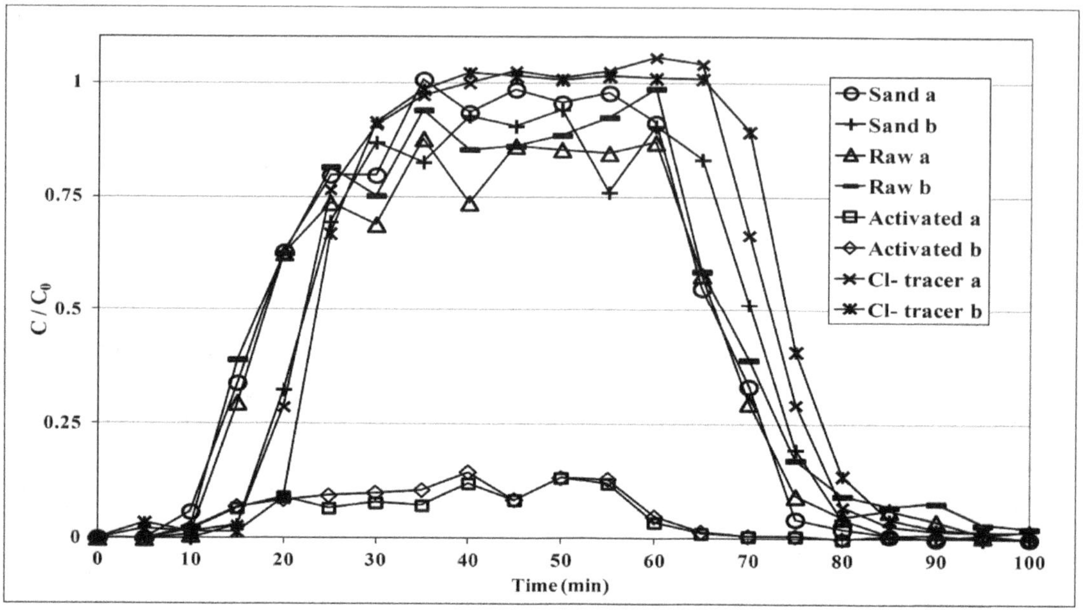

Figure 5. Breakthrough curves of *E. coli* from small column experiments carried out at a flow rate of 1 mL (0.20 cm)/min.

3.2.2. *E. coli* Removal Efficiency in Large Column Experiments

Figure 6 illustrates the *E. coli* removal efficiencies of the sand and raw and activated hydrochar-amended columns during the 30 days of experiments with intermittent operation. Similar to the results from the small column experiments, only the activated hydrochar amendment was effective at *E. coli* removal. The removal efficiency of the activated hydrochar-amended column was the highest (99.7%) on the first day, and declined gently until the last day of the experiment (78.9%). The sand-only and raw hydrochar-amended columns showed a comparable *E. coli* removal performance, in the range of 11.4–57.2%. Table 2 summarizes the overall *E. coli* removal efficiencies and the effect of the idle time during the large column experiments. Under all experimental conditions applied, the activated hydrochar-amended columns showed greater *E. coli* removal performances than the other columns. During 30 days of operation, the sand columns with activated hydrochar supplements showed an average total *E. coli* removal efficiency of 91.2%. In contrast, the sand-only and raw hydrochar-supplemented columns showed an average total *E. coli* removal efficiency of only 24.4% and 36.5%, respectively. The effect of the idle time on the *E. coli* removal was only observed in the sand-only columns; throughout the experimental period, the removal in the first PV was significantly greater than that of the second PV. The *E. coli* removal efficiency in the second PV, which represents the direct removal of *E. coli* when passing through the column media, was 17.2%. In contrast, the removal efficiencies in the first PV (stored in the sand media for 24 or 72 h and thus representing the effect of the idle time) were 52.1% and 66.9%, respectively. Throughout the experiments, the idle time had no clear effect on the *E. coli* removal performance of both types of hydrochar amendments, except for the 72 h idle time applied to the raw hydrochar-amended columns. Such an extended idle time increased the *E. coli* removal efficiency by 14.8%. The direct removal of *E. coli* (without idle time) was similar, at ~19% in the second PV in the sand-only and raw hydrochar-amended columns. This indicates a close correspondence with the results from the small column experiments.

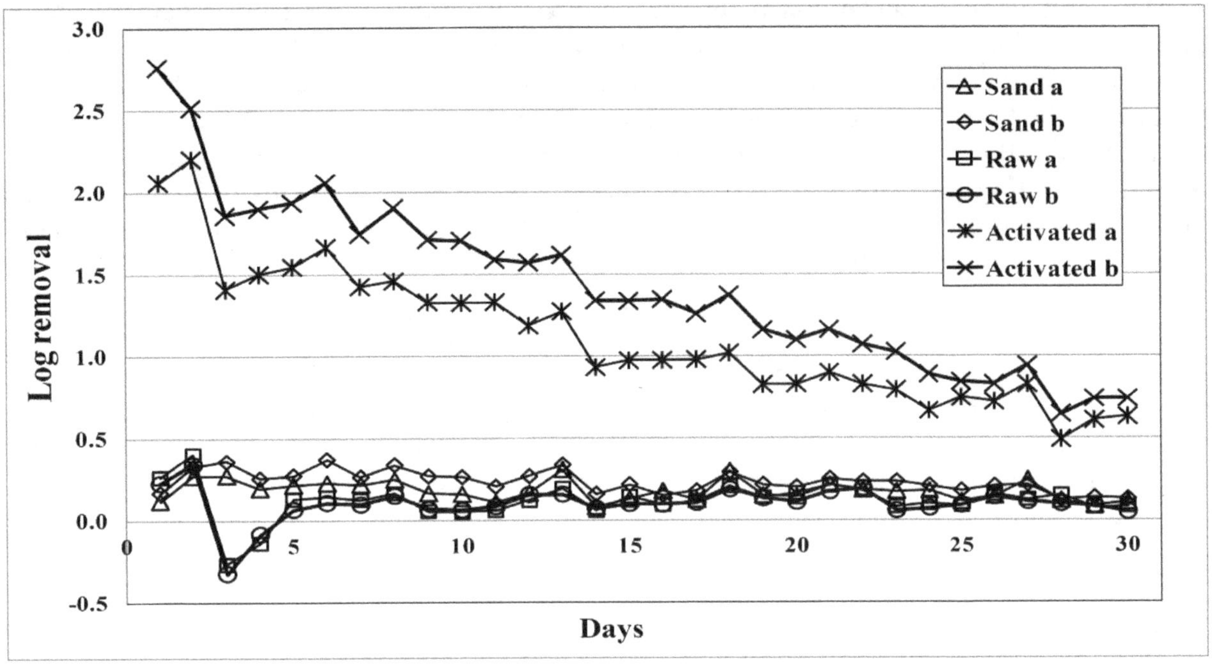

Figure 6. *E. coli* removal efficiencies from large columns during 30 days of flushing with a flow rate of 33.3 mL (1.35 cm)/min.

Table 2. Removal efficiency (average% ± standard deviation) of *E. coli* in large column experiments during 30 days of intermittent flushing in duplicate columns of each treatment.

Content	Pore Volume			Total
	First		Second	
	24 h idle	72 h idle		
Sand	52.1 ± 13.3 (n = 46)	66.9 ± 14.7 (n = 12)	17.2 ± 8.64 § (n = 60)	36.5 ± 10.1 (n = 60)
Raw	23.0 ± 17.3 † (n = 44)	35.1 ± 12.9 (n = 10)	20.0 ± 12.5 § † (n = 56)	24.4 ± 10.5 † (n = 56)
Activated	92.8 ± 5.0 ‡ (n = 46)	92.6 ± 8.0 ‡ (n = 12)	90.0 ± 9.1 ‡ (n = 60)	± 7.5 ‡ (n = 60)

Within each column, values followed by the symbol § are not significantly different using Tukey's test at $p < 0.05$. Within each row, values followed by the same symbol † or ‡ are not significantly different using Tukey's test at $p < 0.05$.

4. Discussion

4.1. Effect of KOH Activation of Hydrochar on E. coli Removal

The performance of an adsorbent mainly depends on its surface, which provides internal pore structure and adsorptive sites [30]. Several attractive or repulsive forces between *E. coli* cells and the adsorbent surface regulate their mutual interaction [31]. Considering the characteristics of *E. coli*, the adsorptive removal will be facilitated when the adsorbent possesses a highly porous surface with a positive (less negative) surface charge and hydrophobic properties. It has been reported that the KOH activation of hydrochars derived from plant materials increased the surface roughness, resulting in enhanced heavy metal or *E. coli* removal from artificially contaminated influents [17,19]. However, the BET and SEM analyses carried out in this research could not explain the improvement in *E. coli* removal efficiency induced by the KOH activation. The modification in the surface morphology of the hydrochar by the KOH treatment was insignificant (Figure 4). Moreover, the specific surface area of the activated hydrochar was 27% less than that of the raw hydrochar (See Section 3.1.4).

A possible explanation for the decrease in the specific surface area derived from the KOH activation and subsequent washing processes can be the collapse of micro and mesopores, resulting in the development of a macroporous structure on the hydrochar surface. However, a large surface area of an adsorbent, mainly consisting of micro (<2 nm) and mesopores (2–50 nm), may not be a direct indicator for efficient *E. coli* removal, considering the size *E. coli* cells (~2 μm). Further investigations on the surface area and surface properties of the macropores are thus recommended in order to obtain a better understanding of the *E. coli* removal mechanisms by (KOH treated) biochar.

Because the investigations carried out to compare the characteristics of raw and activated hydrochar showed only minor differences, the improved *E. coli* removal performance obtained from the KOH activation could result from the increase in hydrophobicity of the hydrochar surface, which enforces the hydrophobic attraction. Hydrochar consists of a hydrophobic core and a hydrophilic outer surface [26,32]. The KOH activation carried out in this research would have brought more hydrophobic surfaces into contact with *E. coli* cells by removing the hydrophilic surface coatings formed by recondensation and repolymerization of water-soluble substances during the HTC process [27]. Another possible explanation for the advantageous effect of KOH activation can be an increase in the surface charge of the hydrochar, which would have weakened the electrostatic repulsion between *E. coli* cells and hydrochar surfaces.

4.2. Effect of Idling Time on E. coli Removal Efficiency

Previous research on slow-sand filtration units under intermittent operation reported a significant contribution of the idle time to the removal efficiency of bacteria [33]. The observations in this research on the effect of the idle time in the large sand column correspond well with these studies (Table 2). During the pause between daily flushing, bacterial surrogates residing in the sand bed were attenuated.

A more recent research [34] reported an important observation that the activities of the microbial community in the sand bed were more responsible for the viral attenuation than the physico-chemical processes. The attenuation of bacteriophages during the idle time increased along with the maturation of the filter bed, while the filters that operated under the suppression of microbial activities did not show any significant attenuation [34]. This may, however, not explain our results, because the bactericidal effect of the idle time observed in this research already existed from the first batch of the intermittent operation, and it was maintained at a comparable level throughout the 30 days of experiments (Table 2). This discrepancy might be due to differences in experimental conditions such as the type of influent/test micro-organisms, the size of sand grains, or the presence of standing water for oxygen infiltration into the filter bed.

Based on our observation, it could be speculated that the extended contact time between *E. coli* cells and sand surfaces provided more chances for bacterial attachment due to the motility of *E. coli* cells [35] or Brownian diffusion [36]. In contrast, the advantageous effect of the idle time on *E. coli* removal was negligible when either raw or activated hydrochar was supplemented in the sand bed. The provision of carbonaceous surfaces in the sand medium could have induced more desirable conditions for *E. coli* survival, by providing extra nutrients and more protection against external stress factors [37–40]. In order to clarify the *E. coli* inactivation mechanism during the idle time, more investigations on *E. coli*–surface interactions under static conditions are recommended.

4.3. Potential of HTC-Sand Filters for Pathogen Removal

Potable water can be provided either by centralized water treatment and supply systems or by decentralized (point-of-use) technologies that are installed on a house. Decentralized technologies have been recognized as appropriate options for poor rural communities. Biosand filters (BSF) are an example of a point-of-use technology that have been applied in developing countries [41]. Since the formation of a biofilm layer (Schmutzdecke) in a BSF plays a key role in removing microbial agents, insufficient pathogen removal during the startup (ripening) period is one of the main limitations of BSF. Previous research on BSF reported only 60–70% of *E. coli* removal during the ripening period, i.e., the first 3 weeks of operation [33,42]. Considering the superior *E. coli* removal efficiency (96.5%) in the same period, amendments of activated hydrochar to the sand bed would be an attractive option to overcome the inferior performance of BSF during the startup period. Since the experimental conditions employed in this research differed from the general BSF design and operational parameters, extrapolation of the results of this research to supplementation of BSF setups with hydrochar needs to be done carefully. Accordingly, it is recommended to perform more experiments employing (i) smaller sand grain sizes (<0.7mm), (ii) standing heads for ensuring sufficient oxygen supplementation, and (iii) influents with a higher microbial heterogeneity than that of the AGW used in this study, e.g., surface water from lakes, rivers, or canals [43,44].

Though it is reported that hydrothermal treatment decreased the environmental risk of abiotic contaminants, such as heavy metals and pharmaceuticals embedded in sewage sludge [45,46], the release of these contaminants from hydrochar has not yet been investigated. In case the hydrochar contains considerable amounts of these undesirable compounds, long-term monitoring on the effluent quality is recommended prior to the practical implementation of BSF supplemented with the hydrochar adsorbent. The properties of hydrochar are largely determined by parameters such as reaction temperature, time, and pressure, as well as catalyst and feedstock composition [1]. Further research on the optimization of these parameters can improve the stability of heavy metals in the hydrochar [45], and completely degrade the pharmaceutical residues during the hydrothermal treatment [46].

5. Conclusions

This research evaluated the use of hydrochar derived from stabilized sewage sludge as a low-cost adsorbent for pathogen removal in water treatment. The activation of hydrochar carried out at ambient temperatures using a strong alkaline solution increased the adsorptive performance of hydrochar by removing hydrophilic substances from the hydrochar surface, resulting in an increase in hydrophobicity of the biochar particles. Supplementation of activated hydrochar in a sand filtration unit (1.5%, *w*/*w*) with a 50 cm bed height yielded an *E. coli* removal efficiency exceeding 90% during 30 days of intermittent operation. Pathogen removal based on the use of hydrochar is a new concept which has not been extensively studied. Its practical implementation in developing countries still requires follow up investigations on the optimization of the process parameters and the durability of the filtration unit.

Acknowledgments: This research was funded by the Korean Church of Brussels, Mangu Jeja Church (Seoul, Korea), and the Netherlands Ministry of Development Cooperation (DGIS) through the UNESCO-IHE Partnership Research Fund. It was carried out in the framework of the research project 'Addressing the Sanitation Crisis in Unsewered Slum Areas of African Mega-cities' (SCUSA).

Author Contributions: Jae Wook Chung, Jan Willem Foppen, and Piet Nicolaas Luc Lens conceived and designed the experiments; Oghosa Charles Edewi performed the experiments; Jae Wook Chung and Oghosa Charles Edewi analyzed the data; Gabriel Gerner and Rolf Krebs contributed materials and analysis tools; Jae Wook Chung put together the initial drafts and finalized the paper.

References

1. Libra, J.A.; Ro, K.S.; Kammann, C.; Funke, A.; Berge, N.D.; Neubauer, Y.; Titirici, M.M.; Fühner, C.; Bens, O.; Kern, J.; et al. Hydrothermal carbonization of biomass residuals: A comparative review of the chemistry, processes and applications of wet and dry pyrolysis. *Biofuels* **2011**, *2*, 71–106. [CrossRef]
2. Titirici, M.-M.; White, R.J.; Falco, C.; Sevilla, M. Black perspectives for a green future: Hydrothermal carbons for environment protection and energy storage. *Energy Environ. Sci.* **2012**, *5*, 6796–6822. [CrossRef]
3. Funke, A.; Ziegler, F. Hydrothermal carbonization of biomass: A summary and discussion of chemical mechanisms for process engineering. *Biofuels Bioprod. Biorefin.* **2010**, *4*, 160–177. [CrossRef]
4. Lu, X.; Jordan, B.; Berge, N.D. Thermal conversion of municipal solid waste via hydrothermal carbonization: Comparison of carbonization products to products from current waste management techniques. *Waste Manag.* **2012**, *32*, 1353–1365. [CrossRef] [PubMed]
5. Nizamuddin, S.; Jayakumar, N.S.; Sahu, J.N.; Ganesan, P.; Bhutto, A.W.; Mubarak, N.M. Hydrothermal carbonization of oil palm shell. *Korean J. Chem. Eng.* **2015**, *32*, 1789–1797. [CrossRef]

6. Erdogan, E.; Atila, B.; Mumme, J.; Reza, M.T.; Toptas, A.; Elibol, M.; Yanik, J. Characterization of products from hydrothermal carbonization of orange pomace including anaerobic digestibility of process liquor. *Bioresour. Technol.* **2015**, *196*, 35–42. [CrossRef] [PubMed]
7. Islam, M.A.; Tan, I.A.W.; Benhouria, A.; Asif, M.; Hameed, B.H. Mesoporous and adsorptive properties of palm date seed activated carbon prepared via sequential hydrothermal carbonization and sodium hydroxide activation. *Chem. Eng. J.* **2015**, *270*, 187–195. [CrossRef]
8. Fytili, D.; Zabaniotou, A. Utilization of sewage sludge in eu application of old and new methods—A review. *Renew. Sustain. Energy Rev.* **2008**, *12*, 116–140. [CrossRef]
9. Escala, M.; Zumbuhl, T.; Koller, C.; Junge, R.; Krebs, R. Hydrothermal carbonization as an energy-efficient alternative to established drying technologies for sewage sludge: A feasibility study on a laboratory scale. *Energy Fuels* **2013**, *27*, 454–460. [CrossRef]
10. Danso-Boateng, E.; Shama, G.; Wheatley, A.D.; Martin, S.J.; Holdich, R.G. Hydrothermal carbonisation of sewage sludge: Effect of process conditions on product characteristics and methane production. *Bioresour. Technol.* **2015**, *177*, 318–327. [CrossRef] [PubMed]
11. Wirth, B.; Reza, T.; Mumme, J. Influence of digestion temperature and organic loading rate on the continuous anaerobic treatment of process liquor from hydrothermal carbonization of sewage sludge. *Bioresour. Technol.* **2015**, *198*, 215–222. [CrossRef] [PubMed]
12. Smith, K.M.; Fowler, G.D.; Pullket, S.; Graham, N.J.D. Sewage sludge-based adsorbents: A review of their production, properties and use in water treatment applications. *Water Res.* **2009**, *43*, 2569–2594. [CrossRef] [PubMed]
13. Alatalo, S.M.; Repo, E.; Makila, E.; Salonen, J.; Vakkilainen, E.; Sillanpaa, M. Adsorption behavior of hydrothermally treated municipal sludge & pulp and paper industry sludge. *Bioresour. Technol.* **2013**, *147*, 71–76. [PubMed]
14. Spataru, A. The Use of Hydrochar as a Low Cost Adsorbent for Heavy Metal and Phosphate Removal from Wastewater. Master's Thesis, UNESCO-IHE, Delft, The Netherlands, 2014.
15. Chung, J.W.; Foppen, J.W.; Gerner, G.; Krebs, R.; Lens, P.N.L. Removal of rotavirus and adenovirus from artificial ground water using hydrochar derived from sewage sludge. *J. Appl. Microbiol.* **2015**, *119*, 876–884. [CrossRef] [PubMed]
16. Lutterodt, G.; Basnet, M.; Foppen, J.W.A.; Uhlenbrook, S. Determining minimum sticking efficiencies of six environmental *Escherichia coli* isolates. *J. Contam. Hydrol.* **2009**, *110*, 110–117. [CrossRef] [PubMed]
17. Chung, J.W.; Foppen, J.W.; Izquierdo, M.; Lens, P.N.L. Removal of *Escherichia coli* from saturated sand columns supplemented with hydrochar produced from maize. *J. Environ. Qual.* **2014**, *43*, 2096–2103. [CrossRef] [PubMed]
18. APHA. *Standard Methods for the Examination of Water and Wastewater*, 20th ed.; American Public Health Association: Washington, DC, USA, 1998.
19. Regmi, P.; Moscoso, J.L.G.; Kumar, S.; Cao, X.Y.; Mao, J.D.; Schafran, G. Removal of copper and cadmium from aqueous solution using switchgrass biochar produced via hydrothermal carbonization process. *J. Environ. Manag.* **2012**, *109*, 61–69. [CrossRef] [PubMed]
20. Sun, K.J.; Tang, J.C.; Gong, Y.Y.; Zhang, H.R. Characterization of potassium hydroxide (koh) modified hydrochars from different feedstocks for enhanced removal of heavy metals from water. *Environ. Sci. Pollut. Res.* **2015**, *22*, 16640–16651. [CrossRef] [PubMed]
21. Neimark, A.V.; Lin, Y.; Ravikovitch, P.I.; Thommes, M. Quenched solid density functional theory and pore size analysis of micro-mesoporous carbons. *Carbon* **2009**, *47*, 1617–1628. [CrossRef]
22. Jeong, S.-B.; Yang, Y.-C.; Chae, Y.-B.; Kim, B.-G. Characteristics of the treated ground calcium carbonate powder with stearic acid using the dry process coating system. *Mater. Trans.* **2009**, *50*, 409–414. [CrossRef]
23. Matthess, G.; Bedbur, E.; Gundermann, K.O.; Loof, M.; Peters, D. Investigation on filtration mechanisms of bacteria and organic particles in porous-media. I. Background and methods. *Zentralblatt Fuer Hygiene Und Umweltmed.* **1991**, *191*, 53–97.
24. Baccile, N.; Weber, J.; Falco, C.; Titirici, M.-M. Characterization of hydrothermal carbonization materials. In *Sustainable Carbon Materials from Hydrothermal Processes*; John Wiley & Sons, Ltd.: Hoboken, NJ, USA, 2013; pp. 151–211.
25. Parshetti, G.K.; Liu, Z.G.; Jain, A.; Srinivasan, M.P.; Balasubramanian, R. Hydrothermal carbonization of sewage sludge for energy production with coal. *Fuel* **2013**, *111*, 201–210. [CrossRef]

26. Sevilla, M.; Fuertes, A.B. Chemical and structural properties of carbonaceous products obtained by hydrothermal carbonization of saccharides. *Chem. Eur. J.* **2009**, *15*, 4195–4203. [CrossRef] [PubMed]
27. Kumar, S.; Loganathan, V.A.; Gupta, R.B.; Barnett, M.O. An assessment of u(vi) removal from groundwater using biochar produced from hydrothermal carbonization. *J. Environ. Manag.* **2011**, *92*, 2504–2512. [CrossRef] [PubMed]
28. Bansal, R.C.; Goyal, M. *Activated Carbon Adsorption*; CRC Press, Taylor & Francis Group: Boca Raton, FL, USA, 2005; Volume 33487-2742, p. 33.
29. Silverstein, R.M.; Webster, F.X.; Kiemle, D.J. *Spectrometric Identification of Organic Compounds*, 7th ed.; John Wiley & Sons: Hoboken, NJ, USA, 2005; Volume 07030-577, pp. 82–88.
30. Unur, E. Functional nanoporous carbons from hydrothermally treated biomass for environmental purification. *Microporous Mesoporous Mater.* **2013**, *168*, 92–101. [CrossRef]
31. Foppen, J.W.A.; Schijven, J.F. Evaluation of data from the literature on the transport and survival of *Escherichia coli* and thermotolerant coliforms in aquifers under saturated conditions. *Water Res.* **2006**, *40*, 401–426. [CrossRef] [PubMed]
32. Jarrah, N.; van Ommen, J.G.; Lefferts, L. Development of monolith with a carbon-nanofiber-washcoat as a structured catalyst support in liquid phase. *Catal. Today* **2003**, *79*, 29–33. [CrossRef]
33. Elliott, M.; Stauber, C.; Koksal, F.; DiGiano, F.; Sobsey, M. Reductions of *E. Coli*, echovirus type 12 and bacteriophages in an intermittently operated household-scale slow sand filter. *Water Res.* **2008**, *42*, 2662–2670. [CrossRef] [PubMed]
34. Elliott, M.A.; DiGiano, F.A.; Sobsey, M.D. Virus attenuation by microbial mechanisms during the idle time of a household slow sand filter. *Water Res.* **2011**, *45*, 4092–4102. [CrossRef] [PubMed]
35. Maeda, K.; Imae, Y.; Shioi, J.I.; Oosawa, F. Effect of temperature on motility and chemotaxis of *Escherichia coli*. *J. Bacteriol.* **1976**, *127*, 1039–1046. [PubMed]
36. Li, G.; Tam, L.K.; Tang, J.X. Amplified effect of brownian motion in bacterial near-surface swimming. *Proc. Natl. Acad. Sci. USA* **2008**, *105*, 18355–18359. [CrossRef] [PubMed]
37. Gerba, C.P.; Schaiberger, G.E. Effect of particulates on virus survival in seawater. *J. Water Pollut. Control Federation* **1975**, *47*, 93–103.
38. Burton, G.A., Jr.; Gunnison, D.; Lanza, G.R. Survival of pathogenic bacteria in various freshwater sediments. *Appl. Environ. Microbiol.* **1987**, *53*, 633–638. [PubMed]
39. Sherer, B.M.; Miner, J.R.; Moore, J.A.; Buckhouse, J.C. Indicator bacterial survival in stream sediments. *J. Environ. Qual.* **1992**, *21*, 591–595. [CrossRef]
40. Howell, J.M.; Coyne, M.S.; Cornelius, P.L. Effect of sediment particle size and temperature on fecal bacteria mortality rates and the fecal coliform/fecal streptococci ratio. *J. Environ. Qual.* **1996**, *25*, 1216–1220. [CrossRef]
41. Sobsey, M.D.; Stauber, C.E.; Casanova, L.M.; Brown, J.M.; Elliott, M.A. Point of use household drinking water filtration: A practical, effective solution for providing sustained access to safe drinking water in the developing world. *Environ. Sci. Technol.* **2008**, *42*, 4261–4267. [CrossRef] [PubMed]
42. Stauber, C.E.; Elliott, M.A.; Koksal, F.; Ortiz, G.M.; DiGiano, F.A.; Sobsey, M.D. Characterisation of the biosand filter for *E. coli* reductions from household drinking water under controlled laboratory and field use conditions. *Water Sci. Technol.* **2006**, *54*, 1–7. [CrossRef] [PubMed]
43. Chan, C.C.V.; Neufeld, K.; Cusworth, D.; Gavrilovic, S.; Ngai, T. Investigation of the effect of grain size, flow rate and diffuser design on the cawst biosand filter performance. *Int. J. Serv. Learn. Eng. Humanit. Eng. Soc. Entrep.* **2015**, *10*, 1–23.
44. Buzunis, B.J. Intermittently operated slow sand filtration: A new water treatment process. In *Civil Engineering*; University of Calgary: Calgary, AB, Canada, 1995.
45. Huang, H.J.; Yuan, X.Z. The migration and transformation behaviors of heavy metals during the hydrothermal treatment of sewage sludge. *Bioresour. Technol.* **2016**, *200*, 991–998. [CrossRef] [PubMed]
46. vom Eyser, C.; Palmu, K.; Schmidt, T.C.; Tuerk, J. Pharmaceutical load in sewage sludge and biochar produced by hydrothermal carbonization. *Sci. Total Environ.* **2015**, *537*, 180–186. [CrossRef] [PubMed]

11

The Performance and Fouling Control of Submerged Hollow Fiber (HF) Systems

Ebrahim Akhondi [1,2], Farhad Zamani [1,3], Keng Han Tng [4,5], Gregory Leslie [4,5], William B. Krantz [1,6], Anthony G. Fane [1] and Jia Wei Chew [1,3,*]

1. Singapore Membrane Technology Center, Nanyang Environment and Water Research Institute, Nanyang Technological University, Singapore 639798, Singapore; EBRA0001@e.ntu.edu.sg (E.A.); FARH0004@e.ntu.edu.sg (F.Z.); krantz@colorado.edu (W.B.K.); a.fane@unsw.edu.au (A.G.F.)
2. Young Researchers and Elite Club, Central Tehran Branch, Islamic Azad University, Tehran 1469669191, Iran
3. School of Chemical and Biomedical Engineering, Nanyang Technological University, Singapore 637459, Singapore
4. UNESCO Centre for Membrane Science and Technology, Sydney, NSW 2052, Australia; k.h.tng@unsw.edu.au (K.H.T.); g.leslie@unsw.edu.au (G.L.)
5. School of Chemical Engineering, The University of New South Wales, Sydney, NSW 2052, Australia
6. Department of Chemical & Biological Engineering, University of Colorado, Boulder, CO 80309-0424, USA

* Correspondence: JChew@ntu.edu.sg.

Abstract: The submerged membrane filtration concept is well-established for low-pressure microfiltration (MF) and ultrafiltration (UF) applications in the water industry, and has become a mainstream technology for surface-water treatment, pretreatment prior to reverse osmosis (RO), and membrane bioreactors (MBRs). Compared to submerged flat sheet (FS) membranes, submerged hollow fiber (HF) membranes are more common due to their advantages of higher packing density, the ability to induce movement by mechanisms such as bubbling, and the feasibility of backwashing. In view of the importance of submerged HF processes, this review aims to provide a comprehensive landscape of the current state-of-the-art systems, to serve as a guide for further improvements in submerged HF membranes and their applications. The topics covered include recent developments in submerged hollow fiber membrane systems, the challenges and developments in fouling-control methods, and treatment protocols for membrane permeability recovery. The highlighted research opportunities include optimizing the various means to manipulate the hydrodynamics for fouling mitigation, developing online monitoring devices, and extending the submerged HF concept beyond filtration.

Keywords: submerged hollow fiber membranes; water treatment; fouling mitigation; critical flux; module design

---

1. Introduction

Low-pressure membrane processes, such as microfiltration (MF) and ultrafiltration (UF), are popular technologies in the water industry due to their proven efficiency in removing particles, colloids, and high molecular weight organics [1,2]. MF and UF membranes either are contained in a closed pressurized module or incorporated in an uncontained module that is submerged (immersed) in a tank. In submerged systems, the feed enters the tank at atmospheric pressure and the permeate is removed by applying suction on the permeate side of the membrane, which limits the transmembrane pressure (TMP) to <1 atmosphere and more typically to <0.5 atmosphere. The submerged concept is now well-established in the water industry, with applications in surface-water treatment, pretreatment prior to reverse osmosis (RO) in desalination and water reclamation, and membrane bioreactors

(MBRs) [3]. Submerged membranes are either in a flat sheet (FS) format (vertically aligned) or hollow fiber (HF) form either horizontally aligned or more typically vertically aligned, often with suction applied at both ends. Submerged FS membranes are used only in MBRs, whereas submerged HFs cover a range of applications and consequently are much more common. The submerged HF concept has the advantages of a higher packing density, the ability to induce movement by mechanisms such as bubbling, and the feasibility of backwashing [4–8]. The focus of this review is the submerged HF concept.

Submerged membrane systems have to deal with fouling, which represents a major drawback that restricts the application of membrane processes in the water industry [8]. In general, fouling is a widespread and costly problem that affects membrane performance and is a complex function of the feed characteristics, membrane properties, and operating conditions [9–11]. Fouling mechanisms include physical and chemical adsorption, precipitation of sparingly soluble salts, the growth of biofilms, and the deposition of suspended matter onto or into the membrane [12]. Inadequate pretreatment, poor fluid management (process hydrodynamics), extreme operating conditions, and improper membrane selection are factors that exacerbate fouling [13,14].

The key parameters influencing fouling deposition in submerged HF membranes are the membrane characteristics (e.g., membrane material, the structure of membranes/fibers, fiber diameter, length, and tautness), feed properties (e.g., foulant characteristics, concentration, viscosity), operating conditions (e.g., temperature, flux), and hydrodynamic conditions (e.g., surface shear, air flowrate) [15,16]; these parameters are discussed in detail in this review. For feeds with a high solids concentration, such as membrane bioreactors (MBRs), cross-flow operation is required for the constant application of surface shear to mitigate concentration polarization and fouling deposition [14,17]. A common practice for submerged membranes in MBRs is two-phase bubbly flow [18–20]; other approaches could include mechanical vibrations [21–23] and particle fluidization [24], all of which are discussed in this review. An important development that coincided with the introduction of submerged HFs was the realization that dead-end filtration was attractive for dilute feeds (surface waters, RO pretreatment). In this case the filtration is operated in cycles, with the dead-end forward flux interrupted by backwashing and periodic surface flushing [25] or relaxation [26]. The submerged HF module makes backwashing feasible for polymeric membranes.

As HF membrane performance continues to improve, submerged HF systems are increasingly becoming more attractive for water treatment, particularly in membrane bioreactor (MBR) applications [27,28]. In spite of the importance of submerged HF processes and the extensive research literature on advancing such systems, a comprehensive review summarizing the current state-of-the-art systems with respect to performance and fouling control remains a gap in the literature. Therefore, this review focuses on recent developments in submerged hollow fiber (HF) membrane systems, the challenges and developments in fouling control methods, and treatment protocols for membrane permeability recovery.

2. Submerged Membrane-Filtration Applications and Benefits

Compared to conventional water-treatment techniques, the most popular of which involves an integrated system consisting of coagulation, flocculation, sedimentation, and disinfection, the production of drinking water via membrane technology is acknowledged to be attractive, especially in terms of higher quality water and ease of implementation [29–34]. The use of submerged hollow fiber membranes can be classified into three main application areas, namely, surface-water treatment for drinking purposes (Section 2.1), pretreatment for RO desalination and reclamation (Section 2.2), and membrane bioreactors (MBRs) (Section 2.3). The former two usually are operated in the dead-end filtration mode with intermittent backwashing, while the third is usually operated as a continuous filtration process with bubbling for inducing tangential shear to mitigate fouling. Table 1 summarizes the submerged membrane-filtration applications and benefits.

Table 1. Submerged membrane-filtration applications and benefits.

Application	Operation Mode	Intermittent Fouling Control	Is Bubbling Implemented?	Advantages
Surface-water treatment	Dead-end with intermittent foulant removal	Backwashing, relaxation, chemical cleaning	With or without bubbling during foulant removal	Less chemical requirements; consistent quality of the filtrate [35–37]
Pretreatment of RO (reverse osmosis)	Dead-end with intermittent foulant removal	Backwashing, relaxation, chemical cleaning	With or without bubbling during foulant removal	Improved water quality; smaller footprint; less chemical requirements; consistent quality of the filtrate; lowered energy cost for RO plants [38–41]
Membrane bioreactors (MBRs)	Cross-flow with tangential shear	Continuous bubbling, sometimes backwash and relaxation	Continuous bubbling	Small footprint; complete solid-liquid separation; high volumetric organic removal rate; higher effluent quality [19,20,42]

2.1. *Surface-Water Treatment*

The submerged membrane process is employed to remove microparticles and macromolecules, which generally includes inorganic particles, microorganisms, and dissolved organic matter (DOM) [34,43,44]. The microparticles and macromolecules present in the feed tend to affect the membrane pores adversely through pore blocking (i.e., sealing off the membrane pore entrance), pore constriction (i.e., narrowing the membrane pore channels), and/or cake-layer formation, all of which result in a decrease in the membrane permeability [35,45,46]. DOM 'particles', whose size approximates that of the membrane pores, can cause pore blocking, while microparticles and macromolecules larger than the size of the membrane pores result in a fouling layer on the membrane surface. Reversible fouling can be removed by hydraulic flushing/backwashing with air bubbles as scouring agents, whereas irreversible fouling binds more stubbornly to the membrane, thereby necessitating chemical cleaning [35–37,47].

2.2. *Pretreatment for RO Desalination and Reclamation*

Adequate pretreatment of the feed to reverse osmosis (RO) systems is essential to ensure optimal performance. Low-pressure membrane pretreatment is increasingly implemented prior to the RO unit operation in seawater reverse osmosis (SWRO) plants, and also RO plants for the treatment of surface water and treated municipal effluent [7,18,44,48–51]. Similar to surface-water treatment, the low solids content in these applications allows the low-pressure membranes to be operated in the dead-end filtration mode with intermittent backwashing. The main advantages of membrane filtration compared to conventional pretreatments such as coagulation, flocculation, and sand filtration are improved water quality, smaller footprint, less chemical requirements, and consistent quality of the filtrate [38,39,52,53]. Higher energy demand and membrane fouling are the main disadvantages of having a membrane-filtration system for RO pretreatment. RO membranes are very sensitive to foulants, so enhanced pretreatment via low-pressure membranes can significantly improve the performance and reduce the energy cost of RO plants [51]. In particular, RO systems with submerged membrane pretreatment have been proven to exhibit a consistently lower silt-density index (SDI) relative to conventional pretreatment [7,40,41,54,55]. Gravity-driven membrane filtration, which was initially developed as a low-energy process for surface water and diluted wastewater treatment, has also shown potential for seawater pretreatment that requires less energy and no chemical cleaning [56,57].

2.3. *Membrane Bioreactors (MBRs)*

Membrane bioreactor (MBR) technology, which combines conventional activated sludge treatment with low-pressure membrane filtration, is widely used for the treatment of wastewater [18,19,58]. The considerable growth of MBR is driven by the high quality of the water produced, increased water scarcity, and decreasing specific energy requirements [28,59]. The anaerobic membrane bioreactor

(AnMBR), a combination of an anaerobic bioreactor and membrane filtration, also is a promising option for anaerobic treatment of wastewater [60–63].

A small footprint, complete solid-liquid separation, high volumetric organic removal rate, and higher effluent quality are some of the key advantages of the MBR and AnMBR [20,64–66]. In the submerged hollow fiber MBR, the membranes are directly immersed in the aeration tank. The results of bench-scale experiments, as well as many industrial and municipal operations, demonstrate high treatment efficiencies for chemical oxygen demand (COD), total suspended solids (TSS), and turbidity [19,42,63,67,68]. Although the MBR technology has been applied in many full-scale plants worldwide for treating municipal and industrial wastewater, membrane fouling and correspondingly increased energy consumption remain chief obstacles, as highlighted in a recent review [28]. Specifically, because membrane fouling diminishes productivity, fouling mitigation measures such as air scouring and frequent cleaning of the membrane are needed to restore the membrane permeability, which increases the energy requirement; furthermore, frequent cleaning shortens the membrane lifespan and results in higher membrane replacement costs. Aeration, bubbling, or gas sparging are the most common methods for mitigating membrane fouling; the important features of the interaction of bubbles with submerged hollow fibers (HF) are discussed in Section 5.2.

3. Fouling and Concentration Polarization in Submerged HF Systems

As noted earlier, submerged HFs can be operated in either a dead-end or cross-flow mode. In dead-end filtration, tangential shear is absent, while in cross-flow, there is shear on the membrane due to bubbling, vibration, or particle scouring. Figure 1 shows a schematic of dead-end and cross-flow filtration in a submerged system.

The operation of submerged HF membrane systems under either dead-end or cross-flow (usually induced by bubbles) conditions involves very different dynamics. Ideally the process is at steady-state with a fixed flux and TMP for cross-flow, while the process is cyclic with repeatable and regular changes in the TMP for dead-end filtration. However, irrespective of the mode of operation, some degree of fouling inevitably occurs, although membrane fouling is relatively less extensive in the cross-flow mode due to the continuous tangential shear on the membrane.

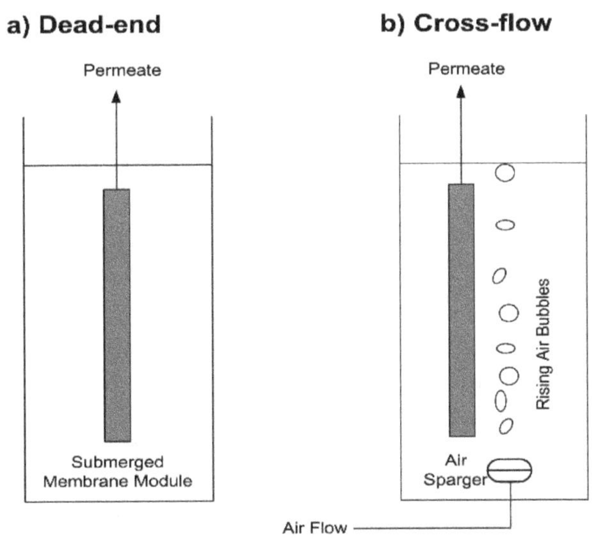

Figure 1. A schematic of dead-end (**a**) and cross-flow; (**b**) submerged filtration.

Accumulation of retained species on the membrane surface is unavoidable in membrane-based separation technologies for liquid feeds. In submerged HF membrane processes, depending on the membrane pore size, the retained species are particulates and macromolecules. The localized accumulation of particles or dissolved species on the membrane leads to concentration polarization

(CP), which is the primary reason for flux decline or TMP rise during the initial period of a membrane-based separation in low-pressure processes [3,69,70]. CP is considered to be reversible and can be controlled in a membrane module by means of optimizing the process hydrodynamics. Unfortunately, membrane fouling resulting from CP can lead to an irreversible loss of membrane permeability [69,71]. Membrane fouling, the process by which foulants, namely colloidal (e.g., clays, flocs), biological (e.g., bacteria, fungi), organic (e.g., oils, polyelectrolytes, humic substances), and scaling (e.g., mineral precipitates in RO systems) foulants, deposit onto the membrane surface or in the membrane pores [72,73], may take different forms, the main mechanisms of which are adsorption (physical and/or chemical), pore blocking, deposition of a cake layer, and gel formation [74–79]. The extent of fouling, which stems from the nature of foulant-membrane interaction, is a complex function of the feed characteristics (e.g., foulant type, foulant concentration, and physicochemical properties of the foulants such as the functional groups, charge, size, and conformation [72,80–82]), operating conditions (e.g., inadequate pretreatment, inadequate control of the hydrodynamics of the system, excessive flux, and low cross-flow velocity (in cross-flow systems) [72,82–84]), and membrane properties (e.g., pore-size distribution, surface roughness, charge properties, and hydrophobicity [70,85–87]).

3.1. Fouling in Submerged Dead-End Filtration

In dead-end filtration, tangential shear is absent and particles are convected by the permeate flow to deposit on the membrane surface, thereby forming a growing cake layer with time. Physical cleaning approaches are typically implemented periodically for the effective removal of the fouling layer in order to prolong the filtration process and membrane lifespan in submerged membrane systems. Such approaches include relaxation (i.e., intermittent cessation of permeation), backwashing (i.e., reversal of permeate flow through the pores), and air backwashing with or without air scouring [25,88–91]. Filtration duration, backwash and relaxation durations, and backwashing flowrate are important parameters in the fouling mitigation of submerged HF membranes [26,90,92,93]. However, a major challenge in the application of these techniques is that the imposed permeate fluxes have to be elevated to maintain a given water production, which in turn could result in a higher fouling rate [94]. More details on backwashing and relaxation can be found in Section 7.

3.2. Fouling in Submerged Cross-Flow Filtration

For filtration with cross-flow, particle back-transport can be caused by the mechanisms of Brownian motion, shear-induced diffusion and/or inertial lift depending on the foulant size and the tangential shear rate [95]. For a submerged HF module, the major hydrodynamic technique to mitigate particle deposition on the fibers is bubbling, which induces unsteady-state shear at the membrane surface through turbulent eddies, fiber oscillations, particle scouring, and the recirculation of the bioreactor liquid [14,15,22,96,97]. The critical flux phenomenon (the flux below which negligible fouling occurs) also is observed in bubbled submerged HFs [98–100]. Bubbling intensity can increase the critical flux [100–103] and eventually reach a plateau beyond which it has little effect. Judd [104] gave typical values of the bubbling intensity (specific aeration demand) in submerged MBRs, defined either as the air flowrate per unit membrane area ($m^3/h\ m^2$) or airflow per unit permeate (m^3/m^3). For complex feeds, such as those encountered in an MBR, the imposed flux is usually 'subcritical' for the biofloc, but could be above critical for the supernatant colloids and macrosolutes. Under these conditions the submerged HF MBR typically initially shows a slow TMP rise that eventually becomes more rapid or displays a TMP 'jump' [105,106]. The TMP jump is not specific to submerged MBRs and can have a number of possible causes [106]. Importantly, earlier TMP jumps occur with higher fluxes and/or inadequate bubbling in the submerged HF MBR. More details on the role of bubbles and the attendant hydrodynamics are in Section 5.2.

4. Blocking and Blocking Mitigation in Submerged HF Modules

Blocking, clogging, or sludging in submerged HF modules obstructs local flows in the fiber bundle and results in an uneven flow distribution; this is detrimental to the performance and can promote membrane fouling [58,107]. It should be noted that this section is targeted at the obstruction in the module rather than membrane fouling. The root cause has been attributed to the accumulation of solids, and the growth and merger of cake layers formed on the individual fibers in the module. In such cases it has been shown that the overall performance of a module packed with HFs is worse than that of a single HF [108]. This aforementioned blocking is distinct from two other forms of blocking that can occur in submerged HF modules: (i) the term 'blocking' can also refer to that of the aerators especially in MBR systems [104]; (ii) it also can refer to the blocking or closure of the pores of the HFs. Both of the latter forms of blocking are distinct from the blocking that occurs due to solids accumulation within the spaces between the fibers in the membrane module that is the focus here.

The hydrodynamics within the module represents the key factor influencing the blocking phenomenon [107,109]. Stagnant regions caused by poor local flow lead to lower shear on the fibers, which in turn results in cake buildup and eventually local blocking [110]. The packing of the fibers within a module has been observed to be very different axially [111], which causes some non-uniformity in the flow. Due to the complexity of the hydrodynamics in the module [111–116], the buildup of cake deposits is not likely to be uniform either among the fibers or along the fiber surface. The misdistribution of fouling deposits, which is acknowledged to be a direct function of the non-uniformity of the flow within the HF module [109,115,117–122], can result in large variations in the performance of fibers at the same position in the module and in poorer performance of the fibers in the middle of the module [108].

The adherence of fibers to one another is traced to the buildup and eventual merging of cake layers, which in turn causes the fibers to foul more rapidly due to hindered local flow; hence, in time it results in blocking. It has been observed that some fibers tend to be held tightly together by the dense cake layers, while other fibers remain freely suspended [108]. The principal difference between constant-pressure and constant-flux operation is that the former is self-limiting whereas the latter is self-accelerating when fouling or blocking occurs [110]. The self-acceleration in the commonly used constant-flux operation is because (i) incipient blocking will reduce the local flow that thereby enhances the deposition [108], and (ii) the flux decline in some fibers has to be matched by a flux increase in other fibers to maintain a constant net flux that thereby accelerates blocking. Either local or global non-uniformities caused by the operating conditions or design parameters [104,109], including high packing densities, low cross-flow velocities, high feed concentrations, high TMP, or lack of means to promote unsteady-state shear [123] (e.g., bubbling [14]), contribute to a greater tendency for blocking.

The control of fouling and blocking in practice is primarily through employing some or all of five strategies [109]: (i) pretreatment of the feed, (ii) physical or chemical cleaning protocols, (iii) flux reduction, (iv) aeration enhancement, and (v) chemical or biochemical modification of the feed. For (i), it is widely acknowledged that upgrading the pretreatment, in particular the screening, is pivotal to the successful retrofitting of an ASP (activated sludge process) or SBR (sequencing batch reactor) with an MBR [109]. Hair, rags, and other debris tend to aggregate at the top of the submerged HF module and become entwined with the filaments, thereby preventing their effective removal by backwashing [124]. Therefore, adequate removal of the solids before the submerged HF module is key to mitigating blocking. Methanogenic [125] and chlorinated [126] pretreatments have also been explored, as well as hybrid processes, both of which add an additional unit operation before the membrane module [127–130]. With respect to (ii), physical cleaning tends to remove reversible fouling while chemical cleaning can remove some irreversible fouling. The primary physical cleaning parameters include duration, frequency, and backwash flux [88,107,109,131,132], while the type and concentration of reagents are the important parameters for effective chemical cleaning [124,131,133–136]. Regarding (iii), flux reduction is not as cost-effective as might be expected, but the attendant extension of membrane lifespan and smoother operation may counterbalance some of the increased cost. Flux reduction is commonly implemented as a sustainable permeability operation (i.e., at a lower flux to maintain stable operation) or as an intermittent operation (i.e., at a higher operating flux with intermittent remedial measures such as relaxation and/or backwashing). As for (iv), aeration is acknowledged to be the main mixing mechanism [137,138] that generates a large amount of momentum that in turn reduces dead-zones and short-circuiting [139]. More details on aeration are given in Section 6.2. In connection with (v), the nature of the biomass in MBRs can be partially controlled through the addition of coagulants or flocculants [124,140–144] and adsorbent agents (e.g., activated carbon [145–151]). In addition, the bioprocess parameters, such as the sludge retention time (SRT), influence the biomass and supernatant as well as the fouling [152].

A suitable mitigation means for blocking has to be determined for each MBR plant. For example, a Mitsubishi Rayon unit failed to resolve the blocking problem via measures such as overnight relaxation and intensive or regular chemical cleaning, but regular backwashing for 30 s during each permeate production cycle achieved a stable low permeation flux [104]. To limit blocking, the PURON® system is designed with free movement of the filaments at the top of the module. This allows for larger solids, such as hair and agglomerated cellulose fibers, to escape without causing clogging in this region. The fibers are reinforced by an inner braid, to withstand the lateral movement of the filaments that subjects them to mechanical stress [153].

A major difficulty in the mitigation of module blocking in submerged HFs is the lack of simple techniques for in situ monitoring of the phenomenon. One approach that has been developed [122] is based on the assumption that blocking within the module is initially localized, which would result in a localized drop in flux relative to the overall average flux. Using a simple array of flow detectors mounted in the permeate header, it is possible to measure local fluxes and identify fiber regions that become less productive. It was shown that shifts in the standard deviation of the local fluxes was much more sensitive to local blocking than shifts in the overall system TMP. In the aforementioned study, the flow detectors were based on constant temperature anemometry strips, but other low cost detectors could be used.

5. Parameters Affecting the Performance of a Submerged Hollow Fiber System

Several factors, including the feed characteristics, membrane and module properties (e.g., fiber length, diameter, and looseness), and hydrodynamic properties (e.g., flowrates), collectively affect the performance of submerged HF systems.

5.1. Membrane Properties and Module Configurations

5.1.1. Membrane Materials and Surface Morphology

Membrane characteristics such as material, surface charge, hydrophilicity, pore size, and pore morphology significantly impact membrane performance and the fouling potential. Membrane surfaces can be modified to combat and mitigate adhesive fouling [25]. The membrane material largely influences the initial rates of deposition of foulants due to the tendency of some materials to adsorb certain solutes or particulates more readily, as quantifiable by the Gibbs free energy of foulant-membrane interaction [154–157]. When the adsorption becomes such that the effective pore size of the membrane is reduced, the flux is adversely affected [158–160].

It is well-known that the severity of fouling increases as the hydrophobicity of the membrane increases, because organic molecules have a higher affinity for hydrophobic materials [47,159,161–165]. Most commercial MF and UF membranes are made from relatively hydrophobic polymers (e.g., polysulfone, polyethersulfone, polypropylene, polyethylene, and polyvinylidene-fluoride (PVDF)), due to their excellent chemical resistance, and thermal and mechanical properties [83,166–170]. In some cases, the membranes are modified by additives to confer increased hydrophilicity for water applications. The charge on the membrane surface also plays a role in either exacerbating or mitigating fouling. Attractive electrostatic forces between a charged surface and the co-ions in the feed solution increase fouling [161], while repulsive electrostatic forces mitigate fouling. In general, a low surface charge or electrically neutral surfaces tend to show better anti-fouling properties during the initial stages of membrane fouling [163,171]. Since most colloidal particles, such as those in natural organic matter (NOM), that tend to deposit are negatively charged, membrane surfaces are usually negatively charged [161,168,169], although positively charged membrane surfaces are also used to repel positively charged colloids and cations [172]. Compared to having just a positively or negatively charged membrane surface, the addition of a zwitterionic charged material, one composed of neutral molecules that have a positive and a negative electrical charge, has been shown to confer better anti-fouling properties [173]. Inorganic nanoparticles (such as SiO_2 [174,175], Al_2O_3 [176,177], clay [178,179], ZrO_2 [180–182], TiO_2 [183,184], and ZnO [185,186]) also have been used to improve polymeric composite membranes, conferring improvements in the mechanical properties, thermal stability, hydrophilicity, permeation, and antifouling performance of membranes [165].

5.1.2. Fiber/Module Arrangement

Experimental efforts have consistently established that a lower HF packing density, either by having fewer fibers [108], widening the HF module [187,188], or by varying the module configuration [97,189], is linked to a lower fouling tendency. Other than the HF packing density, modules that are designed to enhance either lateral flow [190] or lateral movement of the fibers [188] are known to further improve the performance of submerged HF systems. In a carefully controlled arrangement with nine fibers in a matrix, a detailed analysis has shown that the performance of individual fibers varied with their position in the module, such that the fibers at the edge performed best, while those in the center surrounded by other fibers performed the worst [108]. The underlying reason for this was tied to the lower cross-flow velocity at the center of the module and the 'flux competition' for the surrounded fiber. Interestingly, whereas Yeo and Fane [108] found that a single fiber outperformed a multi-fiber HF module in a single-phase (no bubbling) system due to module blocking, Berube and Lei [191] found that a multi-fiber module performed better than a single fiber in the presence of bubbles (i.e., two-phase flow) due to inter-fiber interactions leading to mechanical erosion of the foulant layer.

Simulation results by Liu et al. [192] revealed that MBRs fitted with hollow fibers in a vertical orientation experienced 25% more membrane surface shear in the filtration zone than horizontally oriented fibers at the same aeration intensity. They also found that the addition of baffles in the membrane modules is a feasible way to promote turbulence and shear in the upper section of the membrane module.

5.1.3. Fiber Looseness

Fiber movement plays an important role in determining the extent of particle deposition on submerged HFs in bubbling cross-flow operation. Greater fiber movement, which can be achieved by using looser fibers, enhances the back-transport of the foulants from the membrane and also results in physical contact between the fibers, both of which reduce fouling [16]. Tight fibers studied by Wicaksana et al. [15] showed a 40% faster rate of TMP rise compared to fibers with a 4% looseness, where the looseness percent is based on the difference between the fiber length and the linear distance between the fixed ends of the fiber relative to the fiber length; this implies that approximately 40% of the fouling mitigation brought about by the bubbles was due to movement of the loose fibers. Simulations by Liu et al. [193] showed that fiber displacement and membrane surface shear are highly variable at different locations along the fiber and with time. In addition, increasing the fiber looseness from 0.5% to 1% increased the average surface shear by 50.4% (0.56–1.13 Pa) for fibers with the same diameter. Yeo et al. [194] found that increasing the looseness from 0% to 1% decreased the fouling tendency regardless of the bubble size, whereas increasing the looseness from 1% to 2% increased the fouling tendency. This suggests an optimum fiber looseness, whereby too much displacement may move the fiber away from the influence of bubbles. This effect will depend on the module geometry relative to the bubbling zone. It is also known that too much looseness can lead to fiber breakage.

5.1.4. Fiber Diameter

A smaller HF diameter has been experimentally shown to improve the performance of submerged HF systems [15,16,187] due to the attendant greater fiber mobility [15], which can be further enhanced by bubbling [16] or turbulence in general [187]. However, theoretical analysis shows that a smaller HF diameter leads to a higher pressure drop along the fiber lumen, such that a higher suction pressure is required for a given average imposed flux [187]. On the other hand, simulations show that the average surface shear was 67% higher for fibers with a diameter of 1.3 mm relative to fibers with a diameter of 1.0 mm that had an identical Young's modulus and looseness [193]. The interplay of these two effects suggests an optimal fiber diameter that balances flux enhancement with flux-distribution and suction-pressure considerations [187].

5.1.5. Fiber Length

Simulation studies have indicated that longer fibers lead to a higher lumen-side axial pressure drop and consequently an uneven permeation flux distribution along the fibers, which can result in the particle deposition being more severe near the suction ends relative to the closed end of the fibers due to a higher local flux [99]. Experimental evidence is in agreement that this non-uniform pattern of particle deposition along the fibers is more apparent for longer fibers (lengths tested were in the range of 0.3–1 m) [100,195–197]. On the other hand, fouling control is facilitated by the greater mobility of longer fibers [98], which can be enhanced by bubbling [15].

5.2. Hydrodynamics in Submerged HF Membranes

Enhanced surface shear, for example caused by an increased cross-flow velocity, is a common strategy to control concentration polarization and fouling in submerged HF systems [138,198,199]. Unsteady-state shear, such as two-phase flow (i.e., gas bubbles or fluidized particles) and vibration, is more energy-efficient than steady-state shear [123]. Air bubbling is particularly attractive in MBRs for aeration, mixing, and augmenting liquid flows [14,82,200,201].

5.2.1. Role of Air Bubbles

The use of bubbly flow has been reviewed by Cui et al. [14] and more recently by Wibisono et al. [82]. The major benefit of rising bubbles is the unsteady or transient shear stress at the membrane surface that causes particle back-transport away from this surface [14,15,22]. Figure 2, based on Cui et al. [14], illustrates the possible interactions of bubbles with the surface of hollow fibers and shows three different effects on the submerged hollow fibers: (i) a shear stress on the surface of the hollow fiber induced by the wake generated by the rising bubbles, (ii) fluctuating liquid flows transverse to the fibers induced by the bubbles, and (iii) lateral fiber movement induced by the bubbles that depends on the looseness of the fiber. Wibisono et al. [82] indicated that aeration intensity not only enhances the hydrodynamics, but also can affect the biomass properties in aerobic membrane bioreactors. Yeo et al. [202] [Yeo, 2017 #321] showed that aeration also influenced the biofilm growth. In addition, Cabassud et al. [203] reported that bubbles seem to alter the structure of the cake or fouling layer such that the specific resistance is reduced. They based this conclusion on the observation that gas sparging applied to the MF of particles increased the fluxes significantly after a period of flux decline at the higher feed concentrations. Wang et al. [204] correlated the bubble hydrodynamics with the critical flux and found that bubble momentum and the bubble-membrane contact area had the most positive correlation with the local critical flux. However, Du et al. [201] found that the shear stress associated with the bubbles was insufficient to mitigate the deposition of fine (1.75 μm) particles.

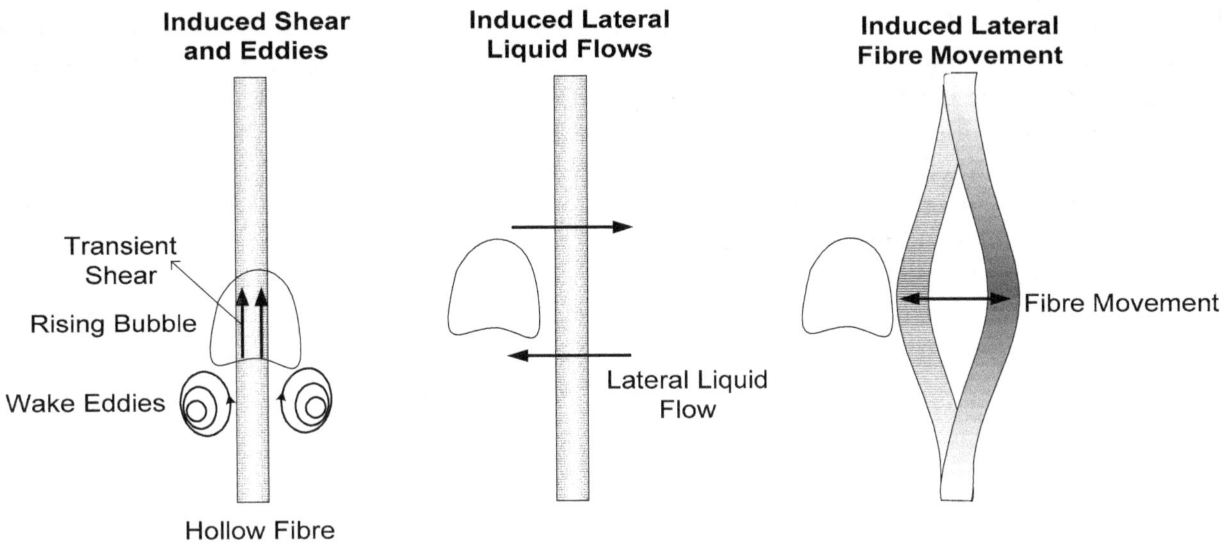

Figure 2. Fouling and cake control mechanisms by bubbles outside fibers [14].

5.2.2. Bubble Characteristics

Several techniques have been used to characterize the hydrodynamic conditions in submerged HF systems. Particle-image velocimetry (PIV) [205] has shown that the bubble size increases with height along the membrane module due to the reduced hydrostatic pressure as the air bubbles move upward, and varies over a wide range of 0.2–50 mm, with a predominant size range of 3–5 mm. A strong sheltering effect attributed to the hydrodynamics was observed within the hollow fiber module that resulted in a 10-fold reduction in the axial velocity relative to the velocity outside the fiber bundle [110]. Nguyen Cong Duc et al. [206] used a bi-optical probe to characterize the bubble velocity, distribution, and size throughout submerged full-scale HF modules. The shear stress was observed to be an important parameter in controlling particle back-transport from the membrane surfaces. Fulton et al. [207] studied the sparged bubble characteristics and the induced shear forces at the surface of submerged hollow fiber membranes using an electrochemical method. The shear stress was observed to be highly unpredictable over time and heterogeneously distributed within the module, ranging from 0.1 to over 10 Pa. Also, no correlation was observed between the shear stress and the bubble frequency or rise velocity. However, this does not corroborate with the general observation of better fouling control with increased aeration intensity (see Section 5.2.3) and also with the results of Yeo et al. [194], who observed an increase in the shear stress with increasing bubble frequency for all bubble types. These disparate observations highlight the challenge in achieving well-distributed two-phase flow in the submerged HF module. The importance of this was shown by Buetenholm et al. [208], who used X-ray computer tomography to detect the instantaneous displacement of fibers in an aerated HF bundle. The data were then incorporated into a computational fluid dynamics (CFD) simulation to allow optimization of the module design and aeration. More recently, Wang et al. [204] used a high speed video camera to characterize the bubble characteristics and direct observation through the membrane (DOTM) to determine the corresponding critical flux of micron-sized polystyrene particles. They found that the local bubble momentum and bubble size had the most positive correlation with the local critical flux.

5.2.3. Effect of Gas Flowrate

It has been reported in many studies that the filtration performance can be improved by controlling the bubbling rate [15,16,209,210]. On the one hand, the critical flux improves in an approximately linear fashion with respect to gas flow in submerged HF systems [100–103,204]. On the other hand, for different sizes of fibers, it has been shown that a modest gas flow can increase the final flux by a factor of 3–6 in a yeast suspension, but the enhancement quickly plateaus at higher gas flowrates such that further increases in the gas flowrate achieve negligible enhancement [187]. This disparate behavior stems from the slugging phenomenon, whereby large changes in the gas flow have a negligible effect on the velocity of the film that is formed adjacent to the surface of the fiber [211]. Similar plateaus have been observed in submerged HF systems for sewage treatment and drinking-water treatment [191,210]. It has been observed that at high gas flowrates (namely, 40 L/h), a maximum flux was observed, after which the flux started to decline [212]. This phenomenon was explained by the relationship between the bubble size and air flowrate; since bubble size tends to increase with the air flowrate, the bubbles become so large after an optimum air flowrate that they start to prevent the liquid from reaching the membrane surface, a phenomenon that is also linked to slugging flow.

5.2.4. Aeration Modes

Armed with the knowledge that the shear stress induced by bubbles is the dominant mechanism in controlling particle back-transport from membrane surfaces, the bubbling mode (e.g., continuous, alternating, pulsed) can be optimized to minimize the energy cost while achieving an optimum shear to improve system performance [213,214]. Yeom et al. [215] carried out a study of the frequency or duration of cycling between filtration and bubbling phases and showed that intermittent aeration is effective for fouling control in a denitrification MBR. Guibert et al. [216] studied the positioning of aeration ports and reported that the injection of air in different zones around the fiber bundles greatly improved the overall system performance. Fulton and Berube [188] studied the effectiveness of continuous, alternating, and pulsed bubbling modes and found that, even though the volume of gas used by pulse sparging was half of that used by the other sparging conditions, relatively similar induced shear stress was observed for all three bubbling modes. Similarly, Tung et al. [217] observed that semi-continuous aeration could suppress the membrane fouling at the same level as at continuous aeration.

5.3. Shear Stress on Membrane Surface by Non-Bubbling Techniques

Submerged HFs also are amenable to fouling control by methods that do not involve bubbling, as discussed below.

5.3.1. Vibrations

Vibrations have been proven to be an effective way to induce shear on a membrane surface and consequently reduce concentration polarization (CP) and fouling [123,218,219]. Various modes of vibration are applicable for different membrane systems. A submerged HF system can be vibrated longitudinally or axially (Figure 3a), transversely (Figure 3b), and rotationally (Figure 3c). Although anaerobic systems are gaining momentum in the wastewater industry due to their potential for energy production, bubbling by recycled biogas has some challenges; hence, the vibration approach is attractive for fouling mitigation in anaerobic MBR (AnMBR) systems. In particular, transverse vibration has been proven to be an effective way to mitigate the fouling in AnMBR applications [220,221].

Many studies have been carried out to probe the effectiveness of the different modes of vibration (i.e., longitudinally, transversely, or rotational). Low et al. [222] found that the use of vibrations slowed the flux decline for the submerged HF system that they investigated, and that longitudinal oscillation outperforms rotational oscillation. Li et al. [223] showed that vibration was more effective for a bentonite suspension compared to a washed yeast solution that may cause internal fouling; hence, vibration is more effective primarily for cake removal but not for the mitigation of internal fouling, which was corroborated by Kola et al. [220]. It was also shown that transverse vibration decreases the fouling rate much more effectively than longitudinal vibration [223]. Genkin et al. [224] reported that adding transverse vibration to longitudinal vibration (by using chess-patterned vanes) resulted in an almost doubling of the critical fluxes at the same frequency; adding coagulants further elevated the critical flux of the vibrating system, although floc breakup at higher frequencies (namely, 10 Hz) tended to reduce the critical flux. Beier and Jonsson [225] also found that vibration facilitates the separation of macromolecules (e.g., BSA) and larger components (e.g., yeast cells) at sub-critical fluxes by loosening and removing the built-up cake in the filtration of a mixed suspension for a membrane with pores larger than the macromolecular components. Fiber spacing and looseness were found to be important parameters to improve the benefits of vibration with respect to the turbulence kinetic energy and eddy length scale [226]. A recent study investigated rotating instead of vibrating the HF membranes and found membrane rotation to be more effective than gas scouring [227].

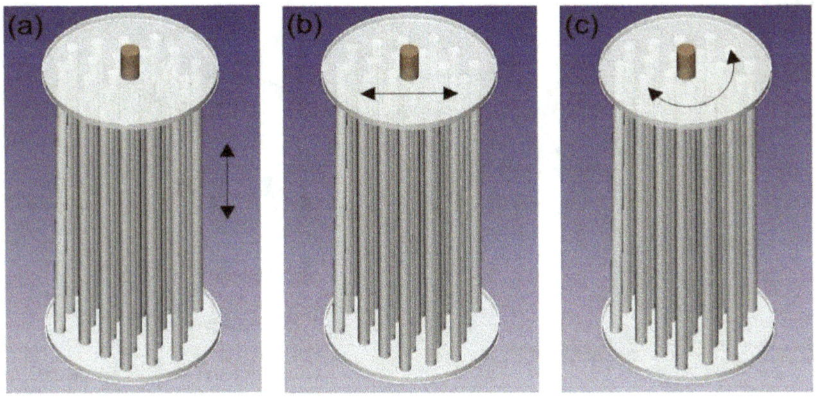

Figure 3. Schematic of an HF (hollow fiber) module with different modes of vibration: (**a**) longitudinal vibration; (**b**) transverse vibration; and (**c**) rotational vibration.

Figure 4 shows that both higher amplitudes and higher frequencies for longitudinal vibration contribute towards fouling mitigation, with an observed reduction in the fouling rate by as much as 90% [23]. Genkin et al. [224] found that the critical flux has a stronger dependency on frequency at higher frequencies but a weaker dependency at lower frequencies, presumably due to a change of the flow regime in the vibrating system. Chatzikonstantinou et al. [228] used high-frequency vibration in a pilot-scale submerged MBR and found it promising with respect to energy savings compared to conventional air-cleaning systems.

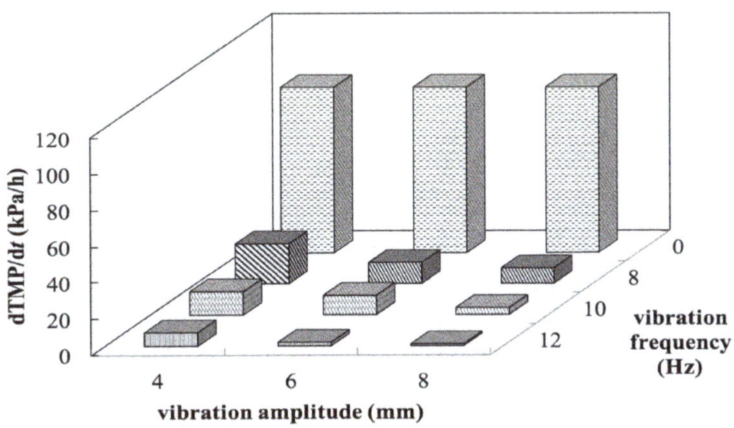

Figure 4. Effect of vibration amplitude and frequency on the fouling rate expressed as the time rate-of-change of the TMP (transmembrane pressure) for filtration of a 4 g/L bentonite suspension [23] for a constant flux of 30 LMH.

The unsteady-state shear induced by vibration can be related to performance enhancement to provide a quantitative assessment of the beneficiation [123]. Beier and coworkers [229–231] applied longitudinal vibration, with frequencies between 0–30 Hz and amplitudes of 0.2, 0.7, and 1.175 mm, on HF bundles with nominal pore diameters of 0.45 μm, and found a general correlation between the critical flux (J_{crit}) and average shear rate (γ_{ave}) induced by vibration on the membrane surface as follows:

$$J_{crit} = a\gamma_{ave}^{n} \tag{1}$$

The correlations based on Equation (1) using different values of a and n for three aqueous suspensions are shown in Figure 5.

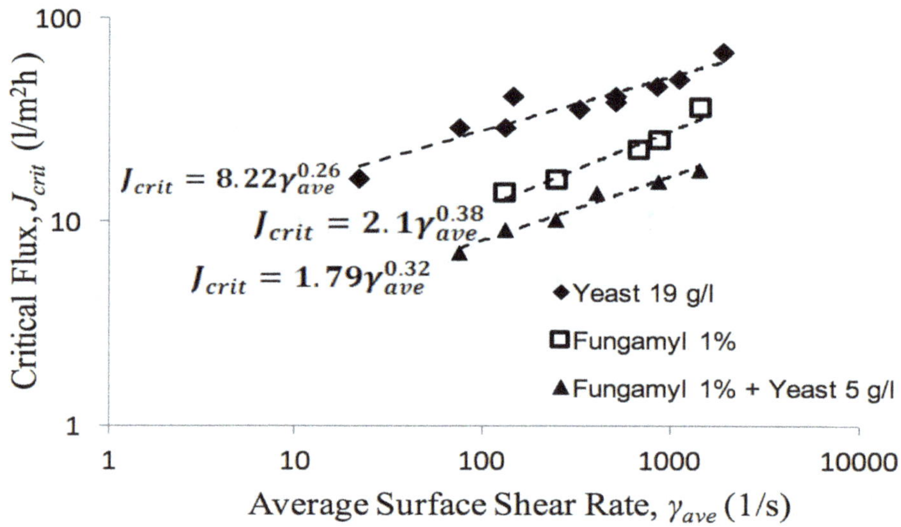

Figure 5. Correlation between the critical flux (J_{crit}) and average shear rate (γ_{ave}) for yeast suspensions with concentration of 19 g/L; 1% Fungymal solutions; and 1% Fungymal + 5 g/L yeast [231].

To calculate the shear rate (γ) on the surface of a vibrating fiber, Beier et al. [229] proposed the following equation:

$$\gamma = A(2\pi f)^{1.5} v^{-0.5} \cos\left(\omega t - \frac{3\pi}{4}\right) \tag{2}$$

where A is the amplitude, f is the frequency of the vibration, v is kinematic viscosity of the fluid, and t is the time. As expected, the shear rate displays a periodic behavior as a function of time since the membrane vibrates with a velocity of:

$$u = 2\pi\ A\ f\ \cos(\omega t) \tag{3}$$

Equation (2) does not depend on the fiber diameter because the Navier-Stokes equation was solved under the assumption that the curvature of the fiber near the membrane surface is negligible [229]. However, recently, Zamani et al. [232] showed that the shear rate of a vibrating fiber is also a function of the fiber diameter. They showed that Equation (2) can have a relative error of 40% for a fiber with a diamter of 1 mm vibrating with an amplitude and frequency of 10 mm and 1 Hz, repectively.

Krantz et al. [233] applied longitudinal vibrations to a bundle of silicon hollow tube membranes to enhance the mass transfer to the liquid on the lumen side of the membrane, and found that the mass-transfer coefficient was increased by a factor of 2.65 relative to that without vibrations. An analytical solution was developed for the velocity profile of the laminar flow within the vibrating tube. However, this solution is not applicable for commonly used submerged HF applications, since the external shear on the fiber is of more interest. However, it could be useful for novel submerged HF processes, such as forward osmosis (FO) and membrane distillation (MD) applications.

5.3.2. Particle Scouring

Particle scouring is one of several unsteady-state shear techniques useful in membrane processes for fouling mitigation and improving the mass-transfer coefficient [123]. In practice, the solid particles are brought into close contact with the membrane surface via fluidization, which is the process whereby the particles are dispersed and suspended by the liquid such that they behave like a fluid [234,235]; thereby, the fouling layer on the membrane surface is mechanically scoured by the particles. A recent review assessed the mechanical cleaning concepts in membrane filtration [27]. Wang et al. [204] concluded that particle fluidization is similar to cleaning via bubbling in terms of (i) the momentum of both bubbles and fluidized granular activated carbon (GAC) correlates more strongly with the critical flux, rather than to the velocities or concentrations; and (ii) an increase in energy input increases the critical flux. In contrast to bubbling, particle fluidization is different with respect to (i) the local critical flux decreasing instead of increasing with height; (ii) its optimization involves a complex interplay of particle size, concentration, and liquid flowrate, instead of simply involving increasing the gas flowrate.

As early as the 1970s, the beneficial impact of the fluidization of particles for membrane processes was recognized as being attributable to both the mixing action of the particles to reduce the solute concentration gradient, and the mechanical action of the particles to both vibrate and clean the membrane surface [236–239]. These studies predate the submerged HF. However, various types of inert solids subsequently have been shown to be beneficial for the mitigation of membrane fouling in submerged HF applications via the scouring mechanism [240–244], although negative effects such as the break-up of sludge flocs [240] and poor filterability of the activated sludge suspension [241] also have been noted. Another potential effect is membrane damage [238,242], which necessitates careful selection of scouring conditions.

The use of powdered activated carbon (PAC) is primarily targeted for mitigating organic accumulation, biological degradation, and reducing the cake-resistance (e.g., [245–253]), all of which contribute towards improving the permeate flux. Almost as an afterthought, PAC also was recognized as being beneficial for inducing fouling-mitigating shear on the membrane [254–257]. Because the increased inertia associated with larger-sized particles can lead to more effective scouring, granular activated carbon (GAC), whose mean diameter is an order-of-magnitude larger than that of PAC, has recently gained interest [24,130,257–264]. Although PAC is more effective than GAC in terms of adsorption capability [257], it has been claimed that GAC is more effective at the higher concentrations encountered in practice and in the longer term [24]. Several reports of the apparent success in the use of GAC in submerged HF membrane systems (namely, the fluidized-bed membrane reactor) for low-cost, sustainable operation have appeared in recent years; hence, a closer look is warranted. The first report was on the use of a two-stage AFBR-AFMBR (i.e., anaerobic fluidized-bed bioreactor-anaerobic fluidized-bed membrane bioreactor) for sustainable control of membrane fouling [24]. Extensive tests subsequently have been carried out [24,27,128,130,244,258–278], especially in view of the potentially lower energy cost than that of bubbling [24,204] and suitability for the anaerobic MBR. Effects of treating different types of wastewater [258,259,261,262,276] (e.g., using municipal versus synthetic wastewater [24,258]), trace organics [128,278], membrane type [273] (including effects on membrane integrity [271,272,275]), screen size [259], fluidized media [236,244,279,280] (including size and packing amount [244,267,268,271,275]), operating conditions [260,271] such as temperature [260–262], scale [261], design [130,261,264,265,276] (e.g., single (AFMBR) versus two-stage (AFBR-AFMBR) systems [130]), which collectively proved the efficacy of GAC in scouring the membranes. Different embodiments of the AFMBR include single (AFMBR) versus two-stage (AFBR-AFMBR) systems [130], as well as simplifications of the two-stage AFBR-AFMBR system termed an IAFMBR (i.e., integrated anaerobic fluidized-bed membrane bioreactor) [277], and hybrids such as the MFC (i.e., microbial fuel cell)-AFMBR [264] and the fluidized bed membrane bioelectrochemical reactor (MBER) [263]. A study on the extent of fouling mitigation by fluidized GAC in an HF module found that larger-sized GAC particles, higher packing densities, and a ratio of hollow fiber spacing to fluidized particle size of approximately 3–5 are beneficial for fouling control [271]. Collectively, these efforts prove the efficacy of particle fluidization in mitigating fouling in submerged HF systems.

The benefits conferred by particle fluidization include low energy cost [24,123,242,258,261,264,281], and amenability for scale-up and continuous operation [24,235], all of which make it an attractive means to improve membrane operations. In particular, the AFMBR energy requirement was only 0.028 kWh/m^3, which is much less than that reported for AMBRs using gas sparging [24]. The momentum, velocity, and concentration of the fluidized GAC particles have been found to play significant roles in membrane fouling mitigation via both experiments and simulation [266–269]. Note that membrane-particle interactions have to be managed to avoid membrane damage. Also, the module geometry must allow movement of the particles to avoid blockages, as highlighted in Section 4.

6. Techniques for Fouling Control in Dead-End Submerged Membrane Systems

As noted earlier, the submerged HF concept is widely used in the dead-end mode in the water-treatment industry. Strategies to mitigate fouling are required to avoid decline in the membrane permeability in dead-end submerged HF systems [6,282,283]. It is possible to minimize fouling both by choosing a suitable membrane material with a reduced tendency to adsorb substances in the feed and by optimizing the operating conditions in the system [5,161,284,285]. The application of relaxation (intermittent cessation of permeation), backwashing (reversal of permeate flow through the pores), and air backwashing with or without air scouring are common physical approaches to remove fouling in submerged systems. It has been shown in many studies that relaxation and backwashing provide an effective removal of the fouling layer, thereby prolonging the filtration process in submerged membrane systems, especially at high imposed fluxes [88,90,131,286,287]. A significant challenge in the application of relaxation and backwashing is that only partial recovery of the permeability is achieved at the end of a filtration cycle, which implies a gradual loss of the effective filtration area due to fouling. Subsequently, in the next filtration cycle the less fouled areas will have to experience increased local fluxes to maintain the overall average flux, which in turn results in a higher fouling rate [94]. However, Figure 6, which shows a plot of the TMP versus time for both continuous and periodic backwashing and filtration, indicates that even with a partial recovery during each cycle, filtration with intermittent backwashing and relaxation outperforms continuous operation. A further discussion of backwashing and relaxation is provided in the following sections.

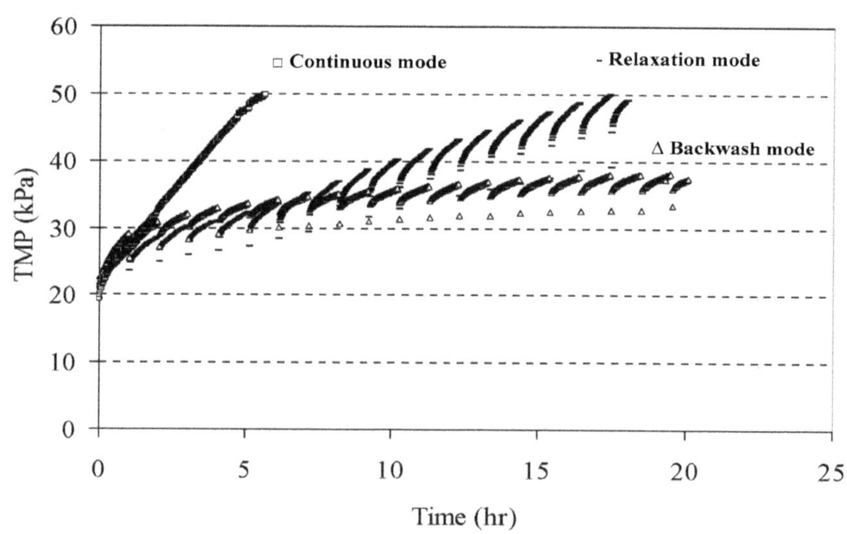

Figure 6. TMP versus time for continuous filtration relative to periodic relaxation and backwashing of real seawater [26].

6.1. *Backwashing*

Backwashing is commonly practiced in most HF filtration systems to limit fouling in both dead-end and cross-flow applications. Typical TMP profiles for dead-end filtration with intermittent backwashing for cake removal are shown in Figure 7. Two modes of operation are usually practiced, either (i) using a fixed cycle time (t_c), whereby backwashing is implemented after a designated filtration time, thereby causing the maximum TMP to increase with each cycle if residual fouling occurs (Figure 7a), or (ii) operating to achieve a fixed TMP_{max}, whereby backwashing is implemented whenever the TMP reaches a predetermined value, thereby requiring the frequency of backwashing to increase with each cycle if residual fouling occurs (Figure 7b). Backwashing is usually effective in reducing the TMP, but some deposits tend to remain attached and contribute an additional residual resistance to the filtration in subsequent cycles. Therefore, other than backwashing alone, cycling between backwashing and chemical cleaning (Section 8) is also a common practice to reduce the minimum TMP (TMP_{min}) attainable. Figure 8 illustrates a typical TMP profile with intermittent backwashing and chemical cleaning with a fixed cycle time.

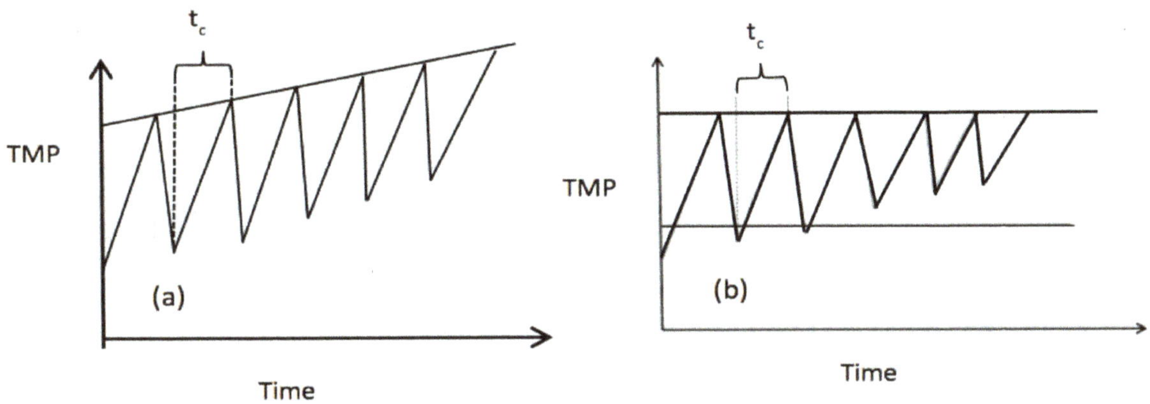

Figure 7. Typical TMP profiles with intermittent backwashing; (**a**) operation to achieve a fixed TMP_{max}; (**b**) operation for a fixed cycle time.

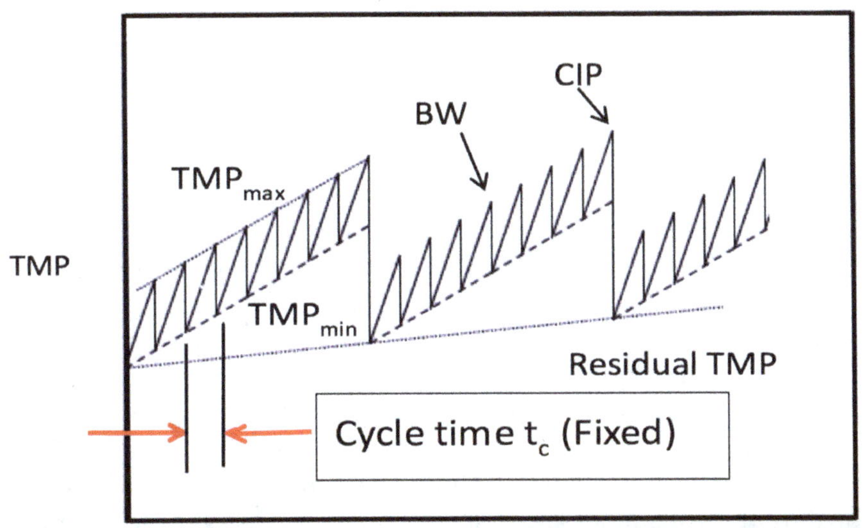

Figure 8. Typical profile of the TMP as a function of time showing the effect of intermittent backwashing and chemical cleaning for a fixed cycle time.

Although backwashing loosens and detaches the fouling cake from the membrane surface so that the foulants can be removed easily by cross-flow or air bubbles [6,25,90,93,288,289], some drawbacks also exist. In cases for which the cake layer serves as a secondary layer to protect the membrane from internal fouling by macromolecular components, over-frequent backwashing can provide more opportunity for macromolecules to enter the membrane pores [290] or change the chemical composition and/or structure of the fouling layer (e.g., from a mixed cake layer of particulates and macromolecules to a fouling structure dominated by the macromolecules after several filtration/cleaning cycles [291]) and consequently the fouling patterns [26]. Generally, the first few cycles of backwashing lead to more significant irreversible fouling, after which the percentage of irreversible fouling with respect to total fouling becomes constant. The reason for the augmented vulnerability to irreversible fouling of new membranes relative to a used membrane is the greater probability of blocking the larger pores in the pore-size distribution, which can be the dominant fouling mechanism in the first few cycles [26,90,91,292].

Overall, an increased backwashing flux was found to be slightly more effective than increased backwash duration when the same amount of backwash volume was used [26,91,93]. Similarly, Akhondi et al. [90] reported that excessive backwash duration and strength resulted in permeate loss, severe pore blocking, and high specific energy consumption. Ye et al. [26] investigated the effect of filtration duration (from 1200 to 5400 s per cycle) on membrane fouling during real seawater filtration while the other operating parameters were kept constant. It was found that the final TMP after 16 h of filtration and the percentage of reversible fouling that can be removed by backwashing did not increase when the filtration duration increased from 1200 to 3600 s, while a further increase in the filtration duration from 3600 to 5400 s promoted membrane fouling due to a more compact cake layer that was more irreversible.

Chua et al. [41] reported that, for a pilot-scale pressurized HF module, prolonging the duration of backwashing was found to be more effective than air scouring in controlling membrane plugging. Studies by Ye et al. [26] showed that increasing the backwash duration from 10 to 30 s led to the final TMP and fouling rate decreasing by more than 50% as well as a slight increase in the percentage of fouling removed by backwashing. However, a further increase in the duration beyond 30 s did not result in any additional improvement, but instead slightly reduced the percentage of fouling removed; this indicates that an excess backwash volume may cause membrane blockage or change the structure of the fouling cake due to impurities in the backwash flux.

Akhondi et al. [160] studied the effect of backwashing on the pore size of hollow fiber ultrafiltration membranes by using the evapoporometry [91,293,294] technique. They reported the following: (i) backwashing can enlarge the pores of a membrane with a greater effect on the larger pores for operation at the same TMP; (ii) pore enlargement due to backwashing was larger for amorphous (PVDF fibers) relative to glassy polymers (PAN fibers) due to the lower modulus-of-elasticity of the former; (iii) cyclic filtration and backwashing at constant flux could more effectively remove foulants both on and within the larger membrane pores compared to the small pores; and (iv) increasing the backwashing flux could remove foulants from smaller pores.

Results for seawater showed that the lowest final TMP and the maximum percentage of fouling removed by backwashing after 16 h of filtration was for the case for which the backwash flux was 1.5 times the filtration flux [26], which suggests the existence of an optimum backwash flux for fouling mitigation. The observation that a further increase in the backwash flux to twice that of the filtration flux led to an increase in the final TMP and a reduction of foulant removal implies that backwashing changes the fouling rate during the filtration cycle. Similar to excessive backwash duration, it seems that an excessive backwash flux also causes convection of impurities to the membrane pores or a residual fouling layer that results in less reversible fouling and a higher fouling rate. The existence of an optimum backwash flux for fouling mitigation was also reported by Chua et al. [41], who found that an increase in the backwash flowrate up to twice that of the permeate flowrate resulted in a process improvement, but no further benefits were observed for a further increase in the backwash flowrate. Compared to the duration or interval of backwashing, the effect of backwashing flux was found to be more significant for fouling mitigation [288].

It has been reported that air scouring during backwashing can assist fouling removal and improve backwash efficiency [289,295]. While the backwashing is expected to detach the cake layer from the fibers, air scouring loosens the deposits and carries them from the membrane surface into the bulk fluid [289,295]. The impact of aeration during backwashing on membrane fouling during seawater filtration was investigated by Ye at al. [26]. Their results showed that backwashing with a moderate air flowrate had a lower final TMP and also slowed down the fouling rate during the filtration. However, a high air flowrate limited the benefits of air scouring and did not improve the reversibility.

6.2. Relaxation

Relaxation, the intermittent cessation of permeation, has been incorporated in many membrane bioreactor (MBR) designs and some other submerged HF systems as a standard operating protocol to control membrane fouling [93,94,288]. For example, the investigation of an MBR found that relaxation was still beneficial even when the relaxation necessitated periods of higher flux to give the same production of permeate [288]. Different relaxation conditions resulted in distinctly different temporal TMP profiles, but all the runs that incorporated relaxation displayed a lower final TMP than the continuous mode. Relaxation was found to be more favorable than backwashing for this MBR application, because, while performances were similar, backwashing may have resulted in membrane pore clogging [288].

The protocol for intermittent filtration/relaxation can be optimized in terms of the ratio of the durations in each cycle (ratios between 0.5–50 were tested) to be more beneficial for fouling removal [94,296], but were not necessarily beneficial for retarding the TMP increase as filtration progressed. This suggests that the relaxation duration and interval should be carefully managed to achieve the best outcome in terms of reducing the fouling resistance during relaxation and retarding the TMP increase during filtration.

Ye et al. [26] investigated the effect of relaxation on the performance of dead-end (i.e., without bubbling) filtration using membranes with two different porosities and a seawater feed. Relaxation was confirmed to limit membrane fouling compared to continuous filtration, but was more effective for the membrane with a higher porosity. The difference in results due to porosity was hypothesized to be that, although the relaxation removed part of the foulant cake for membranes with lower porosities, when the filtration flux resumed the cake was reorganized to a more compact structure. Table 2 lists fouling control methods for submerged HF systems.

Table 2. Fouling control methods in submerged HF systems [18,297–299].

Operation Mode	Fouling Control Technique	Important Parameters	Applications	Benefits
Cross-flow	Bubbling	Bubble characteristics, gas flowrate, bubbling modes (intermittent or continuous)	MBR, AnMBR	Unsteady or transient shear stress; changes biomass properties
	Vibration	Vibration amplitude and frequency	AnMBR, MF, UF, MD, FO	Low energy cost; surface shear; effective cake removal; facilitates separation of macromolecules
	Particle Scouring	Size, fluidization rate	AFBR-AFMBR, IAFMBR	Reduced fouling; low energy cost; amenability for scale-up (disadvantages: membrane damage, blockage)
Dead-end	Backwashing	Backwash flux, backwash duration, backwash frequency	All HF systems	Internal fouling control; can be applied with air scouring
	Relaxation	Relaxation duration, relaxation frequency	All HF systems, especially MBR	-

AnMBR: anaerobic membrane bioreactor; MF: microfiltration; UF: ultrafiltration; MD: membrane distillation; FO: forward osmosis; AFBR-AFMBR: anaerobic fluidized-bed bioreactor-anaerobic fluidized-bed membrane bioreactor; IAFMBR: integrated anaerobic fluidized-bed membrane bioreactor.

7. Chemical Cleaning in Submerged HF Membranes—Procedure, Effect on Membrane Performance

Other than the physical cleaning means discussed in the previous section, chemical cleaning, which involves the use of acids, bases, oxidants, and surfactants, also aids in mitigating membrane fouling. Typically, physical cleaning is followed by chemical cleaning in membrane applications to effectively mitigate fouling [283]. Chemical cleaning was classified by Lin et al. [283] into four categories: (i) clean-in-place (CIP), which involves directly adding chemicals to the submerged HF system; (ii) clean-out-off-place (COP), which involves cleaning the membrane in a separate tank with a higher concentration of chemicals; (iii) chemical washing (CW), which involves adding chemicals to the feed stream; and (iv) chemically enhanced backwashing (CEB), which involves combining chemical and physical cleaning means.

For chemical cleaning, the key factors affecting efficiency in mitigating fouling are the type of chemical agents, cleaning duration and interval, concentration of chemicals, cleaning temperature, and flux [300,301]. The type of chemical used depends mainly on the application, feed characteristics (e.g., pH, ionic strength, and temperature), and membrane materials (e.g., compatibility of the membrane with the chemicals) [302]. Sodium hypochlorite (NaOCl) and citric acid are the most common chemical cleaning agents provided by the main MBR suppliers [18], although they are reported to be ineffective for removing iron species [303], and less effective than the coupling of NaOCl and caustic soda for removing natural organic matter (NOM) [304].

While the robust nature of most submerged HF membranes allows the use of relatively aggressive cleaning, some changes may occur. Kweon et al. [304] evaluated the effectiveness and changes in the membrane surface properties by acidic and alkali cleaning of PVDF HF membranes during the microfiltration of two feed waters. The results indicated that the feed-water quality played an important role in the cleaning efficiency; hence, experiments with the actual feed are necessary for the selection of cleaning procedures. In addition, chemical cleaning leads to changes in the surface properties of the membranes, which may lead to a gradual decrease in the recoverable flux.

8. Submerged HF Membrane Integrity and Failure

Given the chemical and physical stresses experienced by submerged HF membranes during operation, the lifespan of the membrane fibers tends to be significantly shortened. The prorated warranty provided by membrane manufacturers can range from 3 to 10 years [305]; however, experience teaches that the effective membrane life can either exceed or fall short of the manufacturer's expectations. In one particular case, a UF plant treating wastewater from a manufacturer of cosmetics experienced rapid membrane failure resulting in an average membrane lifespan of less than 6 months (note that the membranes were cleaned once a week with an alkaline bleach product, and backwashed monthly) [306], and the main cause was found to be high local shear forces due to fibrous material in the wastewater. In another study performed by De Wilde et al., the lifespan of the membranes was determined to be 13 years by extrapolating data based on 3 years of operation [307].

Given that the paramount operating objective is to avoid any failure that could compromise quality and restrict capacity, the development of a strategy that relies exclusively on the manufacturer's warranty to estimate the membrane lifespan and replacement schedules is fraught with uncertainty. With the prevalent variability in the integrity and productivity of membrane modules, operators of full-scale plants would need to manage an inventory of several thousand membranes; thus, anticipating and scheduling activities for the replacement of these membranes in service becomes a unique challenge for drinking-water plants utilizing membrane technology.

Even though membrane ageing and failure are closely related, a distinction should be made between these two factors. Membrane degradation is the result of ageing and the onset of its adverse effects, which in turn leads to membrane failure. Membrane failure, on the other hand, results in a loss of process removal efficiency, and a reduction in product-water throughput as well as product-water non-compliancy.

8.1. Ageing

Membrane ageing of commonly used composite membranes is defined as the deterioration of the surface layer and sub-layers of membranes due to the irreversible deposition of foulants or by frequent exposure to chemical cleaning agents, which leads to the deterioration of the membrane performance [308,309]. While the active layer of the membrane has been found to be chemically modified, pore reduction has been found in the intermediate sublayer (i.e., between the active layer and the porous support) [308]. Membrane material also affects ageing; for example, polyethylsulfone (PES) membranes were found to be more resistant to acid than alkali [310].

To control membrane fouling caused by the retention of dissolved salts, organics, microorganisms, and suspended solids after extended operation, the industry employs routine chemical cleaning protocols involving specific concentrations, temperatures, and extended cleaning times; in some cases, with submerged HFs, strong oxidants such as sodium hypochlorite (NaOCl) are used to control fouling, causing membrane ageing to be exacerbated after repeated cleanings. Prolonged filtration and cleaning cycles not only have an adverse effect on membrane integrity, but can also lead to the internal fouling of membranes, which is irreversible, detrimental to membrane performance, and also reduces the lifespan of the membrane by increasing the likelihood of membrane failure. For submerged HFs, physical cleaning by backwashing (Section 6.1) can also cause changes in the membrane properties. For example, it has been shown that backwashing can cause a change in the pore-size distribution by increasing the diameters of the largest pores [311]. This could make the membranes more susceptible to fouling as these larger pores become blocked.

8.2. Failure

Membrane failure is defined as the loss of mechanical integrity leading to the inability to achieve the rated log-removal values (LRV) of pathogens [312–314]. Membrane failure can occur during two phases of the operational lifespan of a membrane, namely, damage during the manufacturing and installation process, and during membrane filtration. Inconsistent manufacturing and fabrication techniques as well as handling error during installation often cause failure in the former phase. This issue is kept in check via the implementation of rigorous product quality-control methods and integrity testing of membrane modules before commissioning. On the other hand, unlike failure during the manufacturing and installation process, membrane failure during the filtration operation can mainly be attributed to operating parameters and maintenance protocols [315]. During operation, the likelihood of damage to the membrane is high given the stringent nature of operating protocols such as vigorous mechanical cleaning, chemical cleaning using strong oxidants, and high-pressure backwashing. Although these measures ensure that the membrane performance is maintained, they indirectly put a strain on the membrane integrity, leading to membrane ageing and failure.

According to Childress et al. [313], fiber failure can occur via four different mechanisms: chemical attack; damage during operation due to improper installation; faulty membrane module design; and punctures and scores due to the presence of foreign bodies. Furthermore, membrane ageing, coupled with excessive fiber movement due to external loads, can also cause submerged HF membranes to fail. This is discussed further in Sections 8.4 and 8.5.

8.3. Chemical Oxidation

Owing to membrane fouling being an inevitable phenomenon, membrane maintenance protocols using chemical cleaning to control fouling and restore the membrane flux are employed. There is a wide variety of chemical cleaning agents utilized by the industry, with the most common being sodium hypochlorite (NaOCl) because of its ready availability, relatively low price, and high cleaning efficiency. Unfortunately, such oxidants are the main causes of deterioration in the membrane integrity [316], whereby prolonged exposure causes oxidative damage to the membrane [317], which accelerates membrane ageing and degradation that in turn not only leads to discoloration of the membrane fibers, but also embrittlement of the fibers that subsequently increases the likelihood of membrane-fiber fracture [316,318,319]. The embrittlement rate for hollow fibers has been found to be four times that of flat-sheet membranes [309].

8.4. Module Design

The optimization of membrane-module designs in terms of the potting of the membrane fibers and the design of the membrane housing is usually performed to reduce membrane fouling and to maintain membrane integrity. In submerged HF modules, the fibers are located in a constrained geometry. For example, the GE-Zenon system has membranes assembled into cartridges and held in a supporting frame that connects to the aeration and permeate suction header. The ends of the fibers are potted into the permeate carrier. These features are typical of most submerged HF modules. More details can be found elsewhere [3,58].

The potting efficacy of the fibers can significantly affect the performance and integrity of the module. Current membrane modules consist of up to 20,000 hollow fibers held together with either an epoxy or urethane resin; depending on the manufacturing process, the resin can be cured under static or dynamic conditions. In the slower static curing method the resin is allowed to cure without heat or external forces acting on it. Membranes that are potted statically would have the resin wick up the edge of the fiber due to capillary forces, which leads to the development of a sharp edge that potentially can cause fiber breakage (see Figure 9a). As seen in Figure 9b, with this method an elastomer overlay is usually added on top of the potting material to minimize the sharp edges. On the other hand, for the dynamic curing method, centrifugal forces are used under elevated temperatures to cure the potting resin, thereby preventing resin wicking and avoiding the development of sharp edges [313] (see Figure 9c). Notably, with the implementation of air-scouring and external loads, the probability of fiber breakage at the potting site increases; thus, proper selection of the potting method and material is important. Moreover, an optimized arrangement of the fibers within the module has been reported to significantly improve process performance by as much as 200% [119,190].

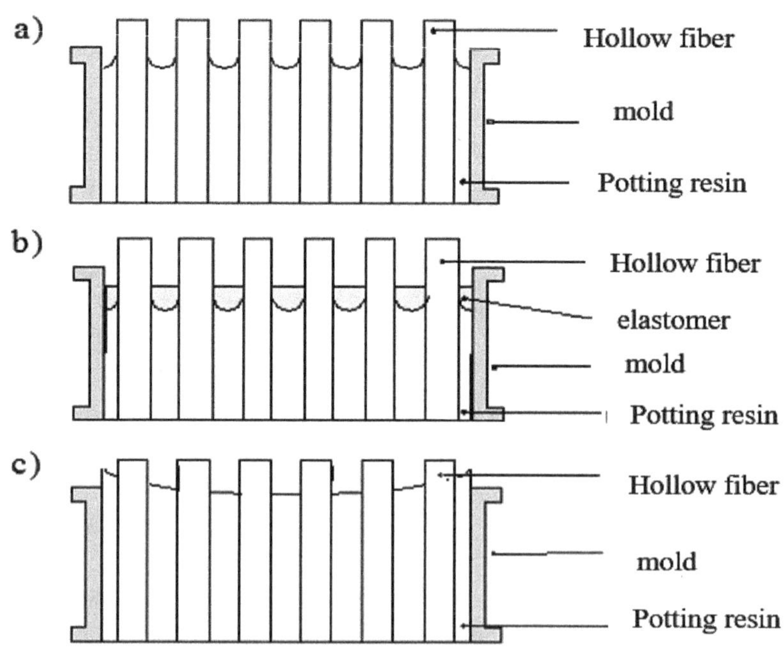

Figure 9. Membrane potting methods employing different conditions: (**a**) Static Conditions, (**b**) Static with Elastomer Overlay, (**c**) Dynamic Conditions.

The design of the membrane module and housing plays an equally important role in maintaining membrane integrity. Although there are many membrane housing designs that try to minimize excessive fiber movement, routine membrane filtration and backwashing carried out at pressures higher than the manufacturer's recommendations can lead to membrane failure via the rupturing of membrane fibers, damage to the membrane module housing, and degradation of the membrane module seals. Consequently, the tolerable stress load on a membrane module depends on the membrane material and structure, and the packing density of the HF membranes in the module. Pilot plant testing indicated that membrane symmetry, which affects the stress at the juncture between the potting material and the fiber, was more important than the potting technique for hollow fiber integrity [313]. At present there is a lack of fundamental data on the stresses experienced by fibers during the filtration and cleaning cycles in the presence of air scouring [320]. This, coupled with the difficulty in accurately measuring and calculating stresses in a multiple fiber system, has significantly limited the development of improved membrane modules. Therefore, for improved membrane-module, design a need exists for both a better understanding of membrane potting methods and the stress-strain forces acting on the membrane fibers.

8.5. Excessive Fiber Movement

The advantage of submerged membranes is that the hydrostatic pressure generated eliminates the need for the membrane modules to be pressurized. For such configurations, air-scouring or bubbling is employed to provide a shear force along the membrane surface to help alleviate the fouling phenomenon [14]. A higher shear force on the membrane surface results in a more efficient removal of foulants. In addition, specifically for bubbly flow around hollow fibers, another mechanism at play is the back and forth movement of the fibers induced by the bubbles, which causes a transverse vibration for loose fibers that leads to enhanced secondary mixing [15]. However, although the bubble-induced shear and fiber movements were able to reduce the fouling rate by up to 10-fold less, the excessive membrane movement due to a higher shear force can also lead to fiber breakage [15]. This phenomenon, coupled with the degradation of the membrane fibers due to ageing, could lead to a higher occurrence of fiber failure. Excessive fiber movement is also constrained if the fiber looseness is limited, for which the practical limits are typically 1–5% [3].

8.6. Foreign Bodies

Membrane damage and integrity compromise also can be caused by unexpected water-quality fluctuations together with the failure of the pretreatment processes, leading to the inadequate removal of foreign material [314]. These foreign bodies, coupled with the effects of strong aeration, can score or puncture the membrane fibers. A membrane autopsy performed by Zappia et al. concluded that the unexpected presence of silicon dioxide spicules (needle-like structures) resulted in multiple membrane occlusions and punctures, leading to a loss in membrane integrity [321]. The erosive effect of fluidized particles [322–325] in the feed stream is also known to compromise the membrane integrity by impacting the membrane surface [123,238,242,326–329]. Patterns of particle scraping was clearly observed [238,279], as well as a decrease in the membrane rejection for the larger fluidized glass beads (3 mm) [242]. As a result, care is needed for submerged HF processes that deliberately introduce scouring by suspended or fluidized media (Section 5.3.2) to control fouling.

8.7. Future Trends for Integrity Assessment

The current state-of-the-art tools for monitoring membrane integrity are limited to detecting compromises via a variety of in situ and ex situ techniques and tools. Despite extensive research performed on membrane-failure mechanisms and their resultant effects, these studies often are based on ex situ or offline analytical techniques, which can only provide information when a serious breach in membrane integrity is detected. Therefore, this underscores the need for the development of non-destructive, computer-aided modeling techniques to predict membrane failure and optimize module design. One possible approach is the prediction of failure in membrane systems via the use of finite element analysis (FEA). FEA is a modeling technique that is widely used by structural, mechanical, and biomedical engineers to perform mechanical analyses on complex structures to determine displacements from applied loads [330,331]. Through FEA, high stress locations along the membrane fiber can be determined that will help identify areas where failure is most likely to occur. Such analyses would aid in membrane and membrane-module design as well as the optimization of operating strategies. FEA can also be used as a diagnostic tool to provide supportive interpretations at similar operating conditions when performing autopsies on failed membranes and modules.

9. Conclusions and Research Opportunities

In a little over two decades, submerged HFs have gone from a curiosity to the mainstream of membrane technology. The major applications are for low-pressure membranes (MF and UF) in the water industry. The submerged HF concept is ideally suited for the dead-end filtration of dilute feeds (surface waters, pretreatment for RO) because effective cake removal can be achieved by backwash flow from the lumen side of the fiber. The submerged HF concept is also well-suited for more concentrated feeds, such as in aerobic MBR, where bubble-induced fiber movement helps to control fouling. The development of the submerged HF system has seen some advances in membranes per se (e.g., improved strength, flexibility, etc.) but the major efforts have been in module design and process optimization, such as fiber geometry, looseness, packing density, bubbling characteristics, backwash protocol, and modifying feed properties (e.g., by adjusting the bioprocess parameters in the MBR). The current generation of submerged HFs is clearly very effective, but there are opportunities for further development that could improve the concept and its applications. These are briefly discussed in the following section.

9.1. Hydrodynamics and Bubbling

Although hydrodynamics and bubbling in submerged HF systems have been actively studied, there may be opportunities for further incremental improvement. Areas for research include identifying the optimal bubble size (somewhere between relatively few large bubbles and many micro-bubbles) and the means for generating these bubble sizes. Also, the potential energy benefits of intermittent bubbling should be further evaluated.

9.2. Non-Bubbled Hydrodynamics

This review has provided an overview of the research activities in vibrations applied to submerged HFs. Further development is required to optimize this strategy in terms of module arrangement (alignment, packing density, use of flow promoters, etc.), vibration frequency and amplitude, etc. An important driver for development is likely to be the anaerobic MBR (AnMBR), where bubbling by recycled biogas has its challenges. Fluidized media to control hydrodynamics may also find application in the AnMBR. Further studies are required to evaluate and compare energy demand for these alternative hydrodynamic control methods.

9.3. Backwashing and Relaxation

Backwashing and relaxation are effective methods for fouling mitigation. However, the chosen protocols are likely to be conservative and suboptimal. Improved performance could involve an adjustable protocol (backwash frequency, duration, flux, etc.) that responds to changes to feed conditions and required production rate. Such a system could use an online monitor for 'fouling propensity' linked to a neural-network-based control system.

9.4. Identifying Sustainable Flux

The concept of critical flux is discussed in Section 5. Due to the limited practical applicability of critical flux, threshold and sustainable flux have been proposed [158,332]. While the threshold flux demarcates a low fouling from a high fouling region, the sustainable flux is one at which moderate fouling is tolerated based on balancing capital and operating costs; some guidelines are available that link measured threshold flux to sustainable flux [158]. However, in practice, it is difficult to measure the threshold flux in an operating plant as it involves flux-stepping involving conditions of high fouling rate that could be detrimental. This could be overcome by the development of a small-scale threshold-flux monitor that simulates the operating plant that could be flux-stepped. The challenge is to design a sufficiently accurate and reliable simulator based on the submerged HF concept.

9.5. Potential Non-Filtration Applications

The submerged HF concept is widely used in membrane filtration (MF and UF) applications. The concept is also amenable to other membrane separations, such as membrane distillation (MD) and forward osmosis (FO). Indeed 'high-retention' MBRs based on MD and FO with submerged HFs have been developed [333,334]. Further work is required to optimize these systems, including development of externally skinned FO hollow fiber membranes to minimize fouling. An interesting extension of FO is pressure-retarded osmosis (PRO), used for harnessing the salinity gradient for the generation of electricity, that can use hollow fibers with pressurized 'draw solute' on the lumen side [335,336]. It would be feasible to use the submerged HF arrangement for PRO with the low salinity feed in the tank.

9.6. Membrane Integrity

The strategies to improve submerged HF membrane integrity have been addressed in Section 9. An additional need is an effective online monitor to detect the loss of integrity. One example of such a device is given in Krantz et al. [337]. The key requirements are reliability, easy implementation, and modest cost for industrial applications.

Acknowledgments: We gratefully acknowledge funding from the Singapore Ministry of Education Academic Research Funds Tier 2 (MOE2014-T2-2-074; ARC16/15) and Tier 1 (2015-T1-001-023; RG7/15), and the Joint Singapore-Germany Research Project Fund (SGP-PROG3-019). We acknowledge support from the Singapore Economic Development Board to the Singapore Membrane Technology Centre.

References

1. Laîné, J.M.; Vial, D.; Moulart, P. Status after 10 years of operation—Overview of uf technology today. *Desalination* **2000**, *131*, 17–25. [CrossRef]
2. Choi, Y.-J.; Oh, H.; Lee, S.; Nam, S.-H.; Hwang, T.-M. Investigation of the filtration characteristics of pilot-scale hollow fiber submerged mf system using cake formation model and artificial neural networks model. *Desalination* **2012**, *297*, 20–29. [CrossRef]
3. Fane, A.G. Submerged membranes. In *Advanced Membrane Technology and Applications*; Li, N.N., Fane, A.G., Ho, W.S.W., Matsuura, T., Eds.; John Wiley & Sons: Hoboken, NJ, USA, 2008; Chapter 10.
4. Ye, Y.; Le Clech, P.; Chen, V.; Fane, A.G.; Jefferson, B. Fouling mechanisms of alginate solutions as model extracellular polymeric substances. *Desalination* **2005**, *175*, 7–20. [CrossRef]
5. Nakatsuka, S.; Nakate, I.; Miyano, T. Drinking water treatment by using ultrafiltration hollow fiber membranes. *Desalination* **1996**, *106*, 55–61. [CrossRef]
6. Günther, J.; Schmitz, P.; Albasi, C.; Lafforgue, C. A numerical approach to study the impact of packing density on fluid flow distribution in hollow fiber module. *J. Membr. Sci.* **2010**, *348*, 277–286. [CrossRef]
7. Di Profio, G.; Ji, X.; Curcio, E.; Drioli, E. Submerged hollow fiber ultrafiltration as seawater pretreatment in the logic of integrated membrane desalination systems. *Desalination* **2011**, *269*, 128–135. [CrossRef]
8. Chellam, S.; Jacangelo, J.G.; Bonacquisti, T.P. Modeling and experimental verification of pilot-scale hollow fiber, direct flow microfiltration with periodic backwashing. *Environ. Sci. Technol.* **1998**, *32*, 75–81. [CrossRef]
9. Raffin, M.; Germain, E.; Judd, S.J. Influence of backwashing, flux and temperature on microfiltration for wastewater reuse. *Sep. Purif. Technol.* **2012**, *96*, 147–153. [CrossRef]
10. Aimar, P. Slow colloidal aggregation and membrane fouling. *J. Membr. Sci.* **2010**, *360*, 70–76. [CrossRef]

11. Tang, C.Y.; Chong, T.H.; Fane, A.G. Colloidal interactions and fouling of nf and ro membranes: A review. *Adv. Colloid Interface Sci.* **2011**, *164*, 126–143. [CrossRef] [PubMed]
12. Fane, A.G.; Chong, T.H.; Le-Clech, P. Fouling in membrane processes. In *Membrane Operations, Innovative Separations and Transformations*; Drioli, E., Giorno, L., Eds.; Wiley-VCH: Weinheim, Germany, 2009; Chapter 6.
13. Vigneswaran, S.; Kwon, D.Y.; Ngo, H.H.; Hu, J.Y. Improvement of microfiltration performance in water treatment: Is critical flux a viable solution? *Water Sci. Technol.* **2000**, *41*, 309–315.
14. Cui, Z.F.; Chang, S.; Fane, A.G. The use of gas bubbling to enhance membrane processes. *J. Membr. Sci.* **2003**, *221*, 1–35. [CrossRef]
15. Wicaksana, F.; Fane, A.G.; Chen, V. Fibre movement induced by bubbling using submerged hollow fibre membranes. *J. Membr. Sci.* **2006**, *271*, 186–195. [CrossRef]
16. Chang, S.; Fane, A.G. Filtration of biomass with laboratory-scale submerged hollow fibre modules—Effect of operating conditions and module configuration. *J. Chem. Technol. Biotechnol.* **2002**, *77*, 1030–1038. [CrossRef]
17. Gomaa, H.G.; Rao, S. Analysis of flux enhancement at oscillating flat surface membranes. *J. Membr. Sci.* **2011**, *374*, 59–66. [CrossRef]
18. Le-Clech, P.; Chen, V.; Fane, T.A.G. Fouling in membrane bioreactors used in wastewater treatment. *J. Membr. Sci.* **2006**, *284*, 17–53. [CrossRef]
19. Radjenović, J.; Matošić, M.; Mijatović, I.; Petrović, M.; Barceló, D. Membrane bioreactor (MBR) as an advanced wastewater treatment technology. In *Emerging Contaminants from Industrial and Municipal Waste*; Springer: Berlin/Heidelberg, Germany, 2008; pp. 37–101.
20. Le-Clech, P. Membrane bioreactors and their uses in wastewater treatments. *Appl. Microbiol. Biotechnol.* **2010**, *88*, 1253–1260. [CrossRef] [PubMed]
21. Lu, Y.; Ding, Z.; Liu, L.; Wang, Z.; Ma, R. The influence of bubble characteristics on the performance of submerged hollow fiber membrane module used in microfiltration. *Sep. Purif. Technol.* **2008**, *61*, 89–95. [CrossRef]
22. Tian, J.-Y.; Xu, Y.-P.; Chen, Z.-L.; Nan, J.; Li, G.-B. Air bubbling for alleviating membrane fouling of immersed hollow-fiber membrane for ultrafiltration of river water. *Desalination* **2010**, *260*, 225–230. [CrossRef]
23. Li, T.; Law, A.W.-K.; Cetin, M.; Fane, A.G. Fouling control of submerged hollow fibre membranes by vibrations. *J. Membr. Sci.* **2013**, *427*, 230–239. [CrossRef]
24. Kim, J.; Kim, K.; Ye, H.; Lee, E.; Shin, C.; McCarty, P.L.; Bae, J. Anaerobic fluidized bed membrane bioreactor for wastewater treatment. *Environ. Sci. Technol.* **2011**, *45*, 576–581. [CrossRef] [PubMed]
25. Hilal, N.; Ogunbiyi, O.O.; Miles, N.J.; Nigmatullin, R. Methods employed for control of fouling in mf and uf membranes: A comprehensive review. *Sep. Sci. Technol.* **2005**, *40*, 1957–2005. [CrossRef]
26. Ye, Y.; Sim, L.N.; Herulah, B.; Chen, V.; Fane, A.G. Effects of operating conditions on submerged hollow fibre membrane systems used as pre-treatment for seawater reverse osmosis. *J. Membr. Sci.* **2010**, *365*, 78–88. [CrossRef]
27. Aslam, M.; Charfi, A.; Lesage, G.; Heran, M.; Kim, J. Membrane bioreactors for wastewater treatment: A review of mechanical cleaning by scouring agents to control membrane fouling. *Chem. Eng. J.* **2017**, *307*, 897–913. [CrossRef]
28. Krzeminski, P.; Leverette, L.; Malamis, S.; Katsou, E. Membrane bioreactors—A review on recent developments in energy reduction, fouling control, novel configurations, lca and market prospects. *J. Membr. Sci.* **2017**, *527*, 207–227. [CrossRef]
29. Arnal, J.M.; Garcia-Fayos, B.; Verdu, G.; Lora, J. Ultrafiltration as an alternative membrane technology to obtain safe drinking water from surface water: 10 years of experience on the scope of the aquapot project. *Desalination* **2009**, *248*, 34–41. [CrossRef]
30. Madaeni, S.S. The application of membrane technology for water disinfection. *Water Res.* **1999**, *33*, 301–308. [CrossRef]
31. Ray, C.; Jain, R. *Low Cost Emergency Water Purification Technologies: Integrated Water Security Series*; Butterworth-Heinemann: Oxford, UK, 2014; pp. 1–205.
32. Loo, S.L.; Fane, A.G.; Krantz, W.B.; Lim, T.T. Emergency water supply: A review of potential technologies and selection criteria. *Water Res.* **2012**, *46*, 3125–3151. [CrossRef] [PubMed]
33. Iannelli, R.; Ripari, S.; Casini, B.; Buzzigoli, A.; Privitera, G.; Verani, M.; Carducci, A. Feasibility assessment of surface water disinfection by ultrafiltration. *Water Sci. Technol. Water Supply* **2014**, *14*, 522–531. [CrossRef]

34. Carvajal, G.; Branch, A.; Sisson, S.A.; Roser, D.J.; van den Akker, B.; Monis, P.; Reeve, P.; Keegan, A.; Regel, R.; Khan, S.J. Virus removal by ultrafiltration: Understanding long-term performance change by application of bayesian analysis. *Water Res.* **2017**, *122*, 269–279. [CrossRef] [PubMed]
35. Zularisam, A.W.; Ismail, A.F.; Sakinah, M. Application and challenges of membrane in surface water treatment. *J. Appl. Sci.* **2010**, *10*, 380–390. [CrossRef]
36. Zularisam, A.W.; Ismail, A.F.; Salim, R. Behaviours of natural organic matter in membrane filtration for surface water treatment—A review. *Desalination* **2006**, *194*, 211–231. [CrossRef]
37. Lin, T.; Shen, B.; Chen, W.; Zhang, X. Interaction mechanisms associated with organic colloid fouling of ultrafiltration membrane in a drinking water treatment system. *Desalination* **2014**, *332*, 100–108. [CrossRef]
38. Guastalli, A.R.; Simon, F.X.; Penru, Y.; de Kerchove, A.; Llorens, J.; Baig, S. Comparison of dmf and uf pre-treatments for particulate material and dissolved organic matter removal in swro desalination. *Desalination* **2013**, *322*, 144–150. [CrossRef]
39. Jamaly, S.; Darwish, N.N.; Ahmed, I.; Hasan, S.W. A short review on reverse osmosis pretreatment technologies. *Desalination* **2014**, *354*, 30–38. [CrossRef]
40. Jeong, S.; Park, Y.; Lee, S.; Kim, J.; Lee, K.; Lee, J.; Chon, H.T. Pre-treatment of swro pilot plant for desalination using submerged mf membrane process: Trouble shooting and optimization. *Desalination* **2011**, *279*, 86–95. [CrossRef]
41. Chua, K.T.; Hawlader, M.N.A.; Malek, A. Pretreatment of seawater: Results of pilot trials in singapore. *Desalination* **2003**, *159*, 225–243. [CrossRef]
42. Wu, B.; Fane, A.G. Microbial relevant fouling in membrane bioreactors: Influencing factors, characterization, and fouling control. *Membranes* **2012**, *2*, 565–584. [CrossRef] [PubMed]
43. Mo, L.; Huanga, X. Fouling characteristics and cleaning strategies in a coagulation-microfiltration combination process for water purification. *Desalination* **2003**, *159*, 1–9. [CrossRef]
44. Huang, H.; Schwab, K.; Jacangelo, J.G. Pretreatment for low pressure membranes in water treatment: A review. *Environ. Sci. Technol.* **2009**, *43*, 3011–3019. [CrossRef] [PubMed]
45. Huang, H.; O'Melia, C.R. Direct-flow microfiltration of aquasols: II. On the role of colloidal natural organic matter. *J. Membr. Sci.* **2008**, *325*, 903–913. [CrossRef]
46. Zhang, M.; Li, C.; Benjamin, M.M.; Chang, Y. Fouling and natural organic matter removal in adsorbent/membrane systems for drinking water treatment. *Environ. Sci. Technol.* **2003**, *37*, 1663–1669. [CrossRef] [PubMed]
47. Jones, K.L.; O'Melia, C.R. Protein and humic acid adsorption onto hydrophilic membrane surfaces: Effects of ph and ionic strength. *J. Membr. Sci.* **2000**, *165*, 31–46. [CrossRef]
48. Lorain, O.; Hersant, B.; Persin, F.; Grasmick, A.; Brunard, N.; Espenan, J.M. Ultrafiltration membrane pre-treatment benefits for reverse osmosis process in seawater desalting. Quantification in terms of capital investment cost and operating cost reduction. *Desalination* **2007**, *203*, 277–285. [CrossRef]
49. Pearce, G.; Talo, S.; Chida, K.; Basha, A.; Gulamhusein, A. Pretreatment options for large scale swro plants: Case studies of of trials at kindasa, saudi arabia, and conventional pretreatment in spain. *Desalination* **2004**, *167*, 175–189. [CrossRef]
50. Pearce, G.K. The case for UF/MF pretreatment to RO in seawater applications. *Desalination* **2007**, *203*, 286–295. [CrossRef]
51. Pearce, G.K. UF/MF pre-treatment to RO in seawater and wastewater reuse applications: A comparison of energy costs. *Desalination* **2008**, *222*, 66–73. [CrossRef]
52. Brehant, A.; Bonnelye, V.; Perez, M. Comparison of MF/UF pretreatment with conventional filtration prior to RO membranes for surface seawater desalination. *Desalination* **2002**, *144*, 353–360. [CrossRef]
53. Bonnélye, V.; Guey, L.; Del Castillo, J. UF/MF as RO pre-treatment: The real benefit. *Desalination* **2008**, *222*, 59–65. [CrossRef]
54. Teng, C.K.; Hawlader, M.N.A.; Malek, A. An experiment with different pretreatment methods. *Desalination* **2003**, *156*, 51–58. [CrossRef]
55. Sohn, J.; Valavala, R.; Han, J.; Her, N.; Yoon, Y. Pretreatment in reverse osmosis seawater desalination: A short review. *Environ. Eng. Res.* **2011**, *16*, 205–212.
56. Akhondi, E.; Wu, B.; Sun, S.; Marxer, B.; Lim, W.; Gu, J.; Liu, L.; Burkhardt, M.; McDougald, D.; Pronk, W.; et al. Gravity-driven membrane filtration as pretreatment for seawater reverse osmosis: Linking biofouling layer morphology with flux stabilization. *Water Res.* **2015**, *70*, 158–173. [CrossRef] [PubMed]

57. Wu, B.; Hochstrasser, F.; Akhondi, E.; Ambauen, N.; Tschirren, L.; Burkhardt, M.; Fane, A.G.; Pronk, W. Optimization of gravity-driven membrane (GDM) filtration process for seawater pretreatment. *Water Res.* **2016**, *93*, 133–140. [CrossRef] [PubMed]
58. Judd, S.; Judd, C. *The MBR Book: Principles and Applications of Membrane Bioreactors for Water and Wastewater Treatment*, 1st ed.; Elsevier: Amsterdam, The Netherlands; Boston, MA, USA; London, UK, 2006; pp. 207–272.
59. Santos, A.; Ma, W.; Judd, S.J. Membrane bioreactors: Two decades of research and implementation. *Desalination* **2011**, *273*, 148–154. [CrossRef]
60. Zhang, Q.; Singh, S.; Stuckey, D.C. Fouling reduction using adsorbents/flocculants in a submerged anaerobic membrane bioreactor. *Bioresour. Technol.* **2017**, *239*, 226–235. [CrossRef] [PubMed]
61. Svojitka, J.; Dvořák, L.; Studer, M.; Straub, J.O.; Frömelt, H.; Wintgens, T. Performance of an anaerobic membrane bioreactor for pharmaceutical wastewater treatment. *Bioresour. Technol.* **2017**, *229*, 180–189. [CrossRef] [PubMed]
62. Chen, C.; Guo, W.; Ngo, H.H.; Chang, S.W.; Duc Nguyen, D.; Dan Nguyen, P.; Bui, X.T.; Wu, Y. Impact of reactor configurations on the performance of a granular anaerobic membrane bioreactor for municipal wastewater treatment. *Int. Biodeterior. Biodegrad.* **2017**, *121*, 131–138. [CrossRef]
63. Ozgun, H.; Dereli, R.K.; Ersahin, M.E.; Kinaci, C.; Spanjers, H.; van Lier, J.B. A review of anaerobic membrane bioreactors for municipal wastewater treatment: Integration options, limitations and expectations. *Sep. Purif. Technol.* **2013**, *118*, 89–104. [CrossRef]
64. Johir, M.A.H.; George, J.; Vigneswaran, S.; Kandasamy, J.; Sathasivan, A.; Grasmick, A. Effect of imposed flux on fouling behavior in high rate membrane bioreactor. *Bioresour. Technol.* **2012**, *122*, 42–49. [CrossRef] [PubMed]
65. Monsalvo, V.M.; McDonald, J.A.; Khan, S.J.; Le-Clech, P. Removal of trace organics by anaerobic membrane bioreactors. *Water Res.* **2014**, *49*, 103–112. [CrossRef] [PubMed]
66. Liao, B.-Q.; Kraemer, J.T.; Bagley, D.M. Anaerobic membrane bioreactors: Applications and research directions. *Crit. Rev. Environ. Sci. Technol.* **2006**, *36*, 489–530. [CrossRef]
67. Ng, H.Y.; Hermanowicz, S.W. Membrane bioreactor operation at short solids retention times: Performance and biomass characteristics. *Water Res.* **2005**, *39*, 981–992. [CrossRef] [PubMed]
68. Chang, I.S.; Kim, S.N. Wastewater treatment using membrane filtration—Effect of biosolids concentration on cake resistance. *Process Biochem.* **2005**, *40*, 1307–1314. [CrossRef]
69. Sablani, S.; Goosen, M.; Al-Belushi, R.; Wilf, M. Concentration polarization in ultrafiltration and reverse osmosis: A critical review. *Desalination* **2001**, *141*, 269–289. [CrossRef]
70. Schäfer, A.I.; Fane, A.G.; Waite, T.D. Fouling effects on rejection in the membrane filtration of natural waters. *Desalination* **2000**, *131*, 215–224. [CrossRef]
71. Zydney, A.L.; Colton, C.K. A concentration polarization model for the filtrate flow in cross-flow microfiltration of particulate suspensions. *Chem. Eng. Commun.* **1986**, *47*, 1–21. [CrossRef]
72. Goosen, M.F.A.; Sablani, S.S.; Al-Hinai, H.; Al-Obeidani, S.; Al-Belushi, R.; Jackson, D. Fouling of reverse osmosis and ultrafiltration membranes: A critical review. *Sep. Sci. Technol.* **2004**, *39*, 2261–2297. [CrossRef]
73. Aoustin, E.; Schäfer, A.I.; Fane, A.G.; Waite, T.D. Ultrafiltration of natural organic matter. *Sep. Purif. Technol.* **2001**, *22*, 63–78. [CrossRef]
74. Peinemann, K.V.; Pereira, N.S. *Membrane Technology: Volume 4: Membranes for Water Treatmen*; Wiley-VCH: Weinheim, Germany, 2010.
75. Huang, H.; Young, T.A.; Jacangelo, J.G. Unified membrane fouling index for low pressure membrane filtration of natural waters: Principles and methodology. *Environ. Sci. Technol.* **2008**, *42*, 714–720. [CrossRef] [PubMed]
76. Cogan, N.G.; Chellam, S. Incorporating pore blocking, cake filtration, and eps production in a model for constant pressure bacterial fouling during dead-end microfiltration. *J. Membr. Sci.* **2009**, *345*, 81–89. [CrossRef]
77. Ho, C.-C.; Zydney, A.L. A combined pore blockage and cake filtration model for protein fouling during microfiltration. *J. Colloid Interface Sci.* **2000**, *232*, 389–399. [CrossRef] [PubMed]
78. Iritani, E.; Katagiri, N.; Takenaka, T.; Yamashita, Y. Membrane pore blocking during cake formation in constant pressure and constant flux dead-end microfiltration of very dilute colloids. *Chem. Eng. Sci.* **2015**, *122*, 465–473. [CrossRef]
79. Abdelrasoul, A.; Doan, H.; Lohi, A.; Cheng, C.H. Modeling of fouling and foulant attachments on heterogeneous membranes in ultrafiltration of latex solution. *Sep. Purif. Technol.* **2014**, *135*, 199–210. [CrossRef]

80. Hong, S.; Elimelech, M. Chemical and physical aspects of natural organic matter (NOM) fouling of nanofiltration membranes. *J. Membr. Sci.* **1997**, *132*, 159–181. [CrossRef]
81. Lee, E.K.; Chen, V.; Fane, A.G. Natural organic matter (NOM) fouling in low pressure membrane filtration—Effect of membranes and operation modes. *Desalination* **2008**, *218*, 257–270. [CrossRef]
82. Wibisono, Y.; Cornelissen, E.R.; Kemperman, A.J.B.; Van Der Meer, W.G.J.; Nijmeijer, K. Two-phase flow in membrane processes: A technology with a future. *J. Membr. Sci.* **2014**, *453*, 566–602. [CrossRef]
83. Fane, A.G.; Fell, C.J.D. A review of fouling and fouling control in ultrafiltration. *Desalination* **1987**, *62*, 117–136. [CrossRef]
84. Pimentel, G.A.; Dalmau, M.; Vargas, A.; Comas, J.; Rodriguez-Roda, I.; Rapaport, A.; Vande Wouwer, A. Validation of a simple fouling model for a submerged membrane bioreactor. *IFAC-PapersOnLine* **2015**, *48*, 737–742. [CrossRef]
85. Tang, C.Y.; Fu, Q.S.; Criddle, C.S.; Leckie, J.O. Effect of flux (transmembrane pressure) and membrane properties on fouling and rejection of reverse osmosis and nanofiltration membranes treating perfluorooctane sulfonate containing wastewater. *Environ. Sci. Technol.* **2007**, *41*, 2008–2014. [CrossRef] [PubMed]
86. Vrijenhoek, E.M.; Hong, S.; Elimelech, M. Influence of membrane surface properties on initial rate of colloidal fouling of reverse osmosis and nanofiltration membranes. *J. Membr. Sci.* **2001**, *188*, 115–128. [CrossRef]
87. Iritani, E. A review on modeling of pore-blocking behaviors of membranes during pressurized membrane filtration. *Dry. Technol.* **2013**, *31*, 146–162. [CrossRef]
88. Vera, L.; González, E.; Díaz, O.; Delgado, S. Application of a backwashing strategy based on transmembrane pressure set-point in a tertiary submerged membrane bioreactor. *J. Membr. Sci.* **2014**, *470*, 504–512. [CrossRef]
89. De Souza, N.P.; Basu, O.D. Relaxation: A beneficial operational step for the reduction of fouling in hollow fiber membranes for drinking water treatment. In Proceedings of the Water Quality Technology Conference and Exposition 2012, Toronto, ON, Canada, 4–8 Novemmmber 2012.
90. Akhondi, E.; Wicaksana, F.; Fane, A.G. Evaluation of fouling deposition, fouling reversibility and energy consumption of submerged hollow fiber membrane systems with periodic backwash. *J. Membr. Sci.* **2014**, *452*, 319–331. [CrossRef]
91. Akhondi, E.; Wicaksana, F.; Krantz, W.B.; Fane, A.G. Influence of dissolved air on the effectiveness of cyclic backwashing in submerged membrane systems. *J. Membr. Sci.* **2014**, *456*, 77–84. [CrossRef]
92. Khirani, S.; Smith, P.J.; Manéro, M.H.; Aim, R.B.; Vigneswaran, S. Effect of periodic backwash in the submerged membrane adsorption hybrid system (SMAHS) for wastewater treatment. *Desalination* **2006**, *191*, 27–34. [CrossRef]
93. Zsirai, T.; Buzatu, P.; Aerts, P.; Judd, S. Efficacy of relaxation, backflushing, chemical cleaning and clogging removal for an immersed hollow fibre membrane bioreactor. *Water Res.* **2012**, *46*, 4499–4507. [CrossRef] [PubMed]
94. Hong, S.P.; Bae, T.H.; Tak, T.M.; Hong, S.; Randall, A. Fouling control in activated sludge submerged hollow fiber membrane bioreactors. *Desalination* **2002**, *143*, 219–228. [CrossRef]
95. Davis, R.H. Modeling of fouling of cross-flow microfiltration membranes. *Sep. Purif. Methods* **1992**, *21*, 75–126. [CrossRef]
96. Braak, E.; Alliet, M.; Schetrite, S.; Albasi, C. Aeration and hydrodynamics in submerged membrane bioreactors. *J. Membr. Sci.* **2011**, *379*, 1–18. [CrossRef]
97. Abdullah, S.Z.; Wray, H.E.; Bérubé, P.R.; Andrews, R.C. Distribution of surface shear stress for a densely packed submerged hollow fiber membrane system. *Desalination* **2015**, *357*, 117–120. [CrossRef]
98. Wicaksana, F.; Fane, A.G.; Chen, V. The relationship between critical flux and fibre movement induced by bubbling in a submerged hollow fibre system. *Water Sci. Technol.* **2005**, *51*, 115–122. [PubMed]
99. Chang, S.; Fane, A.G.; Vigneswaran, S. Modeling and optimizing submerged hollow fiber membrane modules. *AIChE J.* **2002**, *48*, 2203–2212. [CrossRef]
100. Kim, J.; DiGiano, F.A. Defining critical flux in submerged membranes: Influence of length-distributed flux. *J. Membr. Sci.* **2006**, *280*, 752–761. [CrossRef]
101. Fane, A.G.; Chang, S.; Chardon, E. Submerged hollow fibre membrane module—Design options and operational considerations. *Desalination* **2002**, *146*, 231–236. [CrossRef]
102. Madec, A.; Buisson, H.; Ben Aim, R. Aeration to enhance membrane critical flux. In Proceedings of the World Filtration Congress, Brighton, UK, 3–7 April 2000; pp. 199–202.

103. Wu, Z.; Wang, Z.; Huang, S.; Mai, S.; Yang, C.; Wang, X.; Zhou, Z. Effects of various factors on critical flux in submerged membrane bioreactors for municipal wastewater treatment. *Sep. Purif. Technol.* **2008**, *62*, 56–63. [CrossRef]
104. Judd, S.; Judd, C. Design. In *The MBR Book: Principles and Applications of Membrane Bioreactors for Water and Wastewater Treatment*, 1st ed.; Elsevier: Amsterdam, The Netherlands; Boston, MA, USA; London, UK, 2006; Chapter 3, pp. 123–162.
105. Cho, B.D.; Fane, A.G. Fouling transients in nominally sub-critical flux operation of a membrane bioreactor. *J. Membr. Sci.* **2002**, *209*, 391–403. [CrossRef]
106. Zhang, J.; Chua, H.C.; Zhou, J.; Fane, A.G. Factors affecting the membrane performance in submerged membrane bioreactors. *J. Membr. Sci.* **2006**, *284*, 54–66. [CrossRef]
107. Judd, S. The status of membrane bioreactor technology. *Trends Biotechnol.* **2008**, *26*, 109–116. [CrossRef] [PubMed]
108. Yeo, A.; Fane, A.G. Performance of individual fibers in a submerged hollow fiber bundle. *Water Sci. Technol.* **2005**, *51*, 165–172. [PubMed]
109. Judd, S.; Judd, C. Fundamentals. In *The Mbr Book: Principles and Applications of Membrane Bioreactors for Water and Wastewater Treatment*, 1st ed.; Elsevier: Amsterdam, The Netherlands; Boston, MA, USA; London, UK, 2006; Chapter 2, pp. 22–121.
110. Yeo, A.P.S.; Law, A.W.K.; Fane, A.G. Factors affecting the performance of a submerged hollow fiber bundle. *J. Membr. Sci.* **2006**, *280*, 969–982. [CrossRef]
111. Costello, M.J.; Fane, A.G.; Hogan, P.A.; Schofield, R.W. The effect of shell side hydrodynamics on the performance of axial-flow hollow-fiber modules. *J. Membr. Sci.* **1993**, *80*, 1–11. [CrossRef]
112. Zheng, J.M.; Xu, Y.Y.; Xu, Z.K. Flow distribution in a randomly packed hollow fiber membrane module. *J. Membr. Sci.* **2003**, *211*, 263–269. [CrossRef]
113. Wu, J.; Chen, V. Shell-side mass transfer performance of randomly packed hollow fiber modules. *J. Membr. Sci.* **2000**, *172*, 59–74. [CrossRef]
114. Wickramasinghe, S.R.; Semmens, M.J.; Cussler, E.L. Mass-transfer in various hollow fiber geometries. *J. Membr. Sci.* **1992**, *69*, 235–250. [CrossRef]
115. Lipnizki, F.; Field, R.W. Mass transfer performance for hollow fibre modules with shell-side axial feed flow: Using an engineering approach to develop a framework. *J. Membr. Sci.* **2001**, *193*, 195–208. [CrossRef]
116. Busch, J.; Cruse, A.; Marquardt, W. Modeling submerged hollow-fiber membrane filtration for wastewater treatment. *J. Membr. Sci.* **2007**, *288*, 94–111. [CrossRef]
117. Ding, Z.W.; Liu, L.Y.; Ma, R.Y. Study on the effect of flow maldistribution on the performance of the hollow fiber modules used in membrane distillation. *J. Membr. Sci.* **2003**, *215*, 11–23.
118. Li, Z.X.; Zhang, L.Z. Flow maldistribution and performance deteriorations in a counter flow hollow fiber membrane module for air humidification/dehumidification. *Int. J. Heat Mass Transf.* **2014**, *74*, 421–430. [CrossRef]
119. Yang, X.; Wang, R.; Fane, A.G. Novel designs for improving the performance of hollow fiber membrane distillation modules. *J. Membr. Sci.* **2011**, *384*, 52–62. [CrossRef]
120. Chang, S.; Fane, A.G.; Waite, T.D.; Yeo, A. Unstable filtration behavior with submerged hollow fiber membranes. *J. Membr. Sci.* **2008**, *308*, 107–114. [CrossRef]
121. Chen, H.; Cao, C.; Xu, L.L.; Xiao, T.H.; Jiang, G.L. Experimental velocity measurements and effect of flow maldistribution on predicted permeator performances. *J. Membr. Sci.* **1998**, *139*, 259–268. [CrossRef]
122. Wicaksana, F.; Fane, A.G.; Law, A.W.K. The use of constant temperature anemometry for permeate flow distribution measurement in a submerged hollow fibre system. *J. Membr. Sci.* **2009**, *339*, 195–203. [CrossRef]
123. Zamani, F.; Chew, J.W.; Akhondi, E.; Krantz, W.B.; Fane, A.G. Unsteady-state shear strategies to enhance mass-transfer for the implementation of ultrapermeable membranes in reverse osmosis: A review. *Desalination* **2015**, *356*, 328–348. [CrossRef]
124. Gabarron, S.; Gomez, M.; Dvorak, L.; Ruzickova, I.; Rodriguez-Roda, I.; Comas, J. Ragging in mbr: Effects of operational conditions, chemical cleaning, and pre-treatment improvements. *Sep. Sci. Technol.* **2014**, *49*, 2115–2123. [CrossRef]
125. Sanchez, A.; Buntner, D.; Garrido, J.M. Impact of methanogenic pre-treatment on the performance of an aerobic mbr system. *Water Res.* **2013**, *47*, 1229–1236. [CrossRef] [PubMed]

126. Yu, W.Z.; Xu, L.; Graham, N.; Qu, J.H. Pre-treatment for ultrafiltration: Effect of pre-chlorination on membrane fouling. *Sci. Rep.* **2014**, *4*, 6513. [CrossRef] [PubMed]
127. Nguyen, T.T.; Ngo, H.H.; Guo, W.S. Pilot scale study on a new membrane bioreactor hybrid system in municipal wastewater treatment. *Bioresour. Technol.* **2013**, *141*, 8–12. [CrossRef] [PubMed]
128. Dutta, K.; Lee, M.Y.; Lai, W.W.P.; Lee, C.H.; Lin, A.Y.C.; Lin, C.F.; Lin, J.G. Removal of pharmaceuticals and organic matter from municipal wastewater using two-stage anaerobic fluidized membrane bioreactor. *Bioresour. Technol.* **2014**, *165*, 42–49. [CrossRef] [PubMed]
129. Sheldon, M.S.; Zeelie, P.J.; Edwards, W. Treatment of paper mill effluent using an anaerobic/aerobic hybrid side-stream membrane bioreactor. *Water Sci. Technol.* **2012**, *65*, 1265–1272. [CrossRef] [PubMed]
130. Bae, J.; Shin, C.; Lee, E.; Kim, J.; Mccarty, P.L. Anaerobic treatment of low-strength wastewater: A comparison between single and staged anaerobic fluidized bed membrane bioreactors. *Bioresour. Technol.* **2014**, *165*, 75–80. [CrossRef] [PubMed]
131. Wang, Z.; Ma, J.; Tang, C.Y.; Kimura, K.; Wang, Q.; Han, X. Membrane cleaning in membrane bioreactors: A review. *J. Membr. Sci.* **2014**, *468*, 276–307. [CrossRef]
132. Yang, W.B.; Cicek, N.; Ilg, J. State-of-the-art of membrane bioreactors: Worldwide research and commercial applications in north america. *J. Membr. Sci.* **2006**, *270*, 201–211. [CrossRef]
133. Choo, K.H.; Kang, I.J.; Yoon, S.H.; Park, H.; Kim, J.H.; Adiya, S.; Lee, C.H. Approaches to membrane fouling control in anaerobic membrane bioreactors. *Water Sci. Technol.* **2000**, *41*, 363–371.
134. Kang, I.J.; Yoon, S.H.; Lee, C.H. Comparison of the filtration characteristics of organic and inorganic membranes in a membrane-coupled anaerobic bioreactor. *Water Res.* **2002**, *36*, 1803–1813. [CrossRef]
135. Lee, S.M.; Jung, J.Y.; Chung, Y.C. Novel method for enhancing permeate flux of submerged membrane system in two-phase anaerobic reactor. *Water Res.* **2001**, *35*, 471–477. [CrossRef]
136. Brepols, C.; Drensla, K.; Janot, A.; Trimborn, M.; Engelhardt, N. Strategies for chemical cleaning in large scale membrane bioreactors. *Water Sci. Technol.* **2008**, *57*, 457–463. [CrossRef] [PubMed]
137. Brannock, M.W.D.; De Wever, H.; Wang, Y.; Leslie, G. Computational fluid dynamics simulations of mbrs: Inside submerged versus outside submerged membranes. *Desalination* **2009**, *236*, 244–251. [CrossRef]
138. Liu, R.; Huang, X.; Wang, C.; Chen, L.; Qian, Y. Study on hydraulic characteristics in a submerged membrane bioreactor process. *Process Biochem.* **2000**, *36*, 249–254. [CrossRef]
139. Brannock, M.; Wang, Y.; Leslie, G. Mixing characterisation of full-scale membrane bioreactors: CFD modelling with experimental validation. *Water Res.* **2010**, *44*, 3181–3191. [CrossRef] [PubMed]
140. Ji, J.; Li, J.F.; Qiu, J.P.; Li, X.D. Polyacrylamide-starch composite flocculant as a membrane fouling reducer: Key factors of fouling reduction. *Sep. Purif. Technol.* **2014**, *131*, 1–7. [CrossRef]
141. Nouri, N.; Mehrnia, M.R.; Sarrafzadeh, M.H.; Nabizadeh, R. Performance of membrane bioreactor in presence of flocculants. *Desalination Water Treat.* **2014**, *52*, 2933–2938. [CrossRef]
142. Zhang, H.F.; Gao, Z.Y.; Zhang, L.H.; Song, L.F. Performance enhancement and fouling mitigation by organic flocculant addition in membrane bioreactor at high salt shock. *Bioresour. Technol.* **2014**, *164*, 34–40. [CrossRef] [PubMed]
143. Melo-Guimaraes, A.; Torner-Morales, F.J.; Duran-Alvarez, J.C.; Jimenez-Cisneros, B.E. Removal and fate of emerging contaminants combining biological, flocculation and membrane treatments. *Water Sci. Technol.* **2013**, *67*, 877–885. [CrossRef] [PubMed]
144. Nguyen, T.T.; Guo, W.S.; Ngo, H.H.; Vigneswaran, S. A new combined inorganic-organic flocculant (CIOF) as a performance enhancer for aerated submerged membrane bioreactor. *Sep. Purif. Technol.* **2010**, *75*, 204–209. [CrossRef]
145. Johir, M.A.; Shanmuganathan, S.; Vigneswaran, S.; Kandasamy, J. Performance of submerged membrane bioreactor (SMBR) with and without the addition of the different particle sizes of gac as suspended medium. *Bioresour. Technol.* **2013**, *141*, 13–18. [CrossRef] [PubMed]
146. Nguyen, L.N.; Hai, F.I.; Kang, J.G.; Price, W.E.; Nghiem, L.D. Removal of trace organic contaminants by a membrane bioreactor-granular activated carbon (MBR-GAC) system. *Bioresour. Technol.* **2012**, *113*, 169–173. [CrossRef] [PubMed]
147. Gur-Reznik, S.; Katz, I.; Dosoretz, C.G. Removal of dissolved organic matter by granular-activated carbon adsorption as a pretreatment to reverse osmosis of membrane bioreactor effluents. *Water Res.* **2008**, *42*, 1595–1605. [CrossRef] [PubMed]

148. Abegglen, C.; Joss, A.; Boehler, M.; Buetzer, S.; Siegrist, H. Reducing the natural color of membrane bioreactor permeate with activated carbon or ozone. *Water Sci. Technol.* **2009**, *60*, 155–165. [CrossRef] [PubMed]
149. Guo, J.F.; Xia, S.Q.; Lu, Y.J. Characteristics of combined submerged membrane bioreactor with granular activated carbon (GAC) in treating lineal alkylbenzene sulphonates (LAS) wastewater. *AIP Conf. Proc.* **2010**, *1251*, 65–68.
150. Hai, F.I.; Yamamoto, K.; Nakajima, F.; Fukushi, K. Bioaugmented membrane bioreactor (MBR) with a gac-packed zone for high rate textile wastewater treatment. *Water Res.* **2011**, *45*, 2199–2206. [CrossRef] [PubMed]
151. Pham, T.T.; Nguyen, V.A.; Van der Bruggen, B. Pilot-scale evaluation of gac adsorption using low-cost, high-performance materials for removal of pesticides and organic matter in drinking water production. *J. Environ. Eng.* **2013**, *139*, 958–965. [CrossRef]
152. Wu, B.; Yi, S.; Fane, A.G. Microbial behaviors involved in cake fouling in membrane bioreactors under different solids retention times. *Bioresour. Technol.* **2011**, *102*, 2511–2516. [CrossRef] [PubMed]
153. Judd, S.; Judd, C. Commercial technologies. In *The Mbr Book: Principles and Applications of Membrane Bioreactors for Water and Wastewater Treatment*, 1st ed.; Elsevier: Amsterdam, The Netherlands; Boston, MA, USA; London, UK, 2006; Chapter 4, pp. 163–205.
154. Zamani, F.; Ullah, A.; Akhondi, E.; Tanudjaja, H.J.; Cornelissen, E.R.; Honciuc, A.; Fane, A.G.; Chew, J.W. Impact of the surface energy of particulate foulants on membrane fouling. *J. Membr. Sci.* **2016**, *510*, 101–111. [CrossRef]
155. Botton, S.; Verliefde, A.R.D.; Quach, N.T.; Cornelissen, E.R. Influence of biofouling on pharmaceuticals rejection in nf membrane filtration. *Water Res.* **2012**, *46*, 5848–5860. [CrossRef] [PubMed]
156. Cornelissen, E.R.; van den Boomgaard, T.; Strathmann, H. Physicochemical aspects of polymer selection for ultrafiltration and microfiltration membranes. *Colloids Surf. A Physicochem. Eng. Asp.* **1998**, *138*, 283–289. [CrossRef]
157. Van Oss, C.J. Introduction. In *Interfacial Forces in Aqueous Media*, 2nd ed.; CRC Press: Boca Raton, FL, USA, 2006; Chapter 1.
158. Field, R.W.; Pearce, G.K. Critical, sustainable and threshold fluxes for membrane filtration with water industry applications. *Adv. Colloid Interface Sci.* **2011**, *164*, 38–44. [CrossRef] [PubMed]
159. Bildyukevich, A.V.; Plisko, T.V.; Liubimova, A.S.; Volkov, V.V.; Usosky, V.V. Hydrophilization of polysulfone hollow fiber membranes via addition of polyvinylpyrrolidone to the bore fluid. *J. Membr. Sci.* **2017**, *524*, 537–549. [CrossRef]
160. Akhondi, E.; Zamani, F.; Law, A.W.K.; Krantz, W.B.; Fane, A.G.; Chew, J.W. Influence of backwashing on the pore size of hollow fiber ultrafiltration membranes. *J. Membr. Sci.* **2017**, *521*, 33–42. [CrossRef]
161. Rana, D.; Matsuura, T. Surface modifications for antifouling membranes. *Chem. Rev.* **2010**, *110*, 2448–2471. [CrossRef] [PubMed]
162. Marshall, A.D.; Munro, P.A.; Trägårdh, G. The effect of protein fouling in microfiltration and ultrafiltration on permeate flux, protein retention and selectivity: A literature review. *Desalination* **1993**, *91*, 65–108. [CrossRef]
163. Pasmore, M.; Todd, P.; Smith, S.; Baker, D.; Silverstein, J.; Coons, D.; Bowman, C.N. Effects of ultrafiltration membrane surface properties on pseudomonas aeruginosa biofilm initiation for the purpose of reducing biofouling. *J. Membr. Sci.* **2001**, *194*, 15–32. [CrossRef]
164. Brant, J.A.; Childress, A.E. Colloidal adhesion to hydrophilic membrane surfaces. *J. Membr. Sci.* **2004**, *241*, 235–248. [CrossRef]
165. Kochkodan, V.; Hilal, N. A comprehensive review on surface modified polymer membranes for biofouling mitigation. *Desalination* **2015**, *356*, 187–207. [CrossRef]
166. Reddy, A.V.R.; Mohan, D.J.; Bhattacharya, A.; Shah, V.J.; Ghosh, P.K. Surface modification of ultrafiltration membranes by preadsorption of a negatively charged polymer: I. Permeation of water soluble polymers and inorganic salt solutions and fouling resistance properties. *J. Membr. Sci.* **2003**, *214*, 211–221. [CrossRef]
167. Wang, D.; Li, K.; Teo, W.K. Preparation and characterization of polyvinylidene fluoride (PVDF) hollow fiber membranes. *J. Membr. Sci.* **1999**, *163*, 211–220. [CrossRef]
168. Kochkodan, V.; Johnson, D.J.; Hilal, N. Polymeric membranes: Surface modification for minimizing (bio) colloidal fouling. *Adv. Colloid Interface Sci.* **2014**, *206*, 116–140. [CrossRef] [PubMed]
169. Bernardes, P.C.; De Andrade, N.J.; Da Silva, L.H.M.; De Carvalho, A.F.; Fernandes, P.E.; Araújo, E.A.;

Lelis, C.A.; Mol, P.C.G.; De Sá, J.P.N. Modification of polysulfone membrane used in the water filtration process to reduce biofouling. *J. Nanosci. Nanotechnol.* **2014**, *14*, 6355–6367. [CrossRef] [PubMed]

170. Zhu, X.; Loo, H.-E.; Bai, R. A novel membrane showing both hydrophilic and oleophobic surface properties and its non-fouling performances for potential water treatment applications. *J. Membr. Sci.* **2013**, *436*, 47–56. [CrossRef]
171. Lee, W.; Ahn, C.H.; Hong, S.; Kim, S.; Lee, S.; Baek, Y.; Yoon, J. Evaluation of surface properties of reverse osmosis membranes on the initial biofouling stages under no filtration condition. *J. Membr. Sci.* **2010**, *351*, 112–122. [CrossRef]
172. Ulbricht, M.; Richau, K.; Kamusewitz, H. Chemically and morphologically defined ultrafiltration membrane surfaces prepared by heterogeneous photo-initiated graft polymerization. *Colloids Surf. A Physicochem. Eng. Asp.* **1998**, *138*, 353–366. [CrossRef]
173. Susanto, H.; Ulbricht, M. Photografted thin polymer hydrogel layers on pes ultrafiltration membranes: Characterization, stability, and influence on separation performance. *Langmuir* **2007**, *23*, 7818–7830. [CrossRef] [PubMed]
174. Yu, L.-Y.; Xu, Z.-L.; Shen, H.-M.; Yang, H. Preparation and characterization of PVDF–SiO_2 composite hollow fiber uf membrane by sol–gel method. *J. Membr. Sci.* **2009**, *337*, 257–265. [CrossRef]
175. Cui, A.; Liu, Z.; Xiao, C.; Zhang, Y. Effect of micro-sized SiO_2-particle on the performance of PVDF blend membranes via tips. *J. Membr. Sci.* **2010**, *360*, 259–264. [CrossRef]
176. Yan, L.; Li, Y.S.; Xiang, C.B. Preparation of poly (vinylidene fluoride)(PVDF) ultrafiltration membrane modified by nano-sized alumina (Al_2O_3) and its antifouling research. *Polymer* **2005**, *46*, 7701–7706. [CrossRef]
177. Yan, L.; Li, Y.S.; Xiang, C.B.; Xianda, S. Effect of nano-sized al2o3-particle addition on PVDF ultrafiltration membrane performance. *J. Membr. Sci.* **2006**, *276*, 162–167. [CrossRef]
178. Koh, M.J.; Hwang, H.Y.; Kim, D.J.; Kim, H.J.; Hong, Y.T.; Nam, S.Y. Preparation and characterization of porous PVDF-hfp/clay nanocomposite membranes. *J. Mater. Sci. Technol.* **2010**, *26*, 633–638. [CrossRef]
179. Rajabi, H.; Ghaemi, N.; Madaeni, S.S.; Daraei, P.; Khadivi, M.A.; Falsafi, M. Nanoclay embedded mixed matrix PVDF nanocomposite membrane: Preparation, characterization and biofouling resistance. *Appl. Surf. Sci.* **2014**, *313*, 207–214. [CrossRef]
180. Shen, X.; Xie, T.; Wang, J.; Liu, P.; Wang, F. An anti-fouling poly (vinylidene fluoride) hybrid membrane blended with functionalized ZrO_2 nanoparticles for efficient oil/water separation. *RSC Adv.* **2017**, *7*, 5262–5271. [CrossRef]
181. Yang, X.; He, Y.; Zeng, G.; Zhan, Y.; Pan, Y.; Shi, H.; Chen, Q. Novel hydrophilic PVDF ultrafiltration membranes based on a ZrO_2. *J. Mater. Sci.* **2016**, *51*, 8965–8976. [CrossRef]
182. Yang, H.C.; Hou, J.; Chen, V.; Xu, Z.K. Surface and interface engineering for organic-inorganic composite membranes. *J. Mater. Chem. A* **2016**, *4*, 9716–9729. [CrossRef]
183. Oh, S.J.; Kim, N.; Lee, Y.T. Preparation and characterization of PVDF/TiO_2 organic–inorganic composite membranes for fouling resistance improvement. *J. Membr. Sci.* **2009**, *345*, 13–20. [CrossRef]
184. Wu, H.; Liu, Y.; Mao, L.; Jiang, C.; Ang, J.; Lu, X. Doping polysulfone ultrafiltration membrane with TiO_2-PDA nanohybrid for simultaneous self-cleaning and self-protection. *J. Membr. Sci.* **2017**, *532*, 20–29. [CrossRef]
185. Hong, J.; He, Y. Effects of nano sized zinc oxide on the performance of PVDF microfiltration membranes. *Desalination* **2012**, *302*, 71–79. [CrossRef]
186. Liang, S.; Xiao, K.; Mo, Y.; Huang, X. A novel zno nanoparticle blended polyvinylidene fluoride membrane for anti-irreversible fouling. *J. Membr. Sci.* **2012**, *394*, 184–192. [CrossRef]
187. Chang, S.; Fane, A.G. The effect of fibre diameter on filtration and flux distribution—Relevance to submerged hollow fibre modules. *J. Membr. Sci.* **2001**, *184*, 221–231. [CrossRef]
188. Fulton, B.G.; Bérubé, P.R. Optimizing the sparging condition and membrane module spacing for a ZW500 submerged hollow fiber membrane system. *Desalination Water Treat.* **2012**, *42*, 8–16. [CrossRef]
189. Shimizu, Y.; Okuno, Y.-I.; Uryu, K.; Ohtsubo, S.; Watanabe, A. Filtration characteristics of hollow fiber microfiltration membranes used in membrane bioreactor for domestic wastewater treatment. *Water Res.* **1996**, *30*, 2385–2392. [CrossRef]
190. Yang, X.; Fridjonsson, E.O.; Johns, M.L.; Wang, R.; Fane, A.G. A non-invasive study of flow dynamics in membrane distillation hollow fiber modules using low-field nuclear magnetic resonance imaging (MRI). *J. Membr. Sci.* **2014**, *451*, 46–54. [CrossRef]

191. Bérubé, P.R.; Lei, E. The effect of hydrodynamic conditions and system configurations on the permeate flux in a submerged hollow fiber membrane system. *J. Membr. Sci.* **2006**, *271*, 29–37. [CrossRef]
192. Liu, X.; Wang, Y.; Waite, T.D.; Leslie, G. Numerical simulations of impact of membrane module design variables on aeration patterns in membrane bioreactors. *J. Membr. Sci.* **2016**, *520*, 201–213. [CrossRef]
193. Liu, X.; Wang, Y.; Waite, T.D.; Leslie, G. Fluid structure interaction analysis of lateral fibre movement in submerged membrane reactors. *J. Membr. Sci.* **2016**, *504*, 240–250. [CrossRef]
194. Yeo, A.P.S.; Law, A.W.K.; Fane, A.G. The relationship between performance of submerged hollow fibers and bubble-induced phenomena examined by particle image velocimetry. *J. Membr. Sci.* **2007**, *304*, 125–137. [CrossRef]
195. Kim, J.; DiGiano, F.A. Particle fouling in submerged microfiltration membranes: Effects of hollow-fiber length and aeration rate. *J. Water Supply Res. Technol.-Aqua* **2006**, *55*, 535–547. [CrossRef]
196. Lee, M.; Kim, J. Analysis of local fouling in a pilot-scale submerged hollow-fiber membrane system for drinking water treatment by membrane autopsy. *Sep. Purif. Technol.* **2012**, *95*, 227–234. [CrossRef]
197. Li, X.; Li, J.; Cui, Z.; Yao, Y. Modeling of filtration characteristics during submerged hollow fiber membrane microfiltration of yeast suspension under aeration condition. *J. Membr. Sci.* **2016**, *510*, 455–465. [CrossRef]
198. Lin, C.-J.; Rao, P.; Shirazi, S. Effect of operating parameters on permeate flux decline caused by cake formation—A model study. *Desalination* **2005**, *171*, 95–105. [CrossRef]
199. Pradhan, M.; Aryal, R.; Vigneswaran, S.; Kandasamy, J. Application of air flow for mitigation of particle deposition in submerged membrane microfiltration. *Desalination Water Treat.* **2011**, *32*, 201–207. [CrossRef]
200. Visvanathan, C.; Ben Aim, R.; Parameshwaran, K. Membrane separation bioreactors for wastewater treatment. *Crit. Rev. Environ. Sci. Technol.* **2000**, *30*, 1–48. [CrossRef]
201. Du, X.; Qu, F.-S.; Liang, H.; Li, K.; Bai, L.-M.; Li, G.-B. Control of submerged hollow fiber membrane fouling caused by fine particles in photocatalytic membrane reactors using bubbly flow: Shear stress and particle forces analysis. *Sep. Purif. Technol.* **2017**, *172*, 130–139. [CrossRef]
202. Yeo, B.J.L.; Goh, S.; Livingston, A.G.; Fane, A.G. Controlling biofilm development in the extractive membrane bioreactor. *Sep. Sci. Technol.* **2017**, *52*, 113–121. [CrossRef]
203. Cabassud, C.; Laborie, S.; Durand-Bourlier, L.; Lainé, J.M. Air sparging in ultrafiltration hollow fibers: Relationship between flux enhancement, cake characteristics and hydrodynamic parameters. *J. Membr. Sci.* **2001**, *181*, 57–69. [CrossRef]
204. Wang, J.; Fane, A.G.; Chew, J.W. Effect of bubble characteristics on critical flux in the microfiltration of particulate foulants. *J. Membr. Sci.* **2017**, *535*, 279–293. [CrossRef]
205. Liu, N.; Zhang, Q.; Chin, G.-L.; Ong, E.-H.; Lou, J.; Kang, C.-W.; Liu, W.; Jordan, E. Experimental investigation of hydrodynamic behavior in a real membrane bio-reactor unit. *J. Membr. Sci.* **2010**, *353*, 122–134. [CrossRef]
206. Nguyen Cong Duc, E.; Fournier, L.; Levecq, C.; Lesjean, B.; Grelier, P.; Tazi-Pain, A. Local hydrodynamic investigation of the aeration in a submerged hollow fibre membranes cassette. *J. Membr. Sci.* **2008**, *321*, 264–271. [CrossRef]
207. Fulton, B.G.; Redwood, J.; Tourais, M.; Bérubé, P.R. Distribution of surface shear forces and bubble characteristics in full-scale gas sparged submerged hollow fiber membrane modules. *Desalination* **2011**, *281*, 128–141. [CrossRef]
208. Buetehorn, S.; Volmering, D.; Vossenkaul, K.; Wintgens, T.; Wessling, M.; Melin, T. CFD simulation of single- and multi-phase flows through submerged membrane units with irregular fiber arrangement. *J. Membr. Sci.* **2011**, *384*, 184–197. [CrossRef]
209. Bouhabila, E.H.; Ben Aïm, R.; Buisson, H. Microfiltration of activated sludge using submerged membrane with air bubbling (application to wastewater treatment). *Desalination* **1998**, *118*, 315–322. [CrossRef]
210. Ueda, T.; Hata, K.; Kikuoka, Y.; Seino, O. Effects of aeration on suction pressure in a submerged membrane bioreactor. *Water Res.* **1997**, *31*, 489–494. [CrossRef]
211. Chang, S.; Fane, A.G. Filtration of biomass with axial inter-fibre upward slug flow: Performance and mechanisms. *J. Membr. Sci.* **2000**, *180*, 57–68. [CrossRef]
212. Qaisrani, T.M.; Samhaber, W.M. Impact of gas bubbling and backflushing on fouling control and membrane cleaning. *Desalination* **2011**, *266*, 154–161. [CrossRef]
213. Chan, C.C.V.; Bérubé, P.R.; Hall, E.R. Relationship between types of surface shear stress profiles and membrane fouling. *Water Res.* **2011**, *45*, 6403–6416. [CrossRef] [PubMed]

214. Xia, L.; Law, A.W.-K.; Fane, A.G. Hydrodynamic effects of air sparging on hollow fiber membranes in a bubble column reactor. *Water Res.* **2013**, *47*, 3762–3772. [CrossRef] [PubMed]
215. Yeom, I.-T.; Nah, Y.-M.; Ahn, K.-H. Treatment of household wastewater using an intermittently aerated membrane bioreactor. *Desalination* **1999**, *124*, 193–203. [CrossRef]
216. Guibert, D.; Aim, R.B.; Rabie, H.; Côté, P. Aeration performance of immersed hollow-fiber membranes in a bentonite suspension. *Desalination* **2002**, *148*, 395–400. [CrossRef]
217. Tung, K.-L.; Damodar, H.-R.; Damodar, R.-A.; Tsai, J.-H.; Chen, C.-H.; You, S.-J.; Huang, M.-S. Imaging the effect of aeration on particle fouling mitigation in a submerged membrane filtration using a photointerrupt sensor array. *Sep. Sci. Technol.* **2017**, *52*, 228–239. [CrossRef]
218. Jaffrin, M.Y. Dynamic shear-enhanced membrane filtration: A review of rotating disks, rotating membranes and vibrating systems. *J. Membr. Sci.* **2008**, *324*, 7–25. [CrossRef]
219. Pourbozorg, M.; Li, T.; Law, A.W.K. Effect of turbulence on fouling control of submerged hollow fibre membrane filtration. *Water Res.* **2016**, *99*, 101–111. [CrossRef] [PubMed]
220. Kola, A.; Ye, Y.; Ho, A.; Le-Clech, P.; Chen, V. Application of low frequency transverse vibration on fouling limitation in submerged hollow fibre membranes. *J. Membr. Sci.* **2012**, *409–410*, 54–65. [CrossRef]
221. Kola, A.; Ye, Y.; Le-Clech, P.; Chen, V. Transverse vibration as novel membrane fouling mitigation strategy in anaerobic membrane bioreactor applications. *J. Membr. Sci.* **2014**, *455*, 320–329. [CrossRef]
222. Low, S.C.; Juan, H.H.; Siong, L.K. A combined VSEP and membrane bioreactor system. *Desalination* **2005**, *183*, 353–362. [CrossRef]
223. Li, T.; Law, A.W.-K.; Fane, A.G. Submerged hollow fibre membrane filtration with transverse and longitudinal vibrations. *J. Membr. Sci.* **2014**, *455*, 83–91. [CrossRef]
224. Genkin, G.; Waite, T.D.; Fane, A.G.; Chang, S. The effect of vibration and coagulant addition on the filtration performance of submerged hollow fibre membranes. *J. Membr. Sci.* **2006**, *281*, 726–734. [CrossRef]
225. Prip Beier, S.; Jonsson, G. A vibrating membrane bioreactor (VMBR): Macromolecular transmission—Influence of extracellular polymeric substances. *Chem. Eng. Sci.* **2009**, *64*, 1436–1444. [CrossRef]
226. Pourbozorg, M.; Li, T.; Law, A.W.K. Fouling of submerged hollow fiber membrane filtration in turbulence: Statistical dependence and cost-benefit analysis. *J. Membr. Sci.* **2017**, *521*, 43–52. [CrossRef]
227. Ruigomez, I.; Gonzalez, E.; Guerra, S.; Rodriguez-Gomez, L.E.; Vera, L. Evaluation of a novel physical cleaning strategy based on hf membrane rotation during the backwashing/relaxation phases for anaerobic submerged mbr. *J. Membr. Sci.* **2017**, *526*, 181–190. [CrossRef]
228. Chatzikonstantinou, K.; Tzamtzis, N.; Aretakis, N.; Pappa, A. The effect of various high-frequency powerful vibration (HFPV) types on fouling control of hollow fiber membrane elements in a small pilot-scale smbr system. *Desalination Water Treat.* **2016**, *57*, 27905–27913. [CrossRef]
229. Beier, S.P.; Guerra, M.; Garde, A.; Jonsson, G. Dynamic microfiltration with a vibrating hollow fiber membrane module: Filtration of yeast suspensions. *J. Membr. Sci.* **2006**, *281*, 281–287. [CrossRef]
230. Beier, S.P.; Jonsson, G. Dynamic microfiltration with a vibrating hollow fiber membrane module. *Desalination* **2006**, *199*, 499–500. [CrossRef]
231. Beier, S.P.; Jonsson, G. Separation of enzymes and yeast cells with a vibrating hollow fiber membrane module. *Sep. Purif. Technol.* **2007**, *53*, 111–118. [CrossRef]
232. Zamani, F.; Law, A.W.K.; Fane, A.G. Hydrodynamic analysis of vibrating hollow fibre membranes. *J. Membr. Sci.* **2013**, *429*, 304–312. [CrossRef]
233. Krantz, W.B.; Bilodeau, R.R.; Voorhees, M.E.; Elgas, R.J. Use of axial membrane vibrations to enhance mass transfer in a hollow tube oxygenator. *J. Membr. Sci.* **1997**, *124*, 283–299. [CrossRef]
234. Epstein, N. Liquid–solids fluidization. In *Handbook of Fluidization and Fluid-Particle Systems*; Yang, W.-C., Ed.; Marcel Dekker: New York, NY, USA, 2003.
235. Kunii, D.; Levenspiel, O. *Fluidization Engineering*; Butterworth-Heinemann: Newton, MA, USA, 1991.
236. Bixler, H.J.; Rappe, G.C. Ultrafiltration Process. U.S. Patent 3,541,006, 17 November 1970.
237. Lowe, E.; Durkee, E.L. Dynamic turbulence promotion in reverse osmosis processing of liquid foods. *J. Food Sci.* **1971**, *36*, 31–32. [CrossRef]
238. Van der Waal, M.J.; van der Velden, P.M.; Koning, J.; Smolders, C.A.; Vanswaay, W.P.M. Use of fluidized-beds as turbulence promotors in tubular membrane systems. *Desalination* **1977**, *22*, 465–483. [CrossRef]
239. Hamer, E.A.G. Semipermeable Membrane Cleaning Means. U.S. Patent 3,425,562, 4 February 1969.

240. Wei, C.H.; Huang, X.; Wang, C.W.; Wen, X.H. Effect of a suspended carrier on membrane fouling in a submerged membrane bioreactor. *Water Sci. Technol.* **2006**, *53*, 211–220. [CrossRef] [PubMed]
241. Yang, Q.Y.; Chen, J.H.; Zhang, F. Membrane fouling control in a submerged membrane bioreactor with porous, flexible suspended carriers. *Desalination* **2006**, *189*, 292–302. [CrossRef]
242. De Boer, R.; Zomerman, J.J.; Hiddink, J.; Aufderheyde, J.; Vanswaay, W.P.M.; Smolders, C.A. Fluidized-beds as turbulence promoters in the concentration of food liquids by reverse-osmosis. *J. Food Sci.* **1980**, *45*, 1522–1528. [CrossRef]
243. Zhong, Z.X.; Liu, X.; Chen, R.Z.; Xing, W.H.; Xu, N.P. Adding microsized silica particles to the catalysis/ultrafiltration system: Catalyst dissolution inhibition and flux enhancement. *Ind. Eng. Chem. Res.* **2009**, *48*, 4933–4938. [CrossRef]
244. Aslam, M.; McCarty, P.L.; Bae, J.; Kim, J. The effect of fluidized media characteristics on membrane fouling and energy consumption in anaerobic fluidized membrane bioreactors. *Sep. Purif. Technol.* **2014**, *132*, 10–15. [CrossRef]
245. Urbain, V.; Benoit, R.; Manem, J. Membrane bioreactor: A new treatment tool. *J. Am. Water Works Assoc.* **1996**, *88*, 75–86.
246. Williams, M.D.; Pirbazari, M. Membrane bioreactor process for removing biodegradable organic matter from water. *Water Res.* **2007**, *41*, 3880–3893. [CrossRef] [PubMed]
247. Li, Y.Z.; He, Y.L.; Liu, Y.H.; Yang, S.C.; Zhang, G.J. Comparison of the filtration characteristics between biological powdered activated carbon sludge and activated sludge in submerged membrane bioreactors. *Desalination* **2005**, *174*, 305–314. [CrossRef]
248. Akram, A.; Stuckey, D.C. Flux and performance improvement in a submerged anaerobic membrane bioreactor (SAMBR) using powdered activated carbon (PAC). *Process Biochem.* **2008**, *43*, 93–102. [CrossRef]
249. Satyawali, Y.; Balakrishnan, M. Effect of pac addition on sludge properties in an mbr treating high strength wastewater. *Water Res.* **2009**, *43*, 1577–1588. [CrossRef] [PubMed]
250. Ng, C.A.; Sun, D.; Fane, A.G. Operation of membrane bioreactor with powdered activated carbon addition. *Sep. Sci. Technol.* **2006**, *41*, 1447–1466.
251. Kim, J.S.; Lee, C.H. Effect of powdered activated carbon on the performance of an aerobic membrane bioreactor: Comparison between cross-flow and submerged membrane systems. *Water Environ. Res.* **2003**, *75*, 300–307. [CrossRef] [PubMed]
252. Munz, G.; Gori, R.; Mori, G.; Lubello, C. Powdered activated carbon and membrane bioreactors (MBRPAC) for tannery wastewater treatment: Long term effect on biological and filtration process performances. *Desalination* **2007**, *207*, 349–360. [CrossRef]
253. Tian, J.Y.; Liang, H.; Yang, Y.L.; Tian, S.; Li, G.B. Membrane adsorption bioreactor (MABR) for treating slightly polluted surface water supplies: As compared to membrane bioreactor (MBR). *J. Membr. Sci.* **2008**, *325*, 262–270. [CrossRef]
254. Park, H.; Choo, K.H.; Lee, C.H. Flux enhancement with powdered activated carbon addition in the membrane anaerobic bioreactor. *Sep. Sci. Technol.* **1999**, *34*, 2781–2792. [CrossRef]
255. Remy, M.; Potier, V.; Temmink, H.; Rulkens, W. Why low powdered activated carbon addition reduces membrane fouling in mbrs. *Water Res.* **2010**, *44*, 861–867. [CrossRef] [PubMed]
256. Ng, C.A.; Sun, D.; Zhang, J.S.; Wu, B.; Fane, A.G. Mechanisms of fouling control in membrane bioreactors by the addition of powdered activated carbon. *Sep. Sci. Technol.* **2010**, *45*, 873–889. [CrossRef]
257. Hu, A.Y.; Stuckey, D.C. Activated carbon addition to a submerged anaerobic membrane bioreactor: Effect on performance, transmembrane pressure, and flux. *J. Environ. Eng.* **2007**, *133*, 73–80. [CrossRef]
258. Yoo, R.; Kim, J.; McCarty, P.L.; Bae, J. Anaerobic treatment of municipal wastewater with a staged anaerobic fluidized membrane bioreactor (SAF-MBR) system. *Bioresour. Technol.* **2012**, *120*, 133–139. [CrossRef] [PubMed]
259. Bae, J.; Yoo, R.; Lee, E.; McCarty, P.L. Two-stage anaerobic fluidized-bed membrane bioreactor treatment of settled domestic wastewater. *Water Sci. Technol.* **2013**, *68*, 394–399. [CrossRef] [PubMed]
260. Yoo, R.H.; Kim, J.H.; McCarty, P.L.; Bae, J.H. Effect of temperature on the treatment of domestic wastewater with a staged anaerobic fluidized membrane bioreactor. *Water Sci. Technol.* **2014**, *69*, 1145–1150. [CrossRef] [PubMed]
261. Shin, C.; McCarty, P.L.; Kim, J.; Bae, J. Pilot-scale temperate-climate treatment of domestic wastewater with a staged anaerobic fluidized membrane bioreactor (SAF-MBR). *Bioresour. Technol.* **2014**, *159*, 95–103. [CrossRef] [PubMed]

262. Gao, D.W.; Hu, Q.; Yao, C.; Ren, N.Q.; Wu, W.M. Integrated anaerobic fluidized-bed membrane bioreactor for domestic wastewater treatment. *Chem. Eng. J.* **2014**, *240*, 362–368. [CrossRef]
263. Li, J.; Ge, Z.; He, Z. A fluidized bed membrane bioelectrochemical reactor for energy-efficient wastewater treatment. *Bioresour. Technol.* **2014**, *167*, 310–315. [CrossRef] [PubMed]
264. Ren, L.J.; Ahn, Y.; Logan, B.E. A two-stage microbial fuel cell and anaerobic fluidized bed membrane bioreactor (MFC-AFMBR) system for effective domestic wastewater treatment. *Environ. Sci. Technol.* **2014**, *48*, 4199–4206. [CrossRef] [PubMed]
265. Li, J.; Luo, S.; He, Z. Cathodic fluidized granular activated carbon assisted-membrane bioelectrochemical reactor for wastewater treatment. *Sep. Purif. Technol.* **2016**, *169*, 241–246. [CrossRef]
266. Cahyadi, A.; Yang, S.; Chew, J.W. CFD study on the hydrodynamics of fluidized granular activated carbon in anfmbr applications. *Sep. Purif. Technol.* **2017**, *178*, 75–89. [CrossRef]
267. Wang, J.; Wu, B.; Yang, S.; Liu, Y.; Fane, A.G.; Chew, J.W. Characterizing the scouring efficiency of granular activated carbon (GAC) particles in membrane fouling mitigation via wavelet decomposition of accelerometer signals. *J. Membr. Sci.* **2016**, *498*, 105–115. [CrossRef]
268. Wang, J.; Zamani, F.; Cahyadi, A.; Toh, J.Y.; Yang, S.; Wu, B.; Liu, Y.; Fane, A.G.; Chew, J.W. Correlating the hydrodynamics of fluidized granular activated carbon (GAC) with membrane-fouling mitigation. *J. Membr. Sci.* **2016**, *510*, 38–49. [CrossRef]
269. Wang, J.; Wu, B.; Liu, Y.; Fane, A.G.; Chew, J.W. Effect of fluidized granular activated carbon (GAC) on critical flux in the microfiltration of particulate foulants. *J. Membr. Sci.* **2017**, *523*, 409–417. [CrossRef]
270. Wu, B.; Zamani, F.; Lim, W.; Liao, D.; Wang, Y.; Liu, Y.; Chew, J.W.; Fane, A.G. Effect of mechanical scouring by granular activated carbon (GAC) on membrane fouling mitigation. *Desalination* **2017**, *403*, 80–87. [CrossRef]
271. Wu, B.; Wang, Y.; Lim, W.; Chew, J.W.; Fane, A.G.; Liu, Y. Enhanced performance of submerged hollow fibre microfiltration by fluidized granular activated carbon. *J. Membr. Sci.* **2016**, *499*, 47–55. [CrossRef]
272. Shin, C.; Kim, K.; McCarty, P.L.; Kim, J.; Bae, J. Integrity of hollow-fiber membranes in a pilot-scale anaerobic fluidized membrane bioreactor (AFMBR) after two-years of operation. *Sep. Purif. Technol.* **2016**, *162*, 101–105. [CrossRef]
273. Aslam, M.; McCarty, P.L.; Shin, C.; Bae, J.; Kim, J. Low energy single-staged anaerobic fluidized bed ceramic membrane bioreactor (AFCMBR) for wastewater treatment. *Bioresour. Technol.* **2017**, *240*, 33–41. [CrossRef] [PubMed]
274. Charfi, A.; Aslam, M.; Lesage, G.; Heran, M.; Kim, J. Macroscopic approach to develop fouling model under gac fluidization in anaerobic fluidized bed membrane bioreactor. *J. Ind. Eng. Chem.* **2017**, *49*, 219–229. [CrossRef]
275. Wu, B.; Wong, P.C.Y.; Fane, A.G. The potential roles of granular activated carbon in anaerobic fluidized membrane bioreactors: Effect on membrane fouling and membrane integrity. *Desalination Water Treat.* **2015**, *53*, 1450–1459. [CrossRef]
276. Wu, B.; Li, Y.; Lim, W.; Lee, S.L.; Guo, Q.; Fane, A.G.; Liu, Y. Single-stage versus two-stage anaerobic fluidized bed bioreactors in treating municipal wastewater: Performance, foulant characteristics, and microbial community. *Chemosphere* **2017**, *171*, 158–167. [CrossRef] [PubMed]
277. Gao, D.W.; Hu, Q.; Yao, C.; Ren, N.Q. Treatment of domestic wastewater by an integrated anaerobic fluidized-bed membrane bioreactor under moderate to low temperature conditions. *Bioresour. Technol.* **2014**, *159*, 193–198. [CrossRef] [PubMed]
278. McCurry, D.L.; Bear, S.E.; Bae, J.; Sedlak, D.L.; McCarty, P.L.; Mitch, W.A. Superior removal of disinfection byproduct precursors and pharmaceuticals from wastewater in a staged anaerobic fluidized membrane bioreactor compared to activated sludge. *Environ. Sci. Technol. Lett.* **2014**, *1*, 459–464. [CrossRef]
279. Düppenbecker, B.; Engelhart, M.; Cornel, P. Fouling mitigation in anaerobic membrane bioreactor using fluidized glass beads: Evaluation fitness for purpose of ceramic membranes. *J. Membr. Sci.* **2017**, *537*, 69–82. [CrossRef]
280. Aslam, M.; Charfi, A.; Kim, J. Membrane scouring to control fouling under fluidization of non-adsorbing media for wastewater treatment. *Environ. Sci. Pollut. Res.* **2017**, 1–11. [CrossRef] [PubMed]
281. McCarty, P.L.; Bae, J.; Kim, J. Domestic wastewater treatment as a net energy producer—Can this be achieved? *Environ. Sci. Technol.* **2011**, *45*, 7100–7106. [CrossRef] [PubMed]
282. Ohn, T.; Jami, M.S.; Iritani, E.; Mukai, Y.; Katagiri, N. Filtration behaviors in constant rate microfiltration with cyclic backwashing of coagulated sewage secondary effluent. *Sep. Sci. Technol.* **2003**, *38*, 951–966. [CrossRef]

283. Lin, J.C.T.; Lee, D.J.; Huang, C. Membrane fouling mitigation: Membrane cleaning. *Sep. Sci. Technol.* **2010**, *45*, 858–872. [CrossRef]
284. Ma, H.; Bowman, C.N.; Davis, R.H. Membrane fouling reduction by backpulsing and surface modification. *J. Membr. Sci.* **2000**, *173*, 191–200. [CrossRef]
285. Ma, H.; Hakim, L.F.; Bowman, C.N.; Davis, R.H. Factors affecting membrane fouling reduction by surface modification and backpulsing. *J. Membr. Sci.* **2001**, *189*, 255–270. [CrossRef]
286. De Souza, N.P.; Basu, O.D. Comparative analysis of physical cleaning operations for fouling control of hollow fiber membranes in drinking water treatment. *J. Membr. Sci.* **2013**, *436*, 28–35. [CrossRef]
287. Ferrer, O.; Lefèvre, B.; Prats, G.; Bernat, X.; Gibert, O.; Paraira, M. Reversibility of fouling on ultrafiltration membrane by backwashing and chemical cleaning: Differences in organic fractions behaviour. *Desalination Water Treat.* **2016**, *57*, 8593–8607. [CrossRef]
288. Wu, J.; Le-Clech, P.; Stuetz, R.M.; Fane, A.G.; Chen, V. Effects of relaxation and backwashing conditions on fouling in membrane bioreactor. *J. Membr. Sci.* **2008**, *324*, 26–32. [CrossRef]
289. Bessiere, Y.; Guigui, C.; Remize, P.J.; Cabassud, C. Coupling air-assisted backwash and rinsing steps: A new way to improve ultrafiltration process operation for inside-out hollow fibre modules. *Desalination* **2009**, *240*, 71–77. [CrossRef]
290. Wu, J.; Le-Clech, P.; Stuetz, R.M.; Fane, A.G.; Chen, V. Novel filtration mode for fouling limitation in membrane bioreactors. *Water Res.* **2008**, *42*, 3677–3684. [CrossRef] [PubMed]
291. Ye, Y.; Chen, V.; Le-Clech, P. Evolution of fouling deposition and removal on hollow fibre membrane during filtration with periodical backwash. *Desalination* **2011**, *283*, 198–205. [CrossRef]
292. Howe, K.J.; Marwah, A.; Chiu, K.-P.; Adham, S.S. Effect of membrane configuration on bench-scale mf and uf fouling experiments. *Water Res.* **2007**, *41*, 3842–3849. [CrossRef] [PubMed]
293. Akhondi, E.; Zamani, F.; Chew, J.W.; Krantz, W.B.; Fane, A.G. Improved design and protocol for evapoporometry determination of the pore-size distribution. *J. Membr. Sci.* **2015**, *496*, 334–343. [CrossRef]
294. Krantz, W.B.; Greenberg, A.R.; Kujundzic, E.; Yeo, A.; Hosseini, S.S. Evapoporometry: A novel technique for determining the pore-size distribution of membranes. *J. Membr. Sci.* **2013**, *438*, 153–166. [CrossRef]
295. Serra, C.; Durand-Bourlier, L.; Clifton, M.J.; Moulin, P.; Rouch, J.-C.; Aptel, P. Use of air sparging to improve backwash efficiency in hollow-fiber modules. *J. Membr. Sci.* **1999**, *161*, 95–113. [CrossRef]
296. Christensen, M.L.; Bugge, T.V.; Hede, B.H.; Nierychlo, M.; Larsen, P.; Jørgensen, M.K. Effects of relaxation time on fouling propensity in membrane bioreactors. *J. Membr. Sci.* **2016**, *504*, 176–184. [CrossRef]
297. Chang, I.S.; Le Clech, P.; Jefferson, B.; Judd, S. Membrane fouling in membrane bioreactors for wastewater treatment. *J. Environ. Eng.* **2002**, *128*, 1018–1029. [CrossRef]
298. Drews, A. Membrane fouling in membrane bioreactors-characterisation, contradictions, cause and cures. *J. Membr. Sci.* **2010**, *363*, 1–28. [CrossRef]
299. Meng, F.G.; Chae, S.R.; Drews, A.; Kraume, M.; Shin, H.S.; Yang, F.L. Recent advances in membrane bioreactors (MBRs): Membrane fouling and membrane material. *Water Res.* **2009**, *43*, 1489–1512. [CrossRef] [PubMed]
300. Chen, J.P.; Kim, S.L.; Ting, Y.P. Optimization of membrane physical and chemical cleaning by a statistically designed approach. *J. Membr. Sci.* **2003**, *219*, 27–45. [CrossRef]
301. Porcelli, N.; Judd, S. Effect of cleaning protocol on membrane permeability recovery: A sensitivity analysis. *J. Am. Water Works Assoc.* **2010**, *102*, 78–86.
302. Trägårdh, G. Membrane cleaning. *Desalination* **1989**, *71*, 325–335. [CrossRef]
303. Zhang, Z.; Bligh, M.W.; Wang, Y.; Leslie, G.L.; Bustamante, H.; Waite, T.D. Cleaning strategies for iron-fouled membranes from submerged membrane bioreactor treatment of wastewaters. *J. Membr. Sci.* **2015**, *475*, 9–21. [CrossRef]
304. Kweon, J.H.; Jung, J.H.; Lee, S.R.; Hur, H.W.; Shin, Y.; Choi, Y.H. Effects of consecutive chemical cleaning on membrane performance and surface properties of microfiltration. *Desalination* **2012**, *286*, 324–331. [CrossRef]
305. Cote, P.; Alam, Z.; Penny, J. Hollow fiber membrane life in membrane bioreactors (MBR). *Desalination* **2012**, *288*, 145–151. [CrossRef]
306. Huisman, I.H.; Williams, K. Autopsy and failure analysis of ultrafiltration membranes from a waste-water treatment system. *Desalination* **2004**, *165*, 161–164. [CrossRef]
307. De Wilde, W.; Thoeye, C.; De Gueldre, G. Membrane life expectancy assessment after 3 years of MBR

operation at WWTP schilde. In Proceedings of the 4th International Water Association Conference on Membranes for Water and Wastewater Treatment, Harrogate, UK, 15–17 May 2007.

308. Benavente, J.; Vázquez, M. Effect of age and chemical treatments on characteristic parameters for active and porous sublayers of polymeric composite membranes. *J. Colloid Interface Sci.* **2004**, *273*, 547–555. [CrossRef] [PubMed]
309. Thominette, F.; Farnault, O.; Gaudichet-Maurin, E.; Machinal, C.; Schrotter, J.-C. Ageing of polyethersulfone ultrafiltration membranes in hypochlorite treatment. *Desalination* **2006**, *200*, 7–8. [CrossRef]
310. Anton, E.; Alvarez, J.R.; Palacio, L.; Pradanos, P.; Hernandez, A.; Pihlajamaki, A.; Luque, S. Ageing of polyethersulfone ultrafiltration membranes under long-term exposures to alkaline and acidic cleaning solutions. *Chem. Eng. Sci.* **2015**, *134*, 178–195. [CrossRef]
311. Akhondi, E. Submerged Hollow Fibre Membrane Fouling: Characterization and Control. Ph.D. Thesis, Nanyang Technological University, Singapore, 2014.
312. Mallevialle, J.; Odendaal, P.E.; Foundation, A.R.; Wiesner, M.R.; eaux-Dumez, L.D.; Commission, S.A.W.R. *Water Treatment Membrane Processes*; McGraw-Hill: New York, NY, USA, 1996.
313. Childress, A.; Le-Clech, P.; Daugherty, J.; Chen, C.; Leslie, G. Mechanical analysis of hollow fiber membrane integrity in water reuse applications. *Desalination* **2005**, *180*, 5–14. [CrossRef]
314. Gijsbertsen-Abrahamse, A.J.; Cornelissen, E.R.; Hofman, J.A.M.H. Fiber failure frequency and causes of hollow fiber integrity loss. *Desalination* **2006**, *194*, 251–258. [CrossRef]
315. Johnson, W.; MacCormick, T. Issues of operational integrity in membrane drinking water plants. *Water Sci. Technol. Water Supply* **2003**, *3*, 73–80.
316. Arkhangelsky, E.; Kuzmenko, D.; Gitis, N.; Vinogradov, M.; Kuiry, S.; Gitis, V. Hypochlorite cleaning causes degradation of polymer membranes. *Tribol. Lett.* **2007**, *28*, 109–116. [CrossRef]
317. Antony, A.; Leslie, G. Degradation of polymeric membranes in water and wastewater treatment. In *Advanced Membrane Science and Technology for Sustainable Energy and Environmental Applications*; Woodhead Publishing Limited: Cambridge, UK, 2011; pp. 718–745.
318. Gaudichet-Maurin, E.; Thominette, F. Ageing of polysulfone ultrafiltration membranes in contact with bleach solutions. *J. Membr. Sci.* **2006**, *282*, 198–204. [CrossRef]
319. Wang, P.; Wang, Z.; Wu, Z.; Zhou, Q.; Yang, D. Effect of hypochlorite cleaning on the physiochemical characteristics of polyvinylidene fluoride membranes. *Chem. Eng. J.* **2010**, *162*, 1050–1056. [CrossRef]
320. Regula, C.; Carretier, E.; Wyart, Y.; Gésan-Guiziou, G.; Vincent, A.; Boudot, D.; Moulin, P. Chemical cleaning/disinfection and ageing of organic uf membranes: A review. *Water Res.* **2014**, *56*, 325–365. [CrossRef] [PubMed]
321. Zappia, L.R.; Hayes, D.; Nolan, P. Source water characterisation and implications for ultrafiltration in the east kimberly, western australia. In Proceedings of the AWA Membranes and Desalination Conference, Brisbane, Australia, 1 July 2013.
322. Chew, J.W.; Hays, R.; Findlay, J.G.; Knowlton, T.M.; Karri, S.B.R.; Cocco, R.A.; Hrenya, C.M. Reverse core-annular flow of geldart group b particles in risers. *Powder Technol.* **2012**, *221*, 1–12. [CrossRef]
323. King, D.H.; Smith, J.W. Wall mass transfer in liquid-fluidized beds. *Can. J. Chem. Eng.* **1967**, *45*, 329–333. [CrossRef]
324. Zhu, J.; Grace, J.R.; Lim, C.J. Tube wear in gas-fluidized beds. 1. Experimental findings. *Chem. Eng. Sci.* **1990**, *45*, 1003–1015. [CrossRef]
325. Zhu, J.; Lim, C.J.; Grace, J.R.; Lund, J.A. Tube wear in gas-fluidized beds. 2. Low velocity impact erosion and semiempirical model for bubbling and slugging fluidized-beds. *Chem. Eng. Sci.* **1991**, *46*, 1151–1156. [CrossRef]
326. Bethune, B. Surface cracking of glassy polymers under a sliding spherical indenter. *J. Mater. Sci.* **1976**, *11*, 199–205. [CrossRef]
327. Cicek, N.; Dionysiou, D.; Suidan, M.T.; Ginestet, P.; Audic, J.M. Performance deterioration and structural changes of a ceramic membrane bioreactor due to inorganic abrasion. *J. Membr. Sci.* **1999**, *163*, 19–28. [CrossRef]
328. Doll, T.E.; Frimmel, F.H. Cross-flow microfiltration with periodical back-washing for photocatalytic degradation of pharmaceutical and diagnostic residues-evaluation of the long-term stability of the photocatalytic activity of TiO_2. *Water Res.* **2005**, *39*, 847–854. [CrossRef] [PubMed]

329. Lawn, B.; Wilshaw, R. Indentation fracture—Principles and applications. *J. Membr. Sci.* **1975**, *10*, 1049–1081. [CrossRef]
330. Nicholson, D.W. *Finite Element Analysis: Thermomechanics of Solids*; CRC Press: Boca Raton, FL, USA, 2003.
331. Xie, Z.; Swain, M.V.; Hoffman, M.J. Structural integrity of enamel: Experimental and modeling. *J. Dent. Res.* **2009**, *88*, 529–533. [CrossRef] [PubMed]
332. Bacchin, P.; Aimar, P.; Field, R.W. Critical and sustainable fluxes: Theory, experiments and applications. *J. Membr. Sci.* **2006**, *281*, 42–69. [CrossRef]
333. Phattaranawik, J.; Fane, A.G.; Pasquier, A.C.S.; Bing, W. A novel membrane bioreactor based on membrane distillation. *Desalination* **2008**, *223*, 386–395. [CrossRef]
334. Yap, W.J.; Zhang, J.S.; Lay, W.C.L.; Cao, B.; Fane, A.G.; Liu, Y. State of the art of osmotic membrane bioreactors for water reclamation. *Bioresour. Technol.* **2012**, *122*, 217–222. [CrossRef] [PubMed]
335. Achilli, A.; Cath, T.Y.; Childress, A.E. Power generation with pressure retarded osmosis: An experimental and theoretical investigation. *J. Membr. Sci.* **2009**, *343*, 42–52. [CrossRef]
336. Cath, T.Y.; Childress, A.E.; Elimelech, M. Forward osmosis: Principles, applications, and recent developments. *J. Membr. Sci.* **2006**, *281*, 70–87. [CrossRef]
337. Krantz, W.B.; Lin, C.S.; Sin, P.C.Y.; Yeo, A.; Fane, A.G. An integrity sensor for assessing the performance of low pressure membrane modules in the water industry. *Desalination* **2011**, *283*, 117–122. [CrossRef]

12

Olive Mill Wastewater: From a Pollutant to Green Fuels, Agricultural Water Source and Bio-Fertilizer

Patrick Dutournié [1], Mejdi Jeguirim [1,*], Besma Khiari [2], Mary-Lorène Goddard [3,4] and Salah Jellali [5]

1 CNRS ISM2 UMR 7361, Université de Haute Alsace, Université de Strasbourg, 68093 Mulhouse, France; Patrick.dutournie@uha.fr
2 National School of Engineers of Carthage, 45 rue des entrepreneurs, Charguia 2–1002, 2035 Tunis, Tunisia; besmakhiari@yahoo.com
3 CNRS, LIMA UMR 7042, Université de Haute-Alsace, Université de Strasbourg, 68093 Mulhouse, France; mary-lorene.goddard@uha.fr
4 LVBE, EA 3991, Université de Haute-Alsace, 68008 Colmar, France
5 Wastewaters and Environment Laboratory, Water research and Technologies Centre, 8020 Soliman, Tunisia; salah.jallali@certe.rnrt.tn
* Correspondence: mejdi.jeguirim@uha.fr.

Abstract: Water shortage is a very concerning issue in the Mediterranean region, menacing the viability of the agriculture sector and in some countries, population wellbeing. At the same time, liquid effluent volumes generated from agro-food industries in general and olive oil industry in particular, are quite huge. Thus, the main aim of this work is to suggest a sustainable solution for the management of olive mill wastewaters (OMWW) with possible reuse in irrigation. This work is a part of a series of papers valorizing all the outputs of a three-phase system of oil mills. It deals with recovery, by condensation, of water from both OMWW and OMWW-impregnated biomasses (sawdust and wood chips), during a convective drying operation (air velocity: 1 m/s and air temperature: 50 °C). The experimental results showed that the water yield recovery reaches about 95%. The condensate waters have low electrical conductivity and salinities but also acidic pH values and slightly high chemical oxygen demand (COD) values. However, they could be returned suitable for reuse in agriculture after additional low-cost treatment.

Keywords: OMWW; drying; water recovery; water characterization; sustainable development

---

1. Introduction

Water scarcity in the Mediterranean region is becoming a growing concern, menacing the viability of agriculture, which represents an important economic sector in many countries [1]. This water shortage is accentuated in the last decades due to the undeniable climate change causing recurrent long periods of drought. The efficient wastewaters treatment and their controlled reuse in irrigation can contribute to water saving and potentially address this water shortage issue [2,3].

Generally, treated municipal wastewater is the main recyclable source for irrigation. This treatment and reuse process contributes strongly to the potable water resources conservation and reduces the wastewater's negative impacts related to the effluents release into the environment [4]. However, the wastewater chemical composition needs to be controlled due to the presence of a variety of pollutants including heavy metals, organic compounds, bacteria, etc. [5,6]. These contaminants may generate problems for agricultural production affecting crop quantity and quality [7,8]. Therefore, the interest turned towards agro-industrial wastewaters, because of the large amounts produced and their relative high contents in nutrients [9].

Still, in the Mediterranean region, the olive oil industry generates large quantities of olive byproducts that should be appropriately managed [10]. Different kinds of extraction processes are commonly used. The three-phase system, mainly used in Tunisia, Greece, and Italy, generates a solid residue (named olive mill solid waste (OMSW)) and a liquid effluent (olive mill wastewater (OMWW)). Approximately, 30 million tons of OMWW are generated each year in the Mediterranean region [11]. The OMWW, which is slightly acidic, contains a large fraction of water (around 80%) and high contents of soluble organic compounds and salts [12]. These wastes are often discharged and stored in natural open-air basins (Figure 1a). Indeed, the OMWW surface rapidly dries (as a result of the heat supplied by air flow and sun heating) and a crust (with a plastic consistency) rapidly covers the surface (Figure 1b–d). Such crust inhibits the mass and heat transfers and, therefore, water evaporation. In these conditions, the aqueous phase could diffuse and percolate in the subsoil, leading to soils infertility and groundwater pollution. Furthermore, it is the origin of bad smells and the development of several varieties of mosquitoes.

Figure 1. Photos of olive mill wastewaters (OMWW) evolution in deposit basins in the region of Mahdia (Tunisia) at different drying stage.

Sustainable OMWW management has been pointed out as an urgent challenge in order to tackle the disadvantages cited above [11–13]. More than 20 applicable procedures are mentioned in scientific publications, including elemental operations such as flocculation, ultrafiltration, and chemical treatments or combined operations such as centrifugation–ultrafiltration. These techniques are generally tested in a laboratory or in a pilot plant, without posterior industrial projections. Some treatment and valorization systems presenting some degree of applicability are summarized in Table 1.

Table 1. Main applied technologies for OMWW treatment.

Technology	Description	Advantages	Drawbacks
Land spreading	OMWW is used as amendment (30–80 m^3 ha^{-1} $year^{-1}$)	- Low cost operation - Soil fertility and crops growth	• Effect on soil is controversial • Possibility of groundwater contamination through infiltration
Composting	Aerobic breakdown of organic content of OMWW	- soil fertility and crops growth	• Location far from urban or transit zones (bad smell) • Possibility of Groundwater contamination by drainage water
Thermal treatment	OMWW mixing with dry olive husk wastes	- Wastes reduction and energy recovery	• High cost of drying • High investment and operating costs
Membrane filtration	Separation of the solid fraction using multilayered filters	- Treatment of high flowrates - Water recovery and solid subsequent reuse	• High energy consumption • High risk of filters clogging
Biological treatment	Biological degradation of the organic matter in anaerobic digesters	- Low energy consumption	• Expensive plantings and machinery • Production of high amounts of activated sludge
Coagulation-flocculation	Aggregation of the suspended matter through polyers addition	- Water recovery	• Expensive plantings and machinery • Production of high amounts of activated sludge
Electrocoagulation	Imposition a current between two electrodes immersed in an electrolyte to produce a coagulant in solution	- Water recovery - No necessity of using expensive machinery, no use of moving parts	• High energy consumption • Production of high amounts of sludge containing heavy metals
Adsorption	Elimination of organic and mineral content by using porous media	- Low cost, availability, and flexibility - Water recovery	• Pollution transfer from liquid to solid phase • Limited treatment yield

The agronomic use of OMWW through its spreading on soils, directly or following a composting process, has several drawbacks, including the modification of soils' physical, chemical, and biological properties. On the contrary, the thermal treatment could be an interesting solution since it permits the recovery of biofuels (for energy production) and biochars (for agronomic and environmental purposes). However, the high moisture content is not economically favorable for the direct OMWW thermochemical conversion [14,15]. In order to face such constraint, the OMWW impregnation on lignocellulosic biomass was proposed to recuperate the organic matter followed by agropellets production [16]. Nevertheless, the pellets' combustion in domestic boilers has some drawbacks, like particulate matters emissions and ash accumulation, due to the high mineral contents in OMWW [17].

Recently, an environmentally friendly strategy for the OMWW valorization, including different procedures, was implemented [18]. The first step was the impregnation of wood sawdust with OMWW, leading to the recovery of the OMWW's organic and nutrient compounds. The second step of drying showed that the impregnation stage led, not only to the oily fraction retention, but also to the acceleration of water evaporation kinetic. The third step consisted of the pyrolysis of these impregnated samples, which led to high yields of chars with attractive nutrients concentrations, especially potassium and phosphorus. The last step concerned the application of the obtained chars as bio-fertilizers. This operation confirmed a beneficial effect on the plant growth [18].

More precisely, and as far as the drying step is concerned, the experiments were carried out in a convective dryer under different operating conditions (velocity, temperature). The kinetic results confirmed that the drying of the impregnated samples was faster than that of OMWW samples. But more importantly, such procedure could probably allow an ecologic recovery of water from OMWW [19]. In this vein, this paper aims to examine the water recovery from the drying step and its characterization for a possible reuse in irrigation. Indeed, water management will be increasingly strategic, especially in the Mediterranean basin. Very few works in the literature deal with the recovery of water for an agricultural purpose after a drying operation [20]. These studies are focused on the recovering of added-value chemicals in water, for example, essential oils extraction [21].

2. Materials and Methods

2.1. Samples Preparation

Olive mill wastewater (OMWW) used in this study was provided by an olive mill located in the city of Touta, North East of Tunisia. This olive mill plant uses the three-phase extraction system for olive oil production. Sawdust was provided by Nollinger sawmill located in Illfurth (France). Wood chips were supplied by the firewood supplier Farmingroad located in Reguisheim (France).

Five samples were prepared during this investigation in order to examine the water recovery process and analyze its quality for the possible reuse in agriculture. The raw OMWW sample (moisture content: 91% wet basis, wb.) was used for the impregnation procedure. This sample was also used without impregnation for the water recovery tests. The impregnation samples were prepared according to the procedure applied in previous investigations [16,17]. Impregnated sawdust (IS) was prepared by mixing 2 kg of OMWW with 0.5 kg of sawdust (moisture content: 24% wb.) for 24 h. The initial moisture content of the IS before drying was 78% wb. In a similar way, impregnated wood chips (IWC) were prepared by mixing 1.5 kg of OMWW with 0.5 kg of wood chips (moisture content: 24% wb.). The initial moisture of IWC before the drying test was 74% wb.

Two complementary samples (blank tests) were required in order to identify the effect of the OMWW impregnation on the water quality. Hence, humidified sawdust (HS) was prepared by emerging 0.5 kg of sawdust in 2 L deionized water respecting the same procedure as for OMWW impregnation. Humidified wood chips (HWC) were prepared by emerging 0.5 kg of wood chips in 1.5 L deionized water.

2.2. Experimental Drying Tests

Experimental drying tests were performed using a laboratory convective dryer (Figure 2) previously described in details [19].

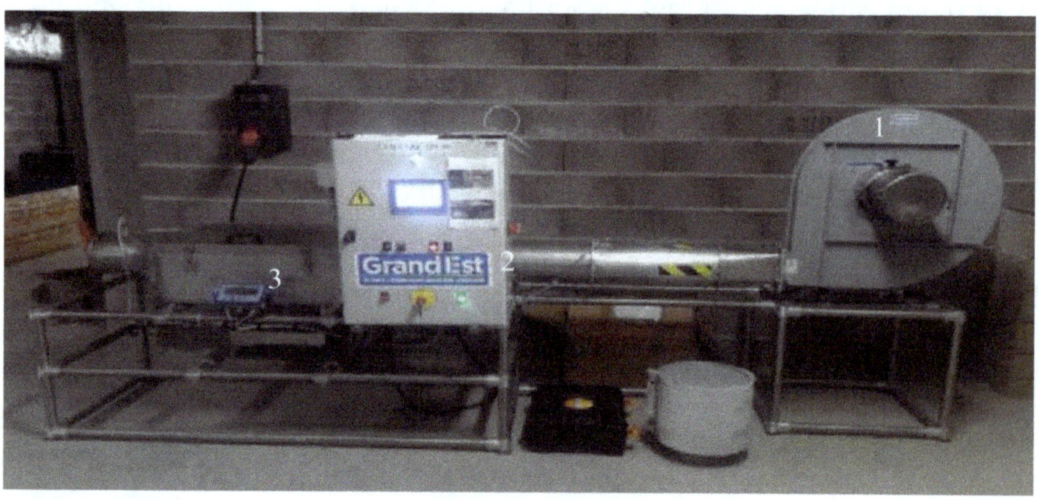

Figure 2. Experimental dryer.

During the drying tests, samples were placed in an open Plexiglas box (19 × 13 × 3 cm). The level of wet samples is accurately adjusted to the upper rim of the box. The mass of the studied sample was recorded in order to measure the moisture content progress versus time. A fan (1) introduces air in the dryer with an air velocity of 1 m/s. The fan is settled for having an air velocity directly onto the material surface. The air heating is achieved by electrical resistances (2) at 50 °C. The air temperature in the drying chamber is controlled by a proportional–integral–derivative (PID) control loop. The operating drying conditions (1 m/s and T = 50 °C) were selected from previous results [19] in order to have a good compromise between the drying time and the energy consumption. Temperature, humidity, and flow rate are continuously measured by several sensors; 1 for fluid velocity in the duct, 3 humidity sensors (room air, hot air at the drying chamber inlet, and at the dryer outlet) and 7 thermocouples (room air, hot air, 4 in the drying chamber, and 1 at the dryer outlet).

From mass measurements, the moisture content X is given by:

$$X(t) = \frac{m(t) - m_f}{m(t)} \tag{1}$$

where m(t) is the sample weight at time t and m_f is the sample weight at the end of the drying.

2.3. Water Condensation and Recovery

In order to examine the water recovery from the impregnated samples, an accessory material was implemented at the dryer outlet. In particular, a boiler condenser (Figure 3) was used for the water condensation at the outlet of the dryer. For this purpose, the moist air flow is introduced in the boiler exchanger. The latter is cooled by an antifreeze fluid circulating into the aluminum jacket.

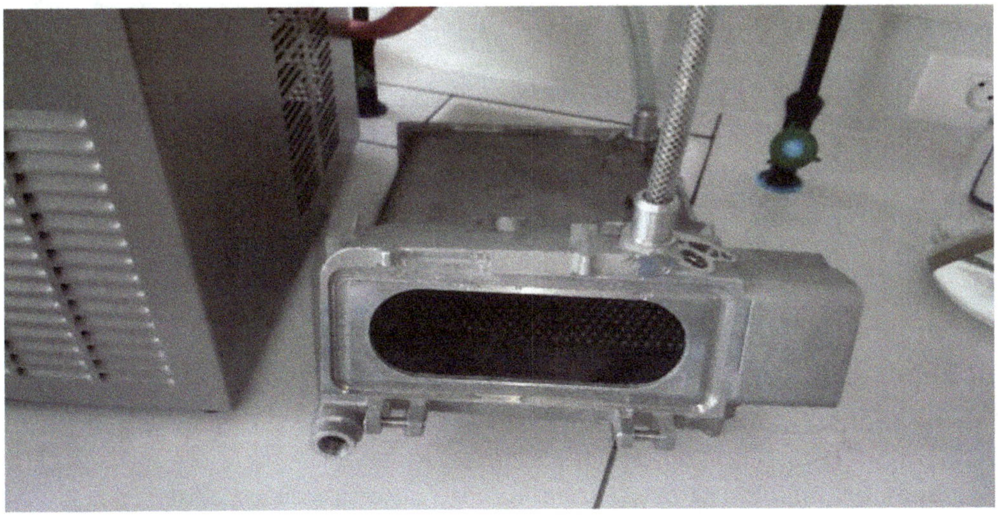

Figure 3. Boiler condenser used for the water recovery experiments.

2.4. Condensed Water Characterization

The condensed water recovered after the drying operation of raw OMWW, IS, IWC, HS, and HWC was physico-chemically characterized by using different analytical techniques in order to assess their potential reuse for irrigation in agriculture. This characterization concerned the assessment of: (i) Their pH by using a HI 2211 (HANNA Instrument, Lingolsheim, France) apparatus, (ii) their electrical conductivity (salinity) through a PC 5000 L (VWR, Radnor, PA, USA) device, (iii) their chemical oxygen demand (COD) via the open reflux using dichromate titrimetric method [22], and (iv) their contents in inorganic anions and cations through ion chromatography analyzer (Metrohm, Herisau, Switzerland).

2.5. *Gas Chromatography–Mass Spectrometry*

2.5.1. Chemicals and Reagents

Reagents were purchased from Sigma–Aldrich (St Quentin Fallavier, France) for methoxyamine hydrochloride and Alfa Aesar (by Thermo Fisher, Karlsruhe, Germany) for anhydrous pyridine and MSTFA). LC-MS grade water was purchased from Fisher Scientific (Illkirch, France).

2.5.2. Derivatization

A volume of 30 μL of each sample was lyophilized and then re-dissolved in 20 μL of 30 mg mL^{-1} methoxyamine hydrochloride in anhydrous pyridine and derivatized at 37 °C for 120 min with mixing at 600 rpm. The samples were incubated for 30 min with mixing at 600 rpm after addition of both 80 μL of N-methyl-N-(trimethylsilyl)trifluoroacetamide (MSTFA). Each derivatized sample was allowed to rest for 30 min prior to injection.

2.5.3. GC–MS Instrument Conditions

Samples (1 μL) were injected into a GC–EI–MS system comprising an AOC-20i Auto-injector, a Shimadzu GC-2010 gas chromatograph, and a Shimadzu GCMS-QP2010 mass spectrometer (Shimadzu, Kyoto, Japan) with an electron impact (EI) ion source and quadrupole analyzer. The GC was operated in constant linear velocity mode (40.3 cm s^{-1}) with helium as the carrier gas and using phenoxyacetic acid as a standard for retention time locking of the method. The MS was adjusted using perfluorotributylamine (PFTBA). An SGE Analytical Science BPX5 column (25 m long, 0.15 mm inner diameter, 0.25 μm film thickness) was used. The injection temperature in splitless mode was set at 310 °C, the MS transfer line at 330 °C, and the ion source adjusted to 200 °C. Helium was used as the carrier gas at a flow rate of 0.97 mL min^{-1}. The following temperature program was used; injection at 110 °C, hold for 4 min, followed by a 10.5 °C min^{-1} oven temperature, ramp to 155 °C, then 11.5 °C min^{-1} oven temperature, ramp to 350 °C, and a final 6 min heating at 350 °C.

2.5.4. Chemical Identification

Gas Chromatography–Mass Spectrometry (GC-MS) results are analyzed from the National Institute of Standards and Technology (NIST) 05 database and identified chemicals are validated by comparison with injection of commercial standards.

3. Results

3.1. *Kinetic Study*

Figure 4 shows the evolution of the drying in time of the OMWW, IS, and IWC in terms of moisture content X versus time.

One can see that for the three samples, the moisture ratio decreased continuously with the drying time, with different constant rate drying period. This is in accordance with other works dealing with the drying of OMWW [23] and of other biomasses impregnated with OWWW [19,23].

During the first phase of drying, the slope of the moisture curve was very dependent on the sample used. Indeed, the highest one was observed for the IWC, followed by IS, and OMWW. As this stage corresponded to the departure of free or/and weakly bound water, one can attribute the difference to the mixture density and its porosity. Indeed, there were more voids within the impregnated wood chips (IWC), which led to a better air infiltration, whereas the impregnated sawdust (IS) was more compact and formed a paste-like mixture. As for the OMWW, the oily layer transformed into a crust at the top of the liquid, which may explain the slow evaporation rate and why the first stage ended earlier (16 h) compared to IWC (19.5 h) and IS (33.8 h).

In the second drying stage (drying stage separation shown by a dotted line in Figure 3), this tendency was inversed and the slopes were in the following order: OMWW > IS > IWC.

The acceleration in OMWW drying, during the second phase, could be the result of the cracking of the crust layer (white patch shown in Figure 5, circled in red), due to the evaporated water under the crust forcing a way out on one hand and to the mechanical action of the convective air on the other hand. The resulting fissures then allow water to escape, and therefore the drying rate is increased.

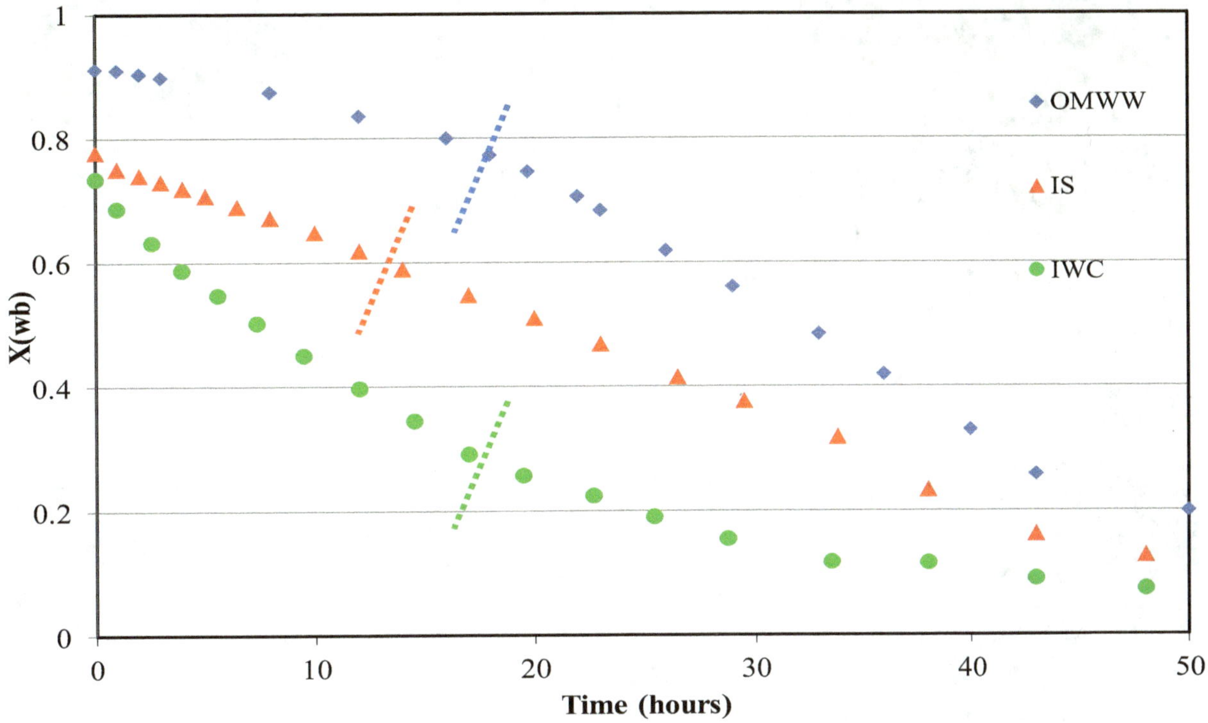

Figure 4. Moisture content profile during the convective drying for raw olive mill wastewater (OMWW), impregnated sawdust (IS), and impregnated wood chips (IWC).

The reason the IS drying rate was higher than that of IWC during the second stage may be related to the impregnation procedure and the material nature. Indeed, as sawdust is a finer and thinner material than the wood chips, its mixture with the oily liquid might have promoted the formation of stronger chemical bonds between the solid matrix and the liquid phase. The phenomenon of water and oil retention on the biomass fibers is probably the reason their slopes were inferior.

For the two impregnated solids, the water loss rate was less in the second stage, compared to the first one. Following the decrease in the concentration gradient between the samples and the surrounding air, mass flow was limited within the solid samples. This is due to bound water, which is more difficult to remove. In fact, the drying process is limited by a diffusion phase within the solid matrix. The vaporization front gradually moves towards the interior of the material. As water vapor has a longer way to cross, the surface pressure decreases. The difference between the latter and the vapor pressure of the surrounding air also decrease, slowing down the exchange and consequently the rate of drying [24].

Finally, it is important to point out that the drying of the impregnated biomasses was faster than that of the OMWW sample (× 2 for IS and × 3.5 for IWC) during only the first stage of drying. Such a conclusion was reached for impregnated olive cake and impregnated sawdust (two times faster than OMWW for both materials) but for the whole operation of drying [19]. This may be linked to the fact that in the mentioned paper, the layer to dry was thin (3 mm) whereas it was 3 cm in the present study. During the second stage, the drying was lower (× 1/2 for IS and × 1/3.5 for IWC) compared to OMWW, which explains why the amount of water evaporated per hour was substantially the same for all the samples.

Figure 5. IWC, OMWW, and IS before and after drying.

3.2. Water Recovery

Figure 6 shows the ratio of recovered water by the initial water during the drying operation for OMWW, IS, and IWC. The final ratios of condensed water are interesting since they were around 80% for OMWW and IWC and as high as 95% for IS. In a context of water shortage, such yields are very promising for further subsequent use.

The impregnation process not only shortened the drying time during the first step, but also increased the contained water recovery ratio by 18% to 35% during the whole condensation operation. Moreover, dried IS samples have the shape of bricks (Figure 7) with good mechanical properties on one hand and with increased high heating values (18 MJ/kg) compared to non-impregnated ones (16.4 MJ/kg) on the other hand. These bricks are homogenous, well-bonded, and can be easily removed from the container, contrarily to the dried OMWW where the oily and plastic phase makes it stick to the box and hard to remove. All these aspects make dried IS easy to handle, transport, and store and, suitable for direct use as bio-fuels or for a densification process by briquetting or pelletizing.

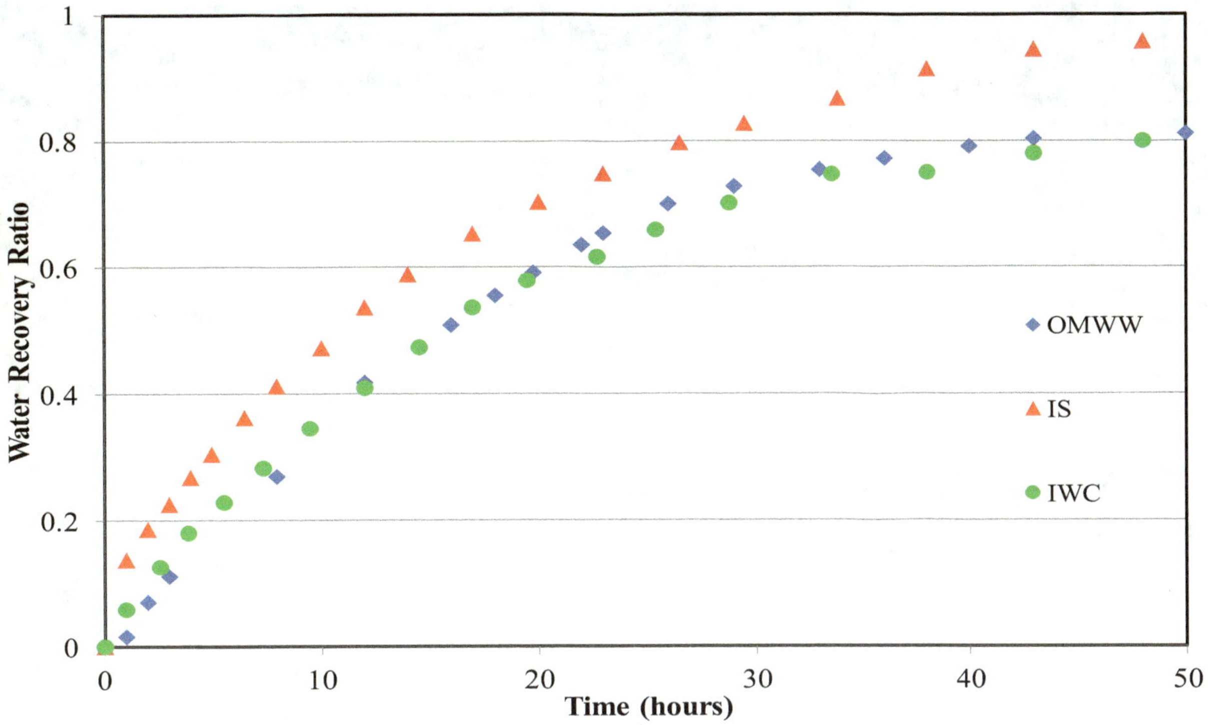

Figure 6. Water recovered ratio during drying for OMWW, IS, and IWC.

Figure 7. Impregnated sawdust at the end of the drying operation.

3.3. *Recovered Water Characterization*

Figure 8 shows the aspect of recuperated condensed water (right) and of the raw OMWW (left). Despite the clearness of the obtained waters, their characterization is essential to ensure their compliance with water irrigation standards.

Figure 8. Aspect of OMWW (left) and recovered-condensed water from raw OMWW (right).

As mentioned in Section 2.4, several parameters were measured for this purpose. Table 2 gives these main physico-chemical properties of the raw OMWW and the recovered water from the drying operation of OMWW, IS, IWC, HS, and HWC, in comparison with the Tunisian standard for both discharging in water bodies and reuse in agriculture (norms NT 4106.002, NT 106.003 and decree-law N°89-1047 of 28 July 1989).

Table 2. Main parameters of raw OMWW and allowable concentrations (standards).

Parameter	Raw OMWW	OMWW	IS	IWC	HS	HWC	Discharging Norm	Reuse in Agriculture Norm
pH	4.8	3.5	3.9	3.8	3.8	3.7	6.5–8.5	6.5–8.5
EC (mS/cm)	9.73	0.25	0.23	0.27	0.05	0.02	5.00	7.00
COD (g/L)	100.00	8.38	2.10	8.40	0.10	0.04	0.13	0.09
Na^+ (mg/L)	2465.0	2.0	3.3	1.2	3.3	2.6	700	-
K^+ (mg/L)	5046.5	5.0	1.9	2.0	2.1	4.5	50	-
Mg^{2+}(mg/L)	1462.5	0.4	1.1	0.4	0.8	1.0	300	-
Ca^{2+} (mg/L)	1907.8	1.7	4.3	1.5	3.0	3.9	500	-
P_{total} (mg/L)	832.1	0.7	1.6	1.4	24.8	1.9	2	-
NO_3^- (mg/L)	1056.8	1.5	1.6	0.4	3.0	5.5	600	-
Cl^- (mg/L)	1912.9	4.8	4.3	2.3	8.5	2.6	700	2000.0

According to Table 2, all the raw OMWW physico-chemical characteristics were much higher than the discharging and reuse requirements. For instance, the OMMW' COD contents and phosphorus concentrations were about 769- and 416-fold higher than the related fixed maximum concentrations, respectively.

At the same time, the drying operation significantly decreased the pH of the recovered water solution compared to the raw OMWW (4.77). Indeed, the pH values of the recovered solutions varied between 3.9 for IS and 3.5 for OMWW. These values are again very low compared to the fixed norms of discharging and reuse (Table 2). However, their adjustment to these norms could be easily performed by filtrating these recovered solutions through low-cost alkaline wastes such as seashell [25] or powdered marble [26]. Furthermore, the drying procedure also significantly decreased the electrical conductivity of the raw OMWW from 9.73 mS cm^{-1} to about 0.25 mS cm^{-1} for the recovered waters. This result was expected since waters that were evaporated during convective drying (low-temperature drying) contain very little inorganic ionic compounds. All these values were significantly lower than the fixed norm for wastewater discharging in water bodies or reuse in agriculture. The blank tests (HS and HWC) showed a low impact of the sawdust and the wood chips since the related recovered waters had very low EC and COD compared to the same solid matrixes when impregnated with OMWW.

Concerning the COD contents, all the recovered water solutions had low values compared to the raw OMWW (100 g L^{-1}). The lowest value (2.1 g L^{-1}) was registered for impregnated sawdust (IS). The highest one (8.4 g L^{-1}) was observed for IWC and could be imputed to the release of supplementary organic matter from the solid matrix into the impregnating liquid phase. Electrical conductivity, pH, and COD tests were duplicated with recovered waters from different drying experiments. While pH and EC were close to each other, COD measurements were sensibly different (for example 2.3 and 14.4 g L^{-1} for OMWW). This difference is attributed to the time spent in open air. Indeed, the more the air was evacuated, the more the COD was decreased (2.3 g L^{-1} with 12 h venting and 14.4 g L^{-1} with no aeration). All the measured COD contents of the recovered waters were higher than the set standards. These values could be significantly decreased through specific treatment (depending on the detailed characterization of the contained dissolved organic matter) and mixing with other waters.

On the other hand, the IC analyses showed that for all the samples, the main anions and cations contents were significantly lower than the fixed norms (Table 2). Moreover, the suggested treatment cited above for the increase of these recovered waters pH by using low-cost alkaline materials such as powdered marble or seashell, will certainly enrich them with mineral elements such as calcium and magnesium.

Moreover, GC-MS investigations showed that the organic chemicals found came from olive oil, residues, and wood. Indeed, classical fatty acids coming from olive oil such as myristic, palmitic, and stearic acids, as well as tyrosol and vanillin were transferred into the recovered waters. Waters from OMWW and impregnated biomasses also contained short-chain acids such as acetic, malonic, butanoic, fumaric, and succinic, which could explain the difference of COD after sample venting.

Wood chips and sawdust significantly released chemicals like glycerol, beta-alanine, D-glucose, urea, and some short-chain organic acids (D-glyceric, succinic, acetic, etc.,) into waters. These chemicals come from both impregnated biomasses and humidified ones.

Therefore, a tertiary treatment is necessary to adjust the pH and decrease the COD in order to use these waters in agriculture. A low-cost solution could be the filtration through low-cost materials such as the cited-above raw/modified mineral wastes (seashell or powdered marble) and a proper ventilation to extract organic volatile compounds.

4. Conclusions

In this paper, a sustainable strategy for recovery of water from olive mill wastewater (OMWW) is proposed. This strategy includes (i) the impregnation of OMWW on lignocellulosic biomass (oak sawdust, wood chips), (ii) the controlled drying of the impregnation samples in a convective dryer, (iii) the water recovery from the drying operation through condensation process, and (iv) the material recovery of the dried impregnated biomasses for thermochemical conversion process (combustion/pyrolysis for energy/biochar production).

The quality of the recovered water was emphasized in this present work. In particular, different characterization techniques were used in order to assess the potential use of the recovered water in agriculture application. It was observed that the condensate waters are attractive for an agricultural purpose, but they do require an additional treatment to adjust the pH and to decrease its COD content.

This study performed at laboratory scale is a useful starting point for scaling up and ensuring sustainability.

Author Contributions: Conceptualization, P.D. and M.J.; methodology, P.D., M.-L.G. and S.J.; software, B.K.; validation, P.D., M.J. and S.J.; formal analysis, M.J. B.K. and M.-L.G.; investigation, P.D. and M.-L.G.; resources, P.D. and S.J.; data curation, P.D., M.J. and B.K.; writing—original draft preparation, B.K. and M.J.; writing—review and editing, P.D. and S.J.; visualization, P.D. and M.-L.G.; supervision, P.D., M.J. and S.J.; project administration, M.J.; funding acquisition, P.D., M.J. and S.J.

References

1. Iglesias, A.; Garrote, L.; Flores, F.; Moneo, M. Challenges to Manage the Risk of Water Scarcity and Climate Change in the Mediterranean. *Water Resour. Manag.* **2007**, *21*, 775–788. [CrossRef]
2. Angelakis, A.N.; Marecos Do Monte, M.H.F.; Bontoux, L.; Asano, T. The status of wastewater reuse practice in the Mediterranean basin: Need for guidelines. *Water Res.* **1999**, *33*, 2201–2217. [CrossRef]
3. Mizyed, N.R. Challenges to treated wastewater reuse in arid and semi-arid areas. *Environ. Sci. Policy* **2013**, *25*, 186–195. [CrossRef]
4. Pedrero, F.; Kalavrouziotis, I.; Alarcón, J.J.; Koukoulakis, P.; Asano, T. Use of treated municipal wastewater in irrigated agriculture—Review of some practices in Spain and Greece. *Agric. Water Manag.* **2010**, *97*, 1233–1241. [CrossRef]
5. Khiari, B.; Wakkel, M.; Abdelmoumen, S.; Jeguirim, M. Dynamics and kinetics of cupric ion removal from wastewaters by Tunisian solid crude olive-oilwaste. *Materials* **2019**, *12*, 365. [CrossRef]
6. Wakkel, M.; Khiari, B.; Zagrouba, F. Textile wastewater treatment by agro-industrial waste: Equilibrium modelling, thermodynamics and mass transfer mechanisms of cationic dyes adsorption onto low-cost lignocellulosic adsorbent. *J. Taiwan Inst. Chem. Eng.* **2019**, *96*, 439–452. [CrossRef]
7. Khan, S.; Cao, Q.; Zheng, Y.M.; Huang, Y.Z.; Zhu, Y.G. Health risks of heavy metals in contaminated soils and food crops irrigated with wastewater in Beijing, China. *Environ. Pollut.* **2008**, *152*, 686–692. [CrossRef] [PubMed]
8. Wakkel, M.; Khiari, B. Basic red 2 and methyl violet adsorption by date pits: Adsorbent characterization, optimization by RSM and CCD, equilibrium and kinetic studies. *Environ. Sci. Pollut. Res.* **2018**, 1–19. [CrossRef] [PubMed]
9. Libutti, A.; Gatta, G.; Gagliardi, A.; Vergine, P.; Pollice, A.; Beneduce, L.; Disciglio, G.; Tarantino, E. Agro-industrial wastewater reuse for irrigation of a vegetable crop succession under Mediterranean conditions. *Agric. Water Manag.* **2018**, *196*, 1–14. [CrossRef]
10. Marrakchi, F.; Kriaa, K.; Hadrich, B.; Kechaou, N. Experimental investigation of processing parameters and effects of degumming, neutralization and bleaching on lampante virgin olive oil's quality. *Food Bioprod. Process.* **2015**, *94*, 124–135. [CrossRef]
11. Souilem, S.; El-Abbassi, A.; Kiai, H.; Hafidi, A.; Sayadi, S.; Galanakis, C.M. Chapter 1—Olive oil production sector: Environmental effects and sustainability challenges. In *Olive Mill Waste*; Galanakis, C.M., Ed.; Academic Press: Cambridge, MA, USA, 2017; pp. 1–28.
12. Negro, M.J.; Manzanares, P.; Ruiz, E.; Castro, E.; Ballesteros, M. Chapter 3—The biorefinery concept for the industrial valorization of residues from olive oil industry. In *Olive Mill Waste*; Galanakis, C.M., Ed.; Academic Press: Cambridge, MA, USA, 2017; pp. 57–78.
13. Kapellakis, I.E.; Tzanakakis, V.A.; Angelakis, A.N. Land application-based management of olive mill wastewater. *Water* **2015**, *7*, 362–376. [CrossRef]
14. Caputo, A.C.; Scacchia, F.; Pelagagge, P.M. Disposal of by-products in olive oil industry: Waste-to-energy solutions. *Appl. Therm. Eng.* **2003**, *23*, 197–214. [CrossRef]
15. Miranda, T.; Esteban, A.; Rojas, S.; Montero, I.; Ruiz, A. Combustion analysis of different olive residues. *Int. J. Mol. Sci.* **2008**, *9*, 512–525. [CrossRef] [PubMed]
16. Jeguirim, M.; Chouchène, A.; Réguillon, A.F.; Trouvé, G.; Le Buzit, G. A new valorisation strategy of olive mill wastewater: Impregnation on sawdust and combustion. *Resour. Conserv. Recycl.* **2012**, *59*, 4–8. [CrossRef]
17. Kraiem, N.; Jeguirim, M.; Limousy, L.; Lajili, M.; Dorge, S.; Michelin, L.; Said, R. Impregnation of olive mill wastewater on dry biomasses: Impact on chemical properties and combustion performances. *Energy* **2014**, *78*, 479–489. [CrossRef]
18. Haddad, K.; Jeguirim, M.; Jerbi, B.; Chouchene, A.; Dutournié, P.; Thevenin, N.; Ruidavets, L.; Jellali, S.; Limousy, L. Olive Mill Wastewater: From a Pollutant to Green Fuels, Agricultural Water Source and Biofertilizer. *ACS Sustain. Chem. Eng.* **2017**, *5*, 8988–8996. [CrossRef]
19. Jeguirim, M.; Dutournié, P.; Zorpas, A.A.; Limousy, L. Olive Mill Wastewater: From a Pollutant to Green Fuels, Agricultural Water Source and Bio-Fertilizer—Part 1. The Drying Kinetics. *Energies* **2017**, *10*, 1423.
20. Amos, W.A. *Report on Biomass Drying Technology*; Technical Report; National Renewable Energy Lab.: Golden, CO, USA, 1999. [CrossRef]

21. Borsato, A.V.; Doni-Filho, L.; Rakocevic, M.; Cocco, L.C.; Paglia, E.C. Chamomile essential oils extracted from flower heads and recovered water during drying process. *J. Food Process. Preserv.* **2009**, *33*, 500–512. [CrossRef]
22. Moore, W.A.; Kroner, R.C.; Ruchhoft, C.C. Dichromate Reflux Method for Determination of Oxygen Consumed. *Anal. Chem.* **1949**, *21*, 953–957. [CrossRef]
23. Montero, I.; Miranda, T.; Arranz, J.I.; Rojas, C.V. Thin Layer Drying Kinetics of By-Products from Olive Oil Processing. *Int. J. Mol. Sci.* **2011**, *12*, 7885–7897. [CrossRef]
24. Khiari, B.; Mihoubi, D.; Mabrouk, S.B.; Sassi, M. Experimental and numerical investigations on water behaviour in a solar tunnel drier. *Desalination* **2004**, *168*, 117–124. [CrossRef]
25. Selimi, A.; Nickelson, R. Effects of liming on soil pH and manganese toxicity in a Goulburn valley pear orchard. *Aust. J. Exp. Agric.* **1972**, *12*, 310–314. [CrossRef]
26. Haddad, K.; Jellali, S.; Jaouadi, S.; Benltifa, M.; Mlayah, A.; Hamzaoui, A.H. Raw and treated marble wastes reuse as low cost materials for phosphorus removal from aqueous solutions: Efficiencies and mechanisms. *C. R. Chim.* **2015**, *18*, 75–87. [CrossRef]

13

Dissolved Oxygen Control in Activated Sludge Process using a Neural Network-Based Adaptive PID Algorithm

Xianjun Du [1,2,3,4], Junlu Wang [1,3,4], Veeriah Jegatheesan [2,*] and Guohua Shi [5]

1 College of Electrical and Information Engineering, Lanzhou University of Technology, Lanzhou 730050, China; 27dxj@163.com (X.D.); wjllanzhou@126.com (J.W.)

2 School of Engineering, Royal Melbourne Institute of Technology (RMIT) University, Melbourne 3000, Australia

3 Key Laboratory of Gansu Advanced Control for Industrial Processes, Lanzhou University of Technology, Lanzhou 730050, China

4 National Demonstration Center for Experimental Electrical and Control Engineering Education, Lanzhou University of Technology, Lanzhou 730050, China

5 Department of Energy and Power Engineering, North China Electric Power University, Baoding 071003, China; ghuashi@outlook.com

* Correspondence: jega.jegatheesan@rmit.edu.au.

Featured Application: This work is currently undergoing field testing at Pingliang Wastewater Treatment Plant situated in Gansu province, China, especially for the control of dissolved oxygen concentration in the activated sludge process of the wastewater treatment. By implementing this control algorithm, we can achieve two goals, namely improving the efficiency of wastewater treatment and reducing the aeration energy. Meanwhile, the method proposed in this work can also be extended to other large- or medium-scale wastewater treatment plants in the future.

Abstract: The concentration of dissolved oxygen (DO) in the aeration tank(s) of an activated sludge system is one of the most important process control parameters. The DO concentration in the aeration tank(s) is maintained at a desired level by using a Proportional-Integral-Derivative (PID) controller. Since the traditional PID parameter adjustment is not adaptive, the unknown disturbances make it difficult to adjust the DO concentration rapidly and precisely to maintain at a desired level. A Radial Basis Function (RBF) neural network (NN)-based adaptive PID (RBFNNPID) algorithm is proposed and simulated in this paper for better control of DO in an activated sludge process-based wastewater treatment. The powerful learning and adaptive ability of the RBF neural network makes the adaptive adjustment of the PID parameters to be realized. Hence, when the wastewater quality and quantity fluctuate, adjustments to some parameters online can be made by RBFNNPID algorithm to improve the performance of the controller. The RBFNNPID algorithm is based on the gradient descent method. Simulation results comparing the performance of traditional PID and RBFNNPID in maintaining the DO concentration show that the RBFNNPID control algorithm can achieve better control performances. The RBFNNPID control algorithm has good tracking, anti-disturbance and strong robustness performances.

Keywords: dissolved oxygen concentration; radial basis function (RBF) neural network; adaptive PID; dynamic simulation

1. Introduction

Currently, the activated sludge process is the most widely used process in wastewater treatment plants to reduce the biochemical oxygen demand (BOD), nutrients and to some extent other micro-pollutants such as pharmaceuticals, personal care products and other household chemicals. The concentration of dissolved oxygen (DO) in the aeration tank(s) in an activated sludge process is an important process control parameter that has a great effect on the treatment efficiency, operational cost and system stability. As the DO drops, the quantity of these filamentous microorganisms increases, adversely affecting the settle-ability of the activated sludge. It is important to recognize these early warning signs and make corrections to dissolved-oxygen levels before the quality of the effluent deteriorates. If dissolved oxygen continues to drop, even low dissolved-oxygen filamentous microorganisms will not be present in the mixed liquor, and treatment efficiencies will be seriously affected. At this point, effluent turbidity will increase and treatment will deteriorate rapidly. Higher dissolved oxygen is often a target, but in reality, this is for the assurance of mixing. If dissolved oxygen is 5.0 or higher there is a good chance that dead zones are minimal since normal currents and mixing will transport the oxygenated mixed liquor throughout the reactor. However, if the dissolved oxygen is excessive then there could be problems in the settling of sludge due to shearing of flocs and re-suspension of inert materials. A high DO concentration also makes the denitrification less efficient. Both the above-mentioned factors will lead to waste of energy. On the other hand, a low DO level cannot supply enough oxygen to the microorganisms in the sludge, so the efficiency of organic matter degradation is reduced [1,2]. Therefore, the premise of how the wastewater treatment process can perform stably will depend on how effectively the concentration of DO is be maintained within a reasonable range [3]. Due to the complex nature of microbial activities that are present in an activated sludge process, even a small change introduced to the system (for example, change in flow rate, water quality of the influent, the temperature of the wastewater in the reactors and so on) can affect the concentration of DO. The air supplied to aeration tanks by blowers allows the oxygen to be transferred from the air to the liquid phase (wastewater). The oxygen transfer is a complex process characterized by large time-delays as well as strong nonlinearity, coupling and disturbance, which further increases the difficulty of controlling the concentration of DO [4,5]. A large number of studies have been carried out and achievements have been made by researchers all over the world to control the concentration of DO level; a series of control methods to control the concentration of DO have been put into practice and they have achieved some good effects.

Currently, the proportional–integral (PI) or proportional–integral–derivative (PID) control strategy is widely used in the process control of wastewater treatment plants. It is well known that the control effect might be affected by the unknown, unexpected disturbances and the great changes of operation conditions while using the PI or PID control strategy. In order to improve the dissolved oxygen control performance of the controller in the wastewater treatment process, various solutions are proposed, such as fuzzy adaptive PID, multivariable robust control and model predictive control (MPC) strategy [6–9]. MPC [2] is an effective way to control DO, not only maintaining the DO concentration at a set value, but also catching up with the real-time changes that occur in the process. Belchior et al. Proposed an adaptive fuzzy control (AFC) strategy for tracking the DO set-points applied to the Benchmark Simulation Model No. 1 (BSM1) [10] that was proposed by International Water Association (IWA) [11]. AFC is a supervised data-driven control method designed with a smooth switching scheme between supervisory and nonsupervisory modes. Results show that it can learn and improve control rules resulting in accurate DO control. Yu et al. simulated intelligent control method and traditional PID control method in combination. Based on their respective advantages, they achieved better control effect when they used the intelligent PID control algorithm into applications of control practice in Haicheng sewage treatment plant, China [12].

Scholars also introduced the neural network into the control of DO in wastewater treatment process, for example, back propagation (BP) neural network [13]. Furthermore, neural network is employed into some control strategies for the wastewater treatment process control. Macnab [14] and

Mirghasemi [15] proposed a robust adaptive neural network control strategy and used it to control the dissolved oxygen in activated sludge process application. The proposed method prevented weight drift and associated bursting, without sacrificing performance. They improved the control performance by using the algorithm, Cerebellar Model Arithmetic Computer (CMAC) to estimate the nonlinear behavior of the system. Results showed that it can effectively avoid state error. Ruan et al. proposed an on-line hybrid intelligent control system based on a genetic algorithm (GA) evolving fuzzy wavelet neural network software sensor to control dissolved oxygen (DO) in an anaerobic/anoxic/oxic (AAO) process for treating papermaking wastewater [16]. The results indicate that the reasonable forecasting and control performances were achieved with optimal DO, and the effluent quality was stable at and below the desired values in real time. It can be an effective control method, attaining not only adequate effluent quality but also minimizing the demand for energy, and is easily integrated into a global monitoring system for purposes of cost management [16]. Qiao Junfei et al. proposed a control method based on self-organizing T-S fuzzy neural network (SO-TSFNN), while using its powerful self-learning, fault-tolerant and adaptive abilities of the environment [17]. It realized the real-time control of dissolved oxygen of the BSM1 and achieved better control effect for DO concentration with good adaptability. Li Minghe et al. proposed a neural network predictive control method for dissolved oxygen based on Levenberg-Marquardt (LM) algorithm [18]. It overcomes the shortages of the BP neural network by combining with the LM algorithm to improve the prediction accuracy of neural network and the tracking performance of dissolved oxygen control. Xu et al. proposed a new control strategy of DO concentration based on fuzzy neural network (FNN). The minimum error of the gradient descent method is used to adjust the parameters of the neural network on-line. Simulation results show that the FNN controller is better than other compared methods [19]. Lin and Luo studied the design approach of a neural adaptive control method based on a disturbance observer. A RBF neural network is employed to approximate the uncertain dynamic model of the wastewater treatment process. The effectiveness of the controller is verified by simulation their study [20]. Han et al. proposed a self-organizing RBF neural network model predictive control (SORBF-MPC) method for controlling DO concentration in WWTP. The hidden nodes in RBF neural network can be added or deleted online on the basis of node activity and mutual information to achieve necessary dynamics of the network. The application results of DO concentration control show that SORBF-MPC can effectively control the process of dissolved oxygen [21]. Zhou Hongbiao proposed a self-organizing fuzzy neural network (SOFNN) control method based on. According to the activation strength and mutual information, the algorithm dynamically adds and reduces the number of neurons in the regular layer to meet the dynamic changes of the actual working conditions. At the same time, the gradient descent algorithm is used to optimize the center, width and output weight of the membership function online to ensure the convergence of SOFNN. Finally, experimental verification was carried out in the international benchmark simulation platform BSM1. Experimental results on the BSM1 show that, compared with control strategies of PID, fuzzy logic control (FLC) and FNN with fixed structure, SOFNN has a better performance on tracking accuracy, control stability and adaptive ability [22].

Although there are many studies on how to control the DO concentration in wastewater treatment system by using neural networks and predictive control methods with great outcomes, these kinds of methods have complicated structures and require large amount of computations. They are difficult to implement in practical engineering applications. Basically, most of the existing wastewater treatment plants (WWTPs) are still using PID, a simple and practical control strategy, to control the process. Unfortunately, since the parameters of the PID control algorithm are difficult to set up in advance which are strongly affected by the nonlinearity and large time-delay characters of the wastewater treatment process the control effect maybe unsatisfactory and the key problem is the parameters are not self-adjusted [23]. Therefore, combining intelligent algorithm with the PID algorithm becomes an effective way to realize simple structures and the control requirements of wastewater treatment process in actual WWTPs.

When we use intelligent algorithm into PID, the parameters can be adjusted real-time according to the control effect of current strategy (such as gradient descent method) to avoid the problem of difficult-to-adjust PID parameters. At the same time, they can be adaptively adjusted according to the change of operation environment and dynamic disturbances. There are two ways to improve the control accuracy: one is to improve the accuracy of the measurement equipment of dissolved oxygen concentration, and another is the selection of the center point and the node width of the neural network.

In this paper, a neural network-based adaptive PID control algorithm is proposed. The radial basis function (RBF) neural network is employed which has good generalization ability besides the strong self-learning and adaptive abilities and has a simple network structure. The proposed network already has research and application basis for the control of practical processes in some other areas [24–26]. Compared with the traditional PID control algorithm, the proposed RBF neural network-based adaptive PID (RBFNNPID) control algorithm comprises the advantages of these two methods. It is simple, easy to implement and has better control accuracy. More importantly, one does not need to set up the best parameters of PID in advance; that is to say, it can solve the problem of traditional PID controller that has difficulty in adjusting parameters online.

Considering the control problem of DO concentration level in the wastewater treatment process, in this paper, the Benchmark model of BSM1 is introduced and the implementation of the RBF neural network-based adaptive PID control algorithm is discussed. It can be seen from the comparison simulation results that RBFNNPID control algorithm can effectively improve the control accuracy of dissolved oxygen concentration under the Benchmark as opposed to traditional PID.

2. Materials and Methods

2.1. Activated Sludge Process (ASP) and Benchmark Simulation Model No. 1 (BSM1)

Activated sludge model No. 1 (ASM1) is a mathematical model that is widely accepted and applied in the research and application of activated sludge process (ASP) used in biological wastewater treatment systems. The typical ASP is shown in Figure 1, which includes two parts, the biological (more accurately biochemical) reaction tanks (or aeration tanks) and the secondary settler [27,28]. In the aeration tanks, the microorganisms are divided into active heterotrophic and autotrophic bacteria. The 13 reaction components and 8 reaction processes of the organic matter present in the influent are incorporated into the ASM1 [28–30]. In each process, all the organic substances and microorganisms have their own reaction rates and stoichiometry. Since the model has been published, researchers have been using the ASM1 model to verify their new proposed control algorithms of the DO concentration of ASP.

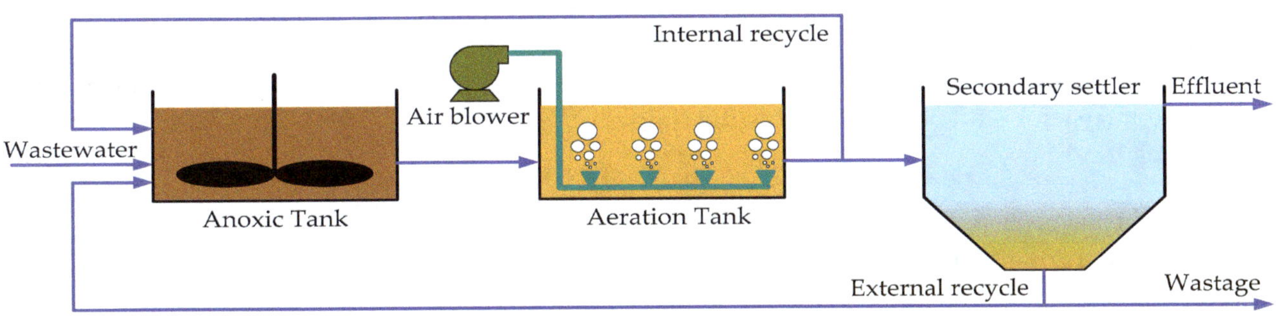

Figure 1. Typical biological (or biochemical) ASP to treat wastewater. ASP: Activated Sludge Process.

The activated sludge process aims to achieve, at minimum cost, sufficiently low concentrations of biodegradable matter and nutrients in the effluent together with minimal sludge production. In order to achieve this, the process has to be controlled [28]. However, it is difficult to predict the performance of the proposed or applied control strategy based on existing reference, process or location.

To enhance the acceptance of innovative control strategies the performance evaluation should be based on a rigorous methodology that includes a simulation model, plant layout, controllers, performance criteria and test procedures.

The first Benchmark Simulation Layout (BSM1), which was based on the ASM1, is relatively a simple layout and is shown in Figure 2. Similar to ASM1, the first part of BSM1 is also a biological (or biochemical) activated sludge reactor, which is comprised of five-compartments, two of them are anoxic tanks and the following three are aerobic tanks; the second part of BSM1 is a secondary settler. Reactors 1 and 2 are unaerated in open-loop, but fully mixed; reactors 3, 4 and 5 are aerated. For the open-loop case, the oxygen transfer coefficients (K_La) are fixed; for reactors 3 and 4 the coefficient (K_La_3 and K_La_4) is set to a constant at 240 d^{-1} (10 h^{-1}), which means the air flow rate of the blower is constant; for reactor 5, the coefficient (K_La_5) is selected as the control variable (or operational variable) in this paper to be manipulated for maintaining the DO concentration at a level of 2mg/L. Thus, the system can achieve biological nitrogen removal through nitrification in the aeration tanks and pre-denitrification in the anoxic tanks. The model equations to be implemented for the proposed layout, the procedure to test the implementation and the performance criteria to be used are described below along with the description of sensors and control handles [28]. For more information, it can be seen in literature [28,29].

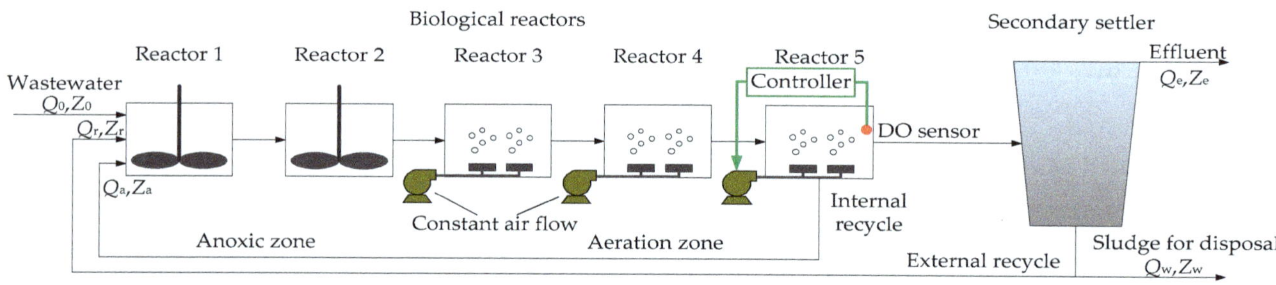

Figure 2. Schematic representation of Benchmark Simulation Model No. 1 (BSM1) model.

The ASM1 [27] has been selected to describe the biological phenomena taking place in the biological reactor and a double-exponential settling velocity function [31] has been selected to describe the secondary settler which is modeled as a 10 layers non-reactive unit (i.e., no biological reaction). In the activated sludge wastewater treatment system, the concentration of DO in the aeration tank is the most important parameter in the process of nitrogen removal [32]. Actually, the DO concentration has a direct impact on the effluent quality with respect to total nitrogen (N_{tot}), nitrate nitrogen (S_{NO}) and ammonia (S_{NH}). Therefore, the study of DO control has its important practical significance and prospect for application.

According to the mass balance of the system, the biochemical reactions that take place in each compartment (reactor) can be described as the follows.

Reactor 1

$$\frac{dZ_1}{dt} = \frac{1}{V_1}(Q_aZ_a + Q_rZ_r + Q_0Z_0 + r_1V_1 - Q_1Z_1) \tag{1}$$

Reactors 2 through 5 (k = 2 to 5)

$$\frac{dZ_k}{dt} = \frac{1}{V_k}(Q_{k-1}Z_{k-1} + r_kV_k - Q_kZ_k) \tag{2}$$

Special case for oxygen ($S_{O,k}$)

$$\frac{dS_{O,k}}{dt} = \frac{1}{V_k}(Q_{k-1}S_{O,k-1} - Q_{k-1}S_{O,k})(K_La)_k(S_O^* - S_{O,k}) + r_k \tag{3}$$

where, Q is the flow rate, Z is the mass concentration of either substrate or bacterial mass, V is the volume of the reactor, r is the reaction rate, K_La is the oxygen transfer coefficient, S_O is the dissolved oxygen concentration. S^* is the saturation concentration for oxygen (S^* = 8 g/m^3 at 15 °C); also $Q_1 = Q_a + Q_r + Q_0$; $Q_k = Q_{k-1}$.

2.2. A Neural Network Based Adaptive PID Algorithm

2.2.1. Radial Basis Function (RBF) Neural Network

Artificial neural network (ANN) is an artificial intelligence system to imitate biological neural networks (BNN). It uses nonlinear processing unit to simulate biological neurons for simulating the behavior of biological synapses among neurons by adjusting the variable weights between connected units. The specific topological structure of the network is organized from each processing unit in a certain connected form. Parallel processing ability and distributed storage are the main features of ANN. Furthermore, it has strong fault tolerance and nonlinear mapping ability with self-organization, self-learning and adaptive reasoning ability [33].

BP (backpropagation) network and RBF network are the most widely used forms of ANN. It is easily to be seen in the widely uses of pattern recognition, prediction, automatic control, etc. [34]. BP algorithm, a supervised learning algorithm, is based on gradient descent algorithm. The drawbacks of BP include an easy fall into local optimum, slow convergence speed, and disunity network structure. RBF network is a feedforward network based on the function approximation theory. It has strong global approximation ability, which can guarantee the network to approximation any kind of nonlinear function with arbitrary accuracy. It can fundamentally overcome the problem of local optimum occurs in BP network. The RBF network has the advantages of simple structure, fast convergence speed and strong generalization ability [35].

Radial basis function (RBF) neural network used in this paper is a three-layer forward network, which is a local approximation method of neural networks. The RBF neural network is composed of three layers, the input layer, the hidden layer and the output layer as shown in Figure 3. The mapping of the input layer to the output layer is nonlinear and the mapping of the space from the hidden layer to the output layer is linear. This kind of mapping configuration itself can speed up the learning rate and avoid the problem of local minima [18].

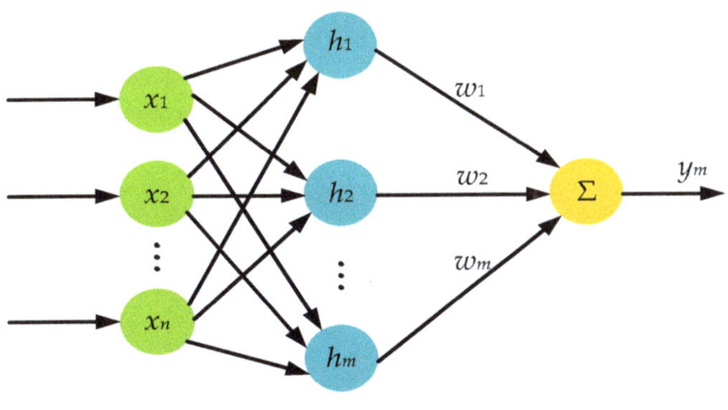

Figure 3. Topology of a radial basis function (RBF) neural network.

In Figure 3, the input vector of the input layer of the neural network is represented as:

$$X = [x_1, x_2, \cdots, x_s, \cdots, x_n]^T \tag{4}$$

where, $x_s = [u_s(k), y_s(k), y_s(k-1)]$, s = 1, 2, ... , n; $u(k)$ is the output of the controller; $y(k)$ is the present (measured) output of the system (or process), that is, the measured value of DO concentration; $y(k-1)$ is the last measured value of DO concentration output from the process.

The middle layer is the hidden layer. The activation function of the hidden layer is composed of radial basis functions. Each array of computing units of hidden layers is called node. The radial basis vector of the nodes in the RBF neural network is shown in Equation (5).

$$T = [h_1, h_2, \cdots, h_j, \cdots, h_m]^T \tag{5}$$

where, h_j is Gaussian function,

$$h_j = \exp(\frac{\|X - C_j\|}{2b_j^2}) \tag{6}$$

where, $j = 1, 2, \ldots, m$. C_j is the central vector of the first j node of the hidden layer of the RBF neural network,

$$C_j = [c_{j1}, c_{j2}, \cdots, c_{ji}, \cdots, c_{jn}]^T \tag{7}$$

where, $i = 1, 2, \ldots, n$.

The basic width vector of the hidden layer node of the RBF neural network is

$$B = [b_1, b_2, \cdots, b_j, \cdots, b_m]^T \tag{8}$$

where, b_j is the parameter of the first j node and $j = 1, 2, \ldots, m$.

The weight vector of RBF neural network W is given by:

$$W = [w_1, w_2, \cdots, w_j, \cdots, w_m]^T \tag{9}$$

Then, the estimated output of the RBF network is defined as:

$$y_m = w_1 h_1 + w_2 h_2 + \cdots + w_m h_m \tag{10}$$

The performance index function of the RBF neural network is set as follows:

$$E_1 = \frac{1}{2}(y(k) - y_m(k))^2 \tag{11}$$

where, $y(k)$ is the system output and $y_m(k)$ is the estimated output of the RBF network.

From the above analysis, the three most important parameters C, W and B of a RBF neural network need to be obtained by the learning algorithm. In this paper, the gradient descent method is employed to obtain those three parameters of the nodes. The iterative algorithm used is as follows:

$$w_j(k) = w_j(k-1) + \eta(y(k) - y_m(k))h_j + \alpha(w_j(k-1) - w_j(k-2)) \tag{12}$$

$$\Delta b_j = (y(k) - y_m(k))w_j h_j \frac{\|X - C_j\|^2}{b_j^3} \tag{13}$$

$$b_j(k) = b_j(k-1) + \eta \Delta b_j + \alpha(b_j(k-1) - b_j(k-2)) \tag{14}$$

$$\Delta c_{ji} = (y(k) - y_m(k))w_j \frac{x_j - c_{ji}}{b_j^2} \tag{15}$$

$$c_{ji}(k) = c_{ji}(k-1) + \eta \Delta c_{ji} + \alpha(c_{ji}(k-1) - c_{ji}(k-2)) \tag{16}$$

and the Jacobian matrix:

$$\frac{\partial y(k)}{\partial u(k)} \approx \frac{\partial y_m(k)}{\partial u(k)} = \sum_{j=1}^{m} w_j h_j \frac{c_{ji} - x_1}{b_j^2} \tag{17}$$

in which, η is the learning rate, α is the momentum factor and $x_1 = \Delta u(k)$ is the control increment which is defined as the first input of the neural network.

2.2.2. Design of the RBF Neural Network Based Adaptive PID (RBFNNPID) Algorithm

In the past decades, Proportional-Integral-Derivative (PID) is the main control method for DO level [36,37]. However, owing to the WWTP's time-varying feature, strong nonlinearity, significant perturbations and large uncertainty, a fixed parameter linear controller is not able to maintain a satisfactory tracking performance under the full range of operating conditions [1,37].

The structure of the RBF neural network-based adaptive PID (RBFNNPID) algorithm is shown in Figure 4. The RBF neural network will adaptively calculate weighting coefficient and the parameter gradient information according to the operating state of the dissolved oxygen control system, by its own great learning ability. These results will be used to update the parameters of the PID controller in real time. Hence, such a repeated execution process realizes the adaptive adjustment of PID parameters and achieves the control of DO concentration.

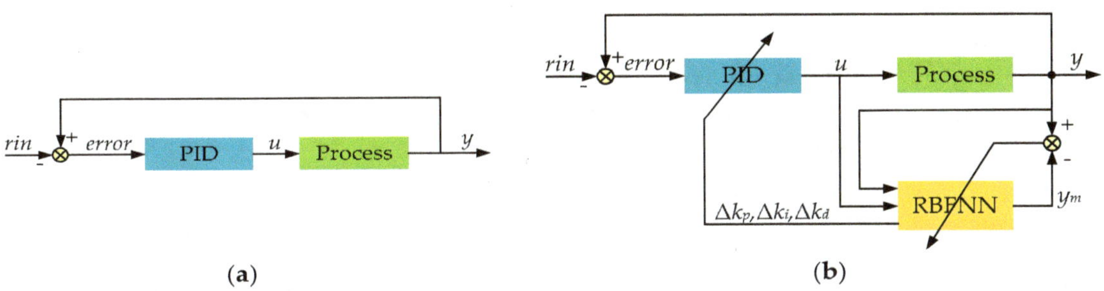

Figure 4. Block diagram comparing two controllers: (**a**) Block diagram of a traditional PID controller in a feedback loop; (**b**) Block diagram of proposed RBF neural network-based adaptive PID (RBFNNPID) controller. PID: Proportional-Integral-Derivative.

We have adopted the incremental PID controller and the control error is:

$$error(k) = rin(k) - y(k) \tag{18}$$

where, *rin* is the desired process value or setpoint of DO concentration; $y(k)$ is the measured process value of DO.

The input of the PID algorithm is three errors, which are defined as:

$$xc(1) = error(k) - error(k-1) \tag{19}$$

$$xc(2) = error(k) \tag{20}$$

$$xc(3) = error(k) - 2error(k-1) + error(k-2) \tag{21}$$

The output of the PID algorithm is:

$$u(k) = u(k-1) + \Delta u(k) \tag{22}$$

$$\Delta u(k) = k_p xc(1) + k_i xc(2) + k_d xc(3) \tag{23}$$

where, k_p, k_i and k_d are the three parameters of the PID controller, which represents the proportion, integration and differentiation. The performance function is defined as:

$$E(k) = \frac{1}{2}(error(k))^2 \tag{24}$$

According to the gradient descent method, the adjustment rules of three parameters are given as:

$$\Delta k_p = -\eta \frac{\partial E}{\partial k_p} = -\eta \frac{\partial E}{\partial y}\frac{\partial y}{\partial u}\frac{\partial u}{\partial k_p} = -\eta error(k)\frac{\partial y}{\partial u} xc(1) \tag{25}$$

$$\Delta k_i = -\eta \frac{\partial E}{\partial k_i} = -\eta \frac{\partial E}{\partial y}\frac{\partial y}{\partial u}\frac{\partial u}{\partial k_i} = -\eta error(k)\frac{\partial y}{\partial u}xc(2) \tag{26}$$

$$\Delta k_d = -\eta \frac{\partial E}{\partial k_d} = -\eta \frac{\partial E}{\partial y}\frac{\partial y}{\partial u}\frac{\partial u}{\partial k_d} = -\eta error(k)\frac{\partial y}{\partial u}xc(3) \tag{27}$$

in which, $\partial y/\partial u$ is the identification information for the Jacobian matrix of the controlled object and it can be obtained through the identification process of neural network. The Jacobian matrix reflects the sensitivity of the output of the controlled object to the change of the input of the control.

The steps of the proposed RBFNNPID control strategy are as follows:

Step 1: Initializing the network parameters, including the number of nodes in input layers and hidden layers, learning rate, inertia coefficient, the base width vector and the weight vector.
Step 2: Sampling to get input *rin* and output y, calculating error in terms of Equation (18).
Step 3: Calculating the output u of regulator according to Equation (22).
Step 4: Calculating network output y_m, adjusting center vector C, base width vector B, weight vector W and the Jacobian matrix in terms of Equations from (10) to (17) to obtain network identification information.
Step 5: Adjusting parameters of regulator in terms of Equations (25)–(27).
Step 6: Back to *Step 2* and repeat the subsequent steps until the end of the simulation time.

The DO control module and the main codes of the S-function module of RBFNNPID can be found in Appendixs A and B. Appendix C describes the stability and convergence analysis of the proposed RBFNNPID algorithm. An example to verify the convergence of the parameters of the neural network is shown in Appendix D.

3. Results

In order to verify the effectiveness and feasibility of the proposed neural network-based adaptive PID (RBFNNPID) algorithm for DO concentration control of the activated sludge wastewater treatment process, comparison simulation of RBFNNPID and traditional PID are designed in this section, including tracking performance and anti- disturbance performance.

We have selected the BSM1 as the simulation model and the dry weather wastewater data provided by IWA as the source data. The dry weather data contains two weeks long actual operational data of a wastewater treatment system, sampled at every 15 min. Figure 5 shows the dynamic influent data and the changes to the concentrations of some of the components between days 7 and 14.

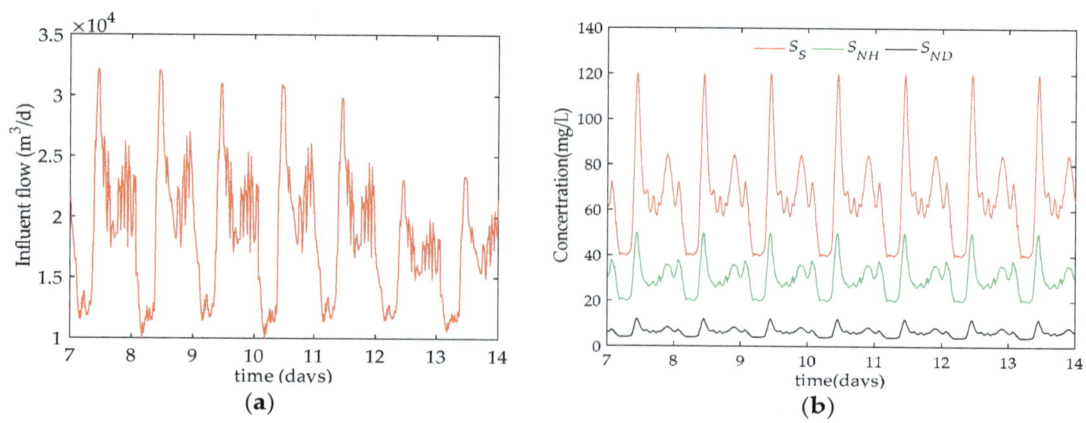

Figure 5. Dynamic data for experimental use: (**a**) Influent flow rate (days 7 to 14); (**b**) Concentration of some of the chemical species in the influent (days 7 to 14) (S_S—readily biodegradable substrate; S_{NH}—ammonium and ammonia nitrogen; S_{ND}—soluble biodegradable organic nitrogen).

In the simulation, the three parameters of both the traditional incremental PID algorithm and the proposed RBFNNPID algorithm are set as: $k_p = 5$, $k_i = 1$, $k_d = 0.5$; the learning rate, η of the three parameters of PID is 0.2; the momentum factor α is 0.05; the network sampling period is 0.001 s; the structure of the RBFNN is defined as "3-6-1", that is to say, the input layer has three nodes, the hidden layer has six nodes and the output layer has one node.

3.1. Tracking Performance Test 1

When the BSM1 wastewater treatment system is operating, due to the dynamic changes in the flow rate and composition of the influent, one need to adjust the oxygen transfer rate (K_La_5) in the fifth tank in real-time to maintain the dissolved oxygen concentration in the appropriate range to ensure the effluent water quality meets the discharge standards. Therefore, how to control the DO concentration around the set point during the process is the aim of control algorithm that is being employed. The tracking performance is one of the deterministic criteria to evaluate whether an algorithm can be applied to an actual wastewater treatment system control.

Generally, the effluent water quality is the best when the DO concentration in the aeration tank is kept between 1~3 mg/L. Therefore, the DO concentration is setup to 2 mg/L in this simulation. The simulation results of last seven days are taken to evaluate the performance of the controller, which is shown in Figure 6.

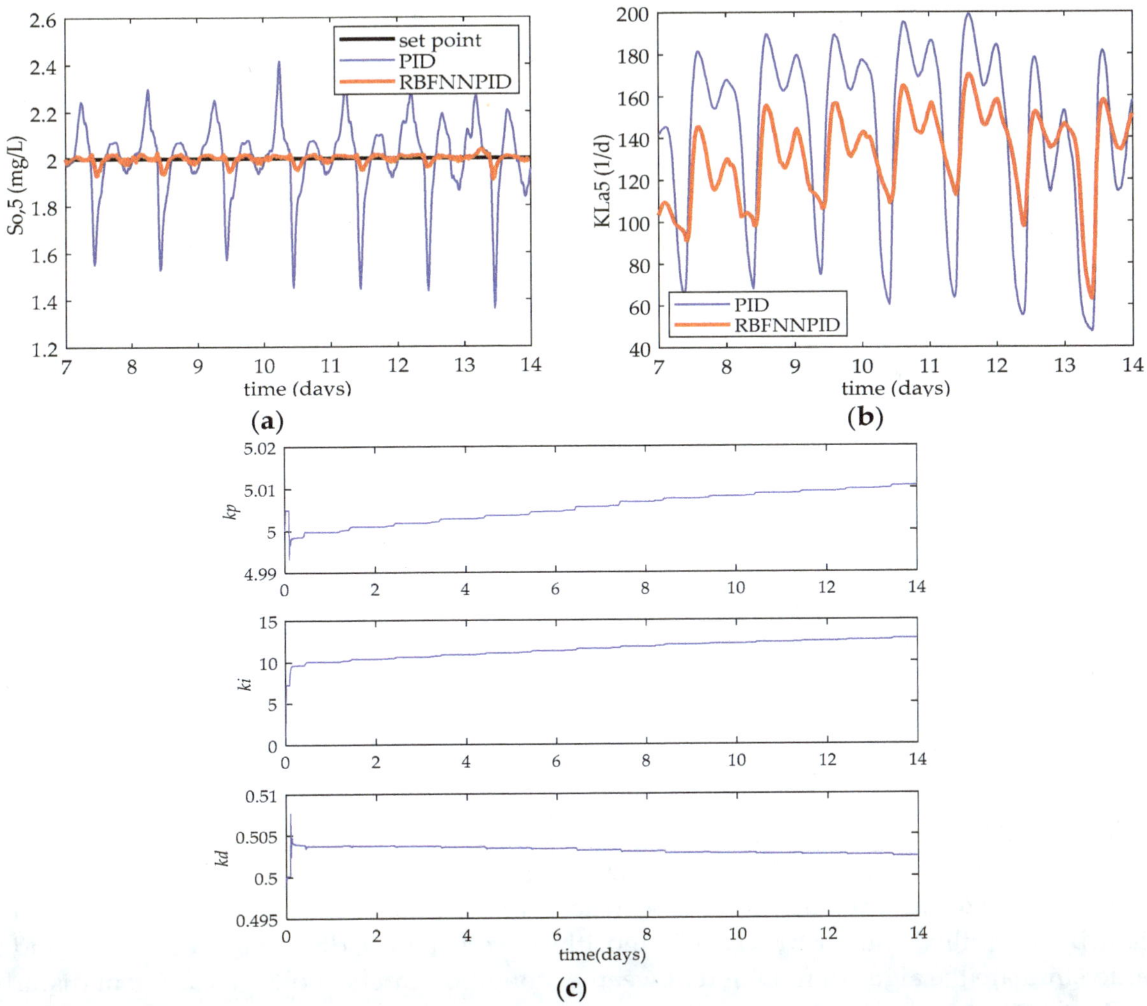

Figure 6. Comparison results of the tracking performance (dry weather): (**a**) DO concentration in the fifth tank; (**b**) Dynamic changes of the manipulated variable K_La_5; (**c**) Dynamic adaptive adjustments of the three parameters *kp*, *ki*, *kd* of RBFNNPID algorithm.

3.2. Tracking Performance Test 2

The set point of DO concentration was changed on days 8, 10 and 12 to 2.5, 1.7 and back to 2 mg/L, respectively to verify the tracking performance of the new proposed RBFNNPID algorithm. Simulation results are shown in Figure 7.

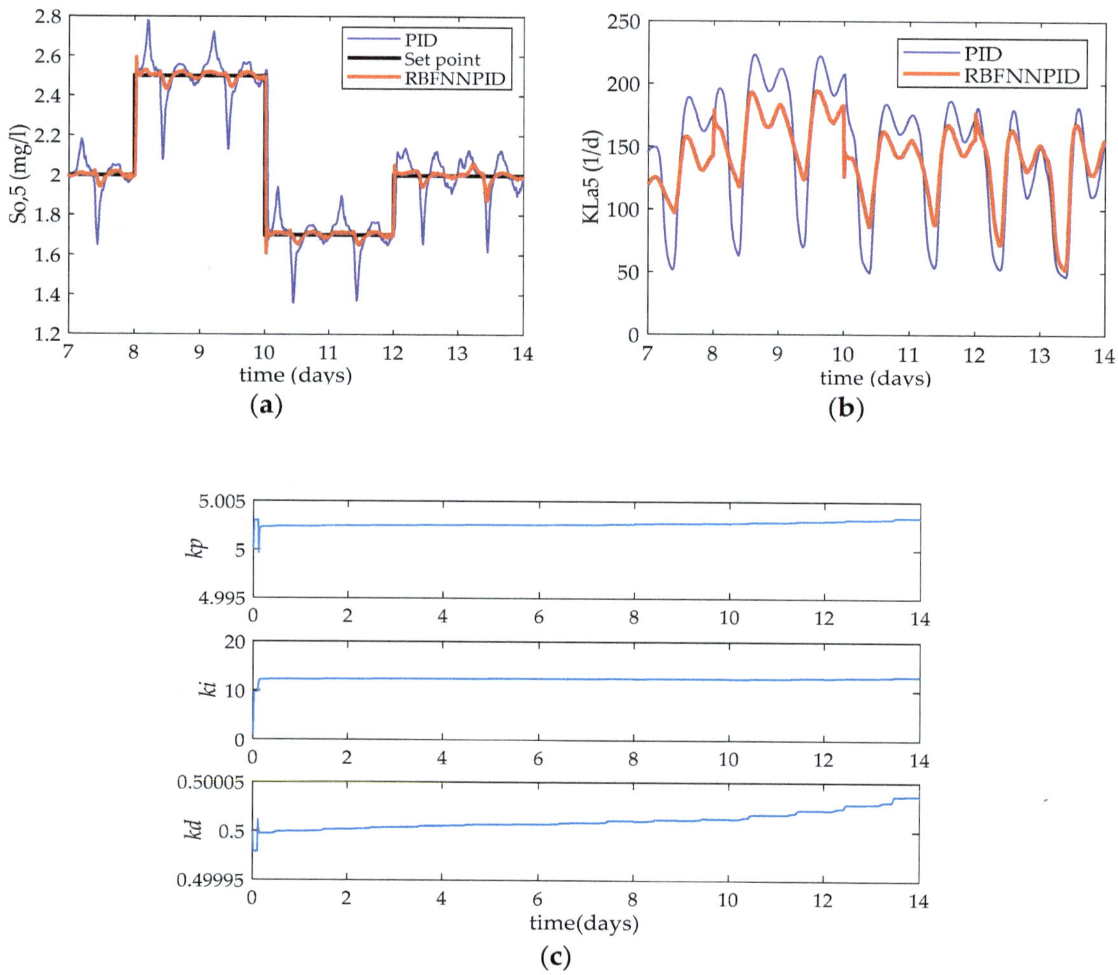

Figure 7. Comparison results of the tracking performance (changing set point of DO): (**a**) DO concentration in the fifth tank; (**b**) Dynamic changes of the manipulated variable K_La_5; (**c**) Dynamic adaptive adjustments of the three parameters *kp*, *ki*, *kd* of RBFNNPID algorithm.

3.3. Anti-Disturbance Performance Test

A good control algorithm should not only have a good tracking performance, but also have a strong anti-disturbance ability. By having these properties, it can be applied to control a complex system such as wastewater treatment process to achieve a precise control effect. To further verify the anti-disturbance ability of the RBFNNPID algorithm, we used the data collected in rain and storm weather to simulate the algorithm. Different weather condition can be looked as different disturbances in the influent. The results are shown in Figures 8 and 9.

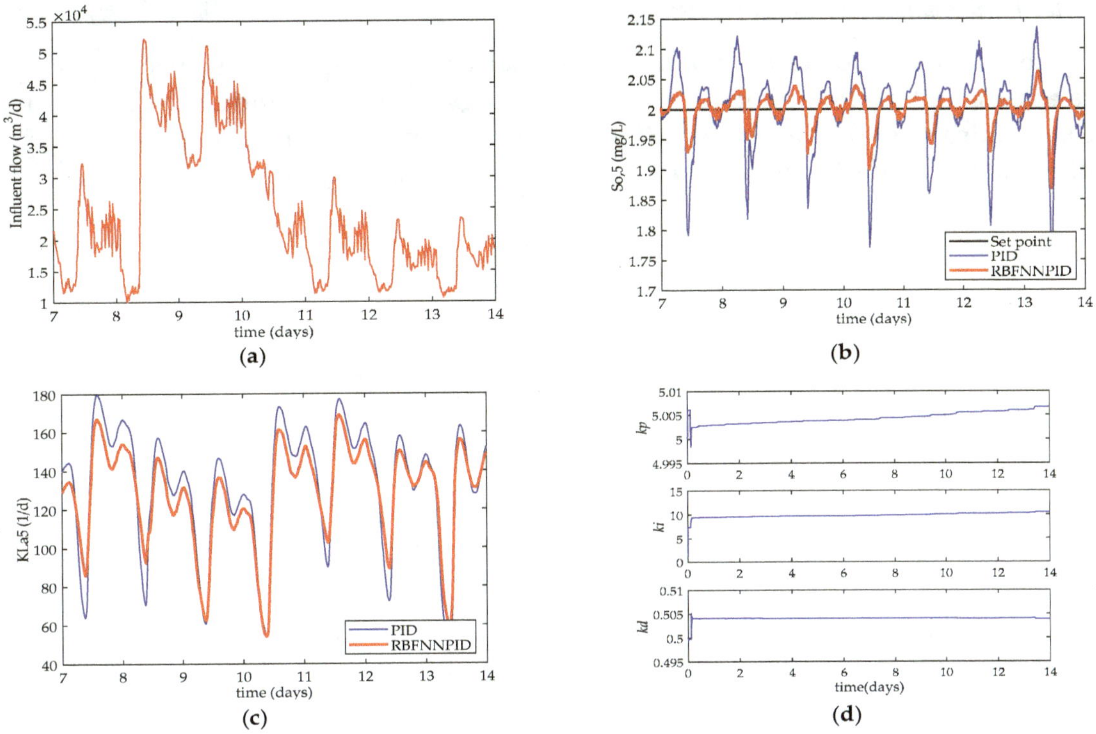

Figure 8. Comparison results of the anti-disturbance performance (rain weather): (**a**) Influent flow rate (days 7 to 14); (**b**) DO concentration in the fifth tank; (**c**) Dynamic changes of the manipulated variable K_La_5; (**d**) Dynamic adaptive adjustments of the three parameters *kp*, *ki*, *kd* of RBFNNPID algorithm.

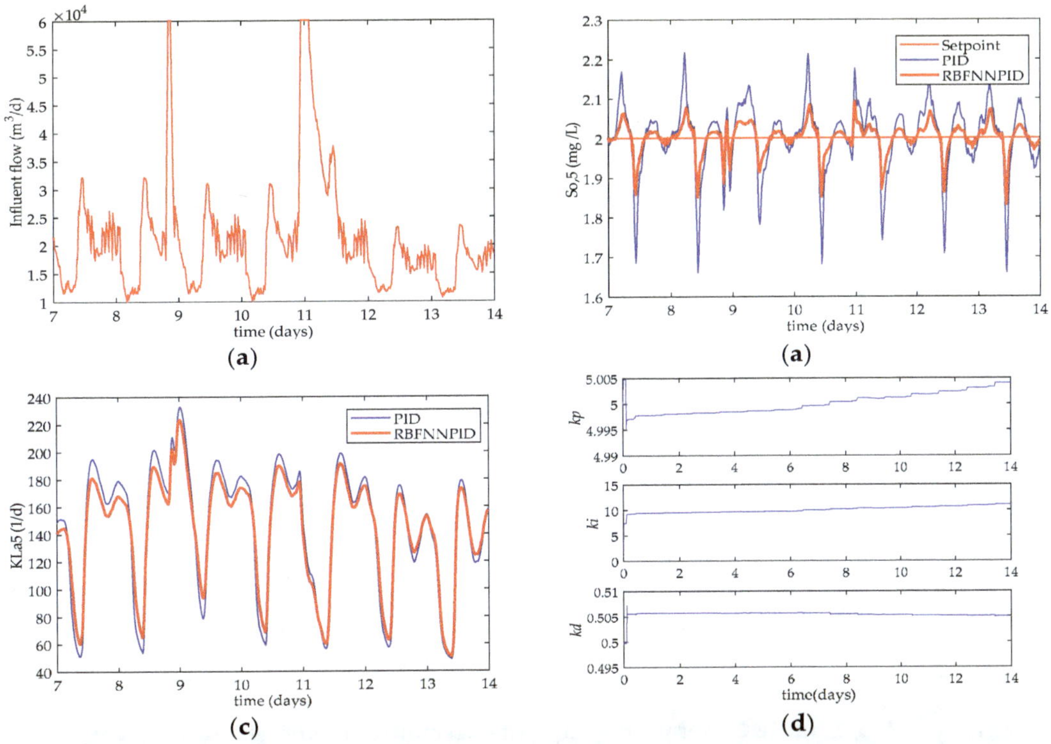

Figure 9. Comparison results of the anti-disturbance performance (storm weather): (**a**) Influent flow rate (days 7 to 14); (**b**) DO concentration in the fifth tank; (**c**) Dynamic changes of the manipulated variable K_La_5; (**d**) Dynamic adaptive adjustments of the three parameters *kp*, *ki*, *kd* of RBFNNPID algorithm.

3.4. Controller Performance Evaluation Index

There are two main indices for evaluating the performance of the dissolved oxygen controller. One is the assessment of the underlying control strategy. Indices include the integral of absolute error (IAE), the integral of squared error (ISE), the maximal deviation from set point ($Dev^{\max}$) and the variance of error (Var_e). The four indices are calculated by Equation (28) through to (31) as shown below.

$$IAE_i = \int_{t=7}^{t=14} |e_i| dt \tag{28}$$

$$ISE_i = \int_{t=7}^{t=14} e_i^2 dt \tag{29}$$

$$Dev_i^{\max} = \max|e_i| \tag{30}$$

$$Var(e_i) = \frac{ISE_i}{T} - (\frac{IAE_i}{T})^2 \tag{31}$$

The aeration cost can be calculated using aeration energy (AE), which will be the economic indicator. AE is mainly used in the last three units of the biochemical reaction tanks. AE can be obtained by using the oxygen transfer function (K_La) of the three units as shown in Equation (32)

$$AE = \frac{S_O^*}{T \cdot 1800} \int_{t=7}^{t=14} \sum_{i=1}^{i=5} V_i \cdot K_L a(t) dt \tag{32}$$

where, S_O^* is the saturation value of dissolved oxygen, V_i is the volume of each unit, and T is the calculation period of AE, in this case T = 7 days.

Generally, the smaller the value of the above evaluation indices, the better the performance of the controller is. Results of the evaluation indices are shown in Tables 1 and 2. We can see that, under the different weather conditions, the RBFNNPID control strategy reduced the values of the above evaluation indices, compared with the traditional PID control strategy, indicating that the control performance of the system has been effectively improved, and the cost of aeration has also been reduced.

Table 1. Performance of two DO control methods.

Weather	Method	ISE	IAE	Dev^{max}	Var_e
Dry	RBFNNPID	1.64×10^{-2}	2.08×10^{-1}	1.89×10^{-1}	2.10×10^{-3}
	PID	4.44×10^{-2}	4.03×10^{-1}	3.43×10^{-1}	6.30×10^{-3}
Rain	RBFNNPID	2.50×10^{-3}	9.47×10^{-2}	6.94×10^{-2}	3.53×10^{-4}
	PID	3.59×10^{-2}	3.61×10^{-1}	2.95×10^{-1}	5.10×10^{-3}
Storm	RBFNNPID	5.70×10^{-3}	1.39×10^{-1}	1.46×10^{-1}	8.16×10^{-4}
	PID	1.38×10^{-2}	2.27×10^{-1}	1.97×10^{-1}	2.00×10^{-3}

Table 2. Aeration energy of two control methods.

Weather	PID (kWh/d)	RBFNNPID (kWh/d)
Dry	7149.9	7032.1
Rain	6955.8	6805.8
Storm	7199.6	6971.8

4. Discussion

PID may fail to achieve the control goal or effect of the process while using the traditional PID control algorithm due to unknown and unexpected disturbances as well as significant changes in operating conditions, such as a significant change in the (i) quality of influent; (ii) weather, etc. Therefore, in general, the parameters of the traditional PID controller need to be adjusted under different operating environments. However, a long period of accumulated experience and several tests are needed in order for the traditional PID to be adjusted to achieve satisfactory results under each operating environment. Clearly, it is not feasible in real time applications and increases the difficulty in applying it in different wastewater treatment plants. Our work will reduce the difficulty of parameter tuning of the traditional PID controller, which is essential to improve the adaptability of the PID control parameters in practice.

The simulation results in Figures 6a, 7a, 8b and 9b show that the DO concentration is difficult to maintain at set point under the control of the conventional incremental PID controller when the influent flow rate and quality changed greatly. On the contrary, RBFNNPID can effectively maintain the DO concentration around the set value with a relatively low error by adjusting the air flow (which can be seen in Figures 6b, 7b, 8c and 9c). It can be seen that the dynamic changes of the manipulated variable $K_L a_5$ is smooth under the control of RBFNNPID. This means we can get a more stable status by using less air supply to the aeration tank. Therefore, using RBFNNPID can reduce the aeration cost which is one of the major electrical costs of the wastewater treatment processes. It also can be verified from the results shown in Tables 1 and 2.

According to the results of the rain weather and storm weather, which can be considered as there has the disturbances of the influent, shown in Figures 8 and 9, compared with the conventional PID controller, the RBFNNPID controller can quickly and accurately track the desired output trajectory values, which means it not only has a good tracking performance, but also has a stronger anti-disturbance ability with the changes to the set points. Figures 6c, 7c, 8d and 9d show the curves of the PID parameters are being adjusted adaptively. Parameters adjusted rapidly at the start of the simulation and small adjustments took place as the simulation goes on.

Applying precise control of the concentration of dissolved oxygen can not only avoid the occurrence of sludge bulking, but also reduce the aeration energy in a wastewater treatment plant. Intermittent aeration can be successfully implemented in a small-scale wastewater treatment plant to reduce the aeration energy while ensuring good effluent water quality [38]; however the same cannot be said in a large-scale wastewater treatment plant such as Pingliang Wastewater Treatment Plant that is situated in Gansu Province, China. In a large-scale wastewater treatment plant, intermittent aeration can reduce only a fraction of aeration energy. However, continuous aeration can effectively reduce the emissions of volatile organic compounds (VOCs) from a wastewater treatment plant, which has been proven as a factor for the increase of haze in some Chinese cities [39].

The characteristics of strong coupling, nonlinearity and large time delay of dissolved oxygen control system in activated sludge wastewater treatment and a control algorithm called RBFNNPID algorithm are discussed in this paper. However, this algorithm has not been applied directly into practice so far as it has certain complexity. We are currently undertaking the following two studies to verify its feasibility, validity, and superiority: (i) Simplifying the algorithm for practical use; (ii) Trialing the algorithm at Pingliang Wastewater Treatment Plant in Gansu Province, China.

5. Conclusions

In this paper, an adaptive PID control algorithm based on RBF neural network is proposed. The RBFNNPID algorithm combines the good learning and adaptive ability of neural networks and the practical advantages of PID algorithm. The gradient descent method is used to adaptively adjust the increment of the three parameters of the PID controller to achieve an optimal control effect on the control of DO concentration. The simulation results show that the RBFNNPID algorithm not only has a better performance of tracking and anti-jamming, but also has a great improvement to the robustness compared to that of the traditional PID. Thus, it can reduce the aeration costs of a wastewater treatment plant employing ASP.

Acknowledgments: China Scholarship Council (CSC) supported the first author to visit RMIT University as visiting scholar while undertaking this work. This work is also supported by the other funding include the National Natural Science Foundation of China (No. 61563032), the Natural Science Foundation of Gansu province (No. 1506RJZA104), University Scientific Research Project of Gansu province (No. 2015B-030), and the Excellent Young Teacher Project of Lanzhou University of Technology (No. Q201408).

Author Contributions: Xianjun Du, Guohua Shi and Junlu Wang conceived and designed the experiments; Junlu Wang performed the experiments; Xianjun Du, Junlu Wang and Veeriah Jegatheesan analyzed the data; Xianjun Du and Veeriah Jegatheesan wrote the paper.

Appendix A

We designed a DO control module in the Matlab Simulink environment which is shown in Figure A1. The output one "kla_out" is used to adjust the air flow of the blower, equipped for reactor 5, to maintain the DO concentration at a level of desired value. The scope is used to monitor the dynamic changes of the three parameters of the controller, which have been shown in Figures 6c, 7c, 8d and 9d. Detail of the inner structure of the controller is shown in Figure A2.

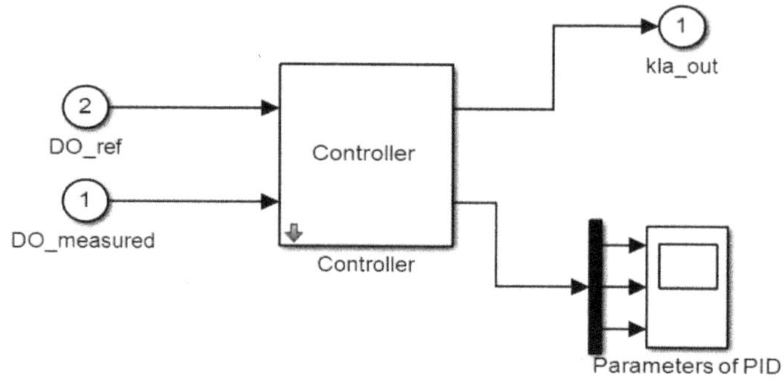

Figure A1. Overview of the DO control module.

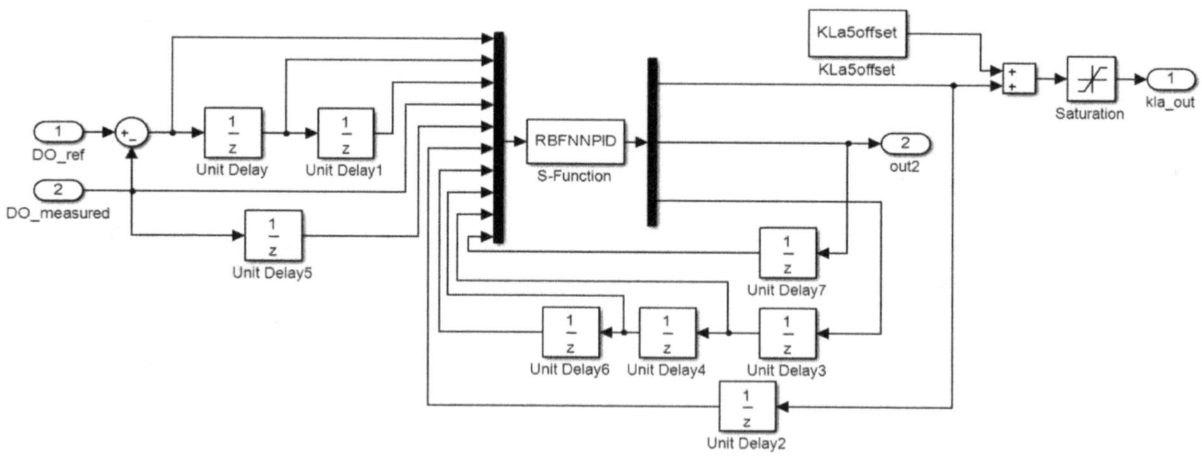

Figure A2. Detailed structure of the RBFNNPID controller.

Appendix B

The main codes of the S-function module, shown in Figure A2, are given below:

```
function [sys, x0, str, ts] = nnrbf_pid(t,x,u,flag,T,nn,K_pid, eta_pid, xite, alfa, beta0, w0)
    switch flag,
        case 0, [sys, x0, str, ts] = mdlInitializeSizes(T,nn);
        case 2, sys = mdlUpdates(u);
        case 3, sys = mdlOutputs(t, x, u, T,nn, K_pid,eta_pid, xite, alfa, beta0, w0);
        case {1, 4, 9}, sys = [];
        otherwise, error (['Unhandled flag = ' , num2str(flag)]);
    end
function [sys,x0,str,ts] = mdlInitializeSizes(T, nn)
    sizes = simsizes;
    sizes. NumContStates = 0;
    sizes.NumDiscStates = 3;
    sizes. NumOutputs = 4 + 5* nn;
    sizes.NumInputs = 9 + 15* nn;
    sizes. DirFeedthrough = 1;
    sizes. NumSampleTimes =1;
    sys = simsizes(sizes) ;
    x0 = zeros(3, 1);
    str = [];
    ts = [T0];
function sys = mdlUpdates(u)
    sys = [ u(1) − u(2); u(1); u(1) + u(3) − 2* u(2)];
function sys = mdlOutputs(t, x, u,T, nn, K_pid, eta_pid, xite, alfa, beta0, w0)
    % Initialization of the radial basis centers
    ci3 = reshape(u(7: 6 + 3* nn), 3, nn);
    ci2 = reshape(u(7 + 5* nn: 6 + 8* nn), 3, nn);
    ci1 = reshape(u(7 + 10* nn: 6 + 13* nn), 3, nn);
    % Initialization of the radial basis width
    bi3 = u(7 + 3* nn: 6 + 4* nn);
    bi2 = u(7 + 8*nn: 6 + 9* nn);
    bi1 = u(7 + 13* nn: 6 + 14* nn);
    % Initialization of the weights
    w3 = u(7 + 4* nn: 6+ 5* nn) ;
    w2 = u(7 + 9* nn: 6+ 10* nn) ;
    w1 = u(7 + 14* nn: 6+ 15* nn) ;
    xx = u([6; 4; 5]);
    if t = 0
        % Initialize the PID parameters
        ci1 = w0(1) * ones(3, nn);
        bi1 = w0(2) *ones(nn, 1);
        w1 = w0(3) * ones(nn, 1);
        K_pid0 = K_pid;
        else
        % Update the PID parameters
        K_pid0 = u(end-2: end);
    end
    for j = 1: nn
        % Gaussian
        h(j, 1) = exp(−norm(xx−ci1( : , j))^2/(2* bi1(j) * bi1(j)));
    end
    % Dynamic of gradient descent method
    dym = u(4) − w1'* h;
    W = w1 + xite* dym* h + alfa* (w1 − w2) + beta0*(w2 − w3) ;
    for j = 1: nn
        dbi(j,1) = xite* dym* w1(j) * h(j) * (bi1(j) ^(−3)) * norm(xx − ci1(:,j))^2;
        dci( : ,j) = xite*dym* w1(j)* h(j) * (xx − ci1(:,j)) * (bi1(j)^(−2));
    end
    bi = bi1 + dbi + alfa* (bi1 − bi2) + beta0*(bi2 − bi3) ;
    ci = ci1 + dci + alfa* (ci1 − ci2) + beta0*(ci2 − ci3) ;
    % Jacobian
    dJac = sum(w.*h.*(−xx (1) + ci (1,:)') ./bi.^2);
    % adjustments of the PID parameters
    KK(1) = K_pid0(1) + u(1) * dJac* eta_pid(1)* x(1);
    KK(2) = K_pid0(2) + u(1) * dJac* eta_pid(2)* x(2);
    KK(3) = K_pid0(3) + u(1) * dJac* eta_pid(3)* x(3);
    sys= [ u(6) + KK* x; KK'; ci( : ) ; bi( : ) ; w( : ) ] ;
```

Appendix C

This section describes the stability and convergence analysis of the proposed RBFNNPID algorithm.

The basic knowledge of the stability and convergence analysis is Lyapunov theorem, also known as Lyapunov stability.

$V(x_1, x_2, K, x_N)$ is an arbitrary function defined in the neighborhood of the origin Ω, where Ω is a state of equilibrium and x_1, x_2, K, x_N are variables, then

$$|x_i| \le H, i = 1, 2, K, N \tag{A1}$$

where, H is a positive constant.

Assuming that V is a continuous differentiable function in Ω and $V(0, 0, K, 0) = 0$. Such that

(i) $V(x) > 0$ is positive definite or $V(x) < 0$ is negative definite, $x \in \Omega$ and $x \neq 0$;
(ii) $V(x) > 0$ is positive semi-definite or $V(x) < 0$ is negative semidefinite, $x \in \Omega$;

Consider an autonomous nonlinear dynamical system

$$\dot{x} = f(x) \tag{A2}$$

where, $f(0) = 0$.

Assuming $x_i = x_i(t)$, $(i = 1, 2, K, N)$ is the solution of the system (A2). We can obtain the derivation

$$\frac{dV}{dt} = \frac{\partial V}{\partial x_1}\frac{\partial x_1}{\partial t} + \frac{\partial V}{\partial x_2}\frac{\partial x_2}{\partial t} + K + \frac{\partial V}{\partial x_N}\frac{\partial x_N}{\partial t} \tag{A3}$$

Introducing the gradient vector (Equation (A4)) into Equation (A3)

$$\nabla V(x) = \left[\frac{\partial V}{\partial x_1}, \frac{\partial V}{\partial x_2}, L, \frac{\partial V}{\partial x_N}\right]^T \tag{A4}$$

We will arrive at the final equation as below:

$$V = [\nabla V(x)]^T f(x) = \omega(x) \tag{A5}$$

The following conclusions can be made from the above analysis:

(i) If $V(x)$ is positive (or negative) definite, and if derivation $V = \omega(x)$ is negative (or positive) semi-definite, the system is said to be Lyapunov stable at the equilibrium of the origin;
(ii) If $V(x)$ is positive (or negative) definite, and if derivation $V = \omega(x)$ is negative (or positive) definite, the system is said to be exponentially stable at the equilibrium of the origin;
(iii) If $V(x)$ is positive (or negative) definite, and if derivation $V = \omega(x)$ is also positive (or negative) definite, the system is said to be unstable at the equilibrium of the origin;

For the adjustment of the weights of the neural network, we need a parameter called learning rate η. If η is too large, NN will be unstable; but if η is too small, the convergence rate will be too slow. Therefore, the selection of the value of learning rate is crucial to the stability and convergence of the system.

Assuming the indicator function of the RBFNNPID controller

$$J(k) = \frac{1}{2}[y(k) - y_m(k)]^2 = \frac{1}{2}e^2(k) \tag{A6}$$

where, $e(k)$ is the learning error of the network.

In order to ensure the adjustment of the weight coefficient is carried out in the direction of the negative gradient relative to $\omega(k)$, there must be

$$\omega(k+1) = \omega(k) - \eta \frac{\partial J(k)}{\partial \omega(k)} \tag{A7}$$

From Equations (A6) and (A7), we can get

$$\Delta\omega(k) = \frac{\partial J(k)}{\partial \omega(k)} = e(k)\frac{\partial e(k)}{\partial \omega(k)} = e(k)\frac{\partial e(k)}{\partial \Delta u(k)}\frac{\partial \Delta u(k)}{\partial \omega(k)} \tag{A8}$$

Defining a Lyapunov function of a discrete-time systems as

$$v(k) = \frac{1}{2}e^2(k) \tag{A9}$$

As we introduced in the paper, the gradient descent method is used as the change of the network-learning algorithm.

$$\Delta v(k) = v(k+1) - v(k) = \frac{1}{2}e^2(k+1) - \frac{1}{2}e^2(k) = \frac{1}{2}\Delta e(k)[2e(k) + \Delta e(k)] \tag{A10}$$

where, $e(0) = 0$, and,

$$\Delta e(k) = \frac{\partial J(k)}{\partial e(k)} = e(k) \tag{A11}$$

It can be obtained from Equation (A8), that

$$\Delta e(k) = e(k) = (\frac{\partial e(k)}{\partial \omega(k)})^T \Delta\omega(k) \tag{A12}$$

and

$$\Delta\omega(k) = -\eta\frac{\partial J(k)}{\partial \omega(k)} = -\eta e(k)\frac{\partial e(k)}{\partial \Delta u(k)}\frac{\partial \Delta u(k)}{\partial \omega(k)} \tag{A13}$$

By substituting Equation (A13) into Equation (A12)

$$\Delta e(k) = -\eta\|\frac{\partial J(k)}{\partial \omega(k)}\|^2 e(k) \tag{A14}$$

Then, substituting Equation (A14) into Equation (A10)

$$\Delta v(k) = -\frac{1}{2}\eta\|\frac{\partial J(k)}{\partial \omega(k)}\|^2 e(k)\left[2e(k) - \eta\|\frac{\partial J(k)}{\partial \omega(k)}\|^2 e(k)\right] = -\frac{1}{2}\eta\|\frac{\partial J(k)}{\partial \omega(k)}\|^2 e^2(k)\left[2 - \eta\|\frac{\partial J(k)}{\partial \omega(k)}\|^2\right] \tag{A15}$$

Knowing from the Lyapunov stability theory, the system is stable when $\Delta v < 0$. In addition, because of $\eta > 0$,

$$2 - \eta\|\frac{\partial J(k)}{\partial \omega(k)}\|^2 > 0 \tag{A16}$$

That is,

$$0 < \eta < \frac{2}{\|\frac{\partial J(k)}{\partial \omega(k)}\|^2} \tag{A17}$$

Therefore, the system is stable.
When $\Delta v < 0$,

$$\frac{1}{2}e^2(k+1) < \frac{1}{2}e^2(k) \tag{A18}$$

$$\lim_{k \to \infty} e(k) = 0 \tag{A19}$$

It means, with the increase of k, $e(k)$ gradually reaches to zero, and it guarantees the convergence of the learning algorithm. Based on the above analysis, if the value of η is according to Equation (A17), the system is stable and the learning algorithm will converge.

Appendix D

To verify the convergence of the parameters of the neural network, the dynamic changes of the centers (Figure A3) as well as weights and widths (Figure A4) of the neural network were computed, taking the first neuron of the hidden layer neurons as the example. From those figures, it can be seen that the parameters are converging with adaptive changes while the control process is going on.

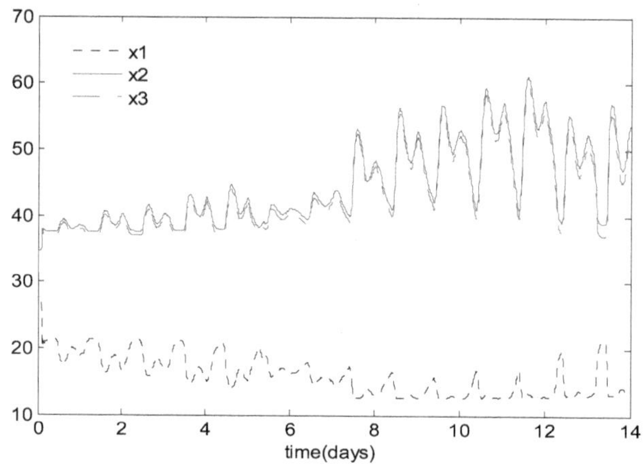

Figure A3. Dynamic changes of the centers of the first neuron (of the hidden layer neurons) of the neural network. (Where, ×1, ×2 and ×3 represent the three centers of the first neuron).

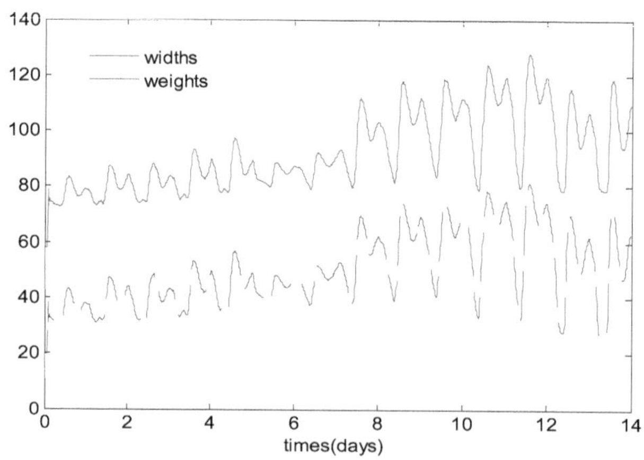

Figure A4. Dynamic changes of the weights and widths of the first neuron (of the hidden layer neurons) of the neural network.

References

1. Bo, Y.C.; Zhang, X. Online adaptive dynamic programming based on echo state networks for dissolved oxygen control. *Appl. Soft Comput.* **2017**. [CrossRef]
2. Holenda, B.; Domokos, E.; Rédey, Á.; Fazakas, J. Dissolved oxygen control of the activated sludge wastewater treatment process using model predictive control. *Comput. Chem. Eng.* **2008**, *32*, 1270–1278. [CrossRef]
3. Zhang, P.; Yuan, M.; Wang, H. Study on Dissolved Oxygen Control Method Based on International Evaluation Benchmark. *Inf. Control* **2007**, *36*, 199–203.

4. Yu, K.; Muhetaer, A.; Wei, L. Evaluation indexes of sewage stabilization from municipal wastewater treatment plant. *China Water Wastewater* **2016**, *5*, 93–97.
5. Chen, C.S. Robust self-organizing neural-fuzzy control with uncertainty observer for MIMO nonlinear systems. *IEEE Trans. Fuzzy Syst.* **2011**, *19*, 694–706. [CrossRef]
6. Nascu, I.; Vlad, G.; Folea, S.; Buzdugan, T. Development and application of a PID auto-tuning method to a wastewater treatment process. In Proceedings of the IEEE International Conference on Automation, Quality and Testing, Robotics (AQTR 2008), Cluj-Napoca, Romania, 22–25 May 2008.
7. Ye, H.T.; Li, Z.Q.; Luo, W.G. Dissolved oxygen control of the activated sludge wastewater treatment process using adaptive fuzzy PID control. In Proceedings of the 32nd Chinese Control Conference (CCC2013), Xi'an, China, 26–28 July 2013.
8. Goldar, A.; Revollar, S.; Lamanna, R.; Vega, P. Neural-MPC for N-removal in activated-sludge plants. In Proceedings of the IEEE European Control Conference (ECC 2014), Strasbourg, France, 24–27 June 2014.
9. Goldar, A.; Revollar, S.; Lamanna, R.; Vega, P. Neural NLMPC schemes for the control of the activated sludge process. In Proceedings of the 11th IFAC Symposium on Dynamics and Control of Process Systems Including Biosystems, Trondheim, Norway, 6–8 June 2016.
10. Belchior, C.A.C.; Araújo, R.A.M.; Landeck, J.A.C. Dissolved oxygen control of the activated sludge wastewater treatment process using stable adaptive fuzzy control. *Comput. Chem. Eng.* **2012**, *37*, 152–162. [CrossRef]
11. Jeppsson, U.; Rosen, C.; Alex, J.; Copp, J.; Gernaey, K.V.; Pons, M.N.; Vanrolleghem, P.A. Towards a benchmark simulation model for plant-wide control strategy performance evaluation of WWTPs. *Water Sci. Technol.* **2006**, *53*, 287–295. [CrossRef] [PubMed]
12. Yu, G.; Zhang, P.; Wei, S.; Fan, M.; Wang, H. Human-simulation intelligent PID control theory and its application for dissolved oxygen in wastewater treatment. *Microcomput. Inf.* **2006**, *22*, 13–15.
13. Syu, M.J.; Chen, B.C. Back-propagation neural network adaptive control of a continuous wastewater treatment process. *Ind. Eng. Chem. Res.* **1998**, *37*, 3625–3630. [CrossRef]
14. Macnab, C.J.B. Stable Neural-Adaptive Control of Activated Sludge Bioreactors. In Proceedings of the 2014 American Control Conference (ACC2014), Portland, OR, USA, 4–6 June 2014.
15. Mirghasemi, S.; Macnab, C.J.B.; Chu, A. Dissolved oxygen control of activated sludge bioreactors using neural-adaptive control. In Proceedings of the IEEE Symposium on Computational Intelligence in Control and Automation (CICA 2014), Orlando, FL, USA, 9–12 December 2014.
16. Ruan, J.; Zhang, C.; Li, Y.; Li, P.; Yang, Z.; Chen, X.; Huang, M.; Zhang, T. Improving the efficiency of dissolved oxygen control using an on-line control system based on a genetic algorithm evolving FWNN software sensor. *J. Environ. Manag.* **2017**, *187*, 550–559. [CrossRef] [PubMed]
17. Qiao, J.; Fu, W.; Han, H. Dissolved oxygen control method based on self-organizing T-S fuzzy neural network. *CIESC J.* **2016**, *67*, 960–966.
18. Li, M.; Zhou, L.; Wang, J. Neural network predictive control for dissolved oxygen based on Levenberg-Marquardt algorithm. *Trans. Chin. Soc. Agric. Mach.* **2016**, *47*, 297–302.
19. Xu, J.; Yang, C.; Qiao, J. A novel dissolve oxygen control method based on fuzzy neural network. In Proceedings of the 36th Chinese Control Conference (CCC2017), Dalian, China, 26–28 July 2017.
20. Lin, M.J.; Luo, F. Adaptive neural control of the dissolved oxygen concentration in WWTPs based on disturbance observer. *Neurocomputing* **2016**, *185*, 133–141. [CrossRef]
21. Han, H.G.; Qiao, J.F.; Chen, Q.L. Model predictive control of dissolved oxygen concentration based on a self-organizing RBF neural network. *Control Eng. Pract.* **2012**, *20*, 465–476. [CrossRef]
22. Zhou, H. Dissolved oxygen control of wastewater treatment process using self-organizing fuzzy neural network. *CIESC J.* **2017**, *68*, 1516–1524.
23. Huang, M.Z.; Han, W.; Wan, J.Q.; Chen, X. Multi-objective optimization for design and operation of anaerobic digestion using GA-ANN and NSGA-II. *J. Chem. Technol. Biotechnol.* **2016**, *91*, 226–233. [CrossRef]
24. Li, Y.; Li, T.; Jiang, Y.; Fan, J.-L. Adaptive PID control of quadrotor based on RBF neural network. *Control Eng. China* **2016**, *23*, 378–382.
25. Chamsai, T.; Jirawattana, P.; Radpukdee, T. Robust adaptive PID controller for a class of uncertain nonlinear systems: An application for speed tracking control of an SI engine. *Math. Probl. Eng.* **2015**, *17*, 1–12. [CrossRef]

26. Lin, C.M.; Chung, C.C. Fuzzy brain emotional learning control system design for nonlinear systems. *Int. J. Fuzzy Syst.* **2015**, *17*, 117–128. [CrossRef]
27. Henze, M.; Grady, C.P.L.; Gujer, W.; Matsuo, T. *Activated Sludge Model No. 1*; IAWPRC Publishing: London, UK, 1987.
28. Gernaey, K.V.; Jeppsson, U.; Vanrolleghem, P.A.; Copp, J.B. *Benchmarking of Control Strategies for Wastewater Treatment Plants*; IWA Publishing: London, UK, 2014.
29. Du, X.J.; Hao, X.H.; Li, H.J.; Ma, Y.W. Study on modelling and simulation of wastewater biochemical treatment activated sludge process. *Asian J. Chem.* **2011**, *23*, 4457–4460.
30. Simsek, H. Mathematical modeling of wastewater-derived biodegradable dissolved organic nitrogen. *Environ. Technol.* **2016**, *37*, 2879–2889. [CrossRef] [PubMed]
31. Takács, I.; Patry, G.G.; Nolasco, D. A dynamic model of the clarification thickening process. *Water Res.* **1991**, *25*, 1263–1271. [CrossRef]
32. Olsson, G.; Newell, B. *Wastewater Treatment Systems: Modelling, Diagnosis and Control*, 1st ed.; IWA Publishing: London, UK, 1999.
33. Maier, H.R.; Dandy, G.C. Neural networks for the prediction and forecasting of water resources variables: A review of modelling issues and applications. *Environ. Model. Softw.* **2000**, *15*, 101–124. [CrossRef]
34. Wang, J.; Shi, P.; Jiang, P.; Xiao, P. Application of BP Neural Network Algorithm in Traditional Hydrological Model for Flood Forecasting. *Water* **2017**, *9*, 48. [CrossRef]
35. Zebardast, B.; Maleki, I. A New Radial Basis Function Artificial Neural Network based Recognition for Kurdish Manuscript. *Int. J. Appl. Evolut. Comput.* **2013**, *4*, 72–87. [CrossRef]
36. Wahab, N.A.; Katebi, R.; Balderud, J. Multivariable PID control design for activated sludge process with nitrification and denitrification. *Biochem. Eng. J.* **2009**, *45*, 239–248. [CrossRef]
37. Luo, F.; Hoang, B.L.; Tien, D.N.; Nguyen, P.H. Hybrid PI controller design and hedge algebras for control problem of dissolved oxygen in the wastewater treatment system using activated sludge method. *Int. Res. J. Eng. Technol.* **2015**, *2*, 733–738.
38. Sánchezmonedero, M.A.; Aguilar, M.I.; Fenoll, R.; Roig, A. Effect of the aeration system on the levels of airborne microorganisms generated at wastewater treatment plants. *Water Res.* **2008**, *42*, 3739–3744. [CrossRef] [PubMed]
39. Chen, W.H. Influences of Aeration and Biological Treatment on the Fates of Aromatic VOCs in Wastewater Treatment Processes. *Aerosol Air Qual. Res.* **2013**, *13*, 225–236. [CrossRef]

14

Formulation of Laccase Nanobiocatalysts Based on Ionic and Covalent Interactions for the Enhanced Oxidation of Phenolic Compounds

Maria Teresa Moreira *, Yolanda Moldes-Diz, Sara Feijoo, Gemma Eibes, Juan M. Lema and Gumersindo Feijoo

Department of Chemical Engineering, Institute of Technology, University of Santiago de Compostela, 15782 Santiago de Compostela, Spain; yolanda.moldes@usc.es (Y.M.-D.); s.feijoomoreira@student.tudelft.nl (S.F.); gemma.eibes@usc.es (G.E.); juan.lema@usc.es (J.M.L.); gumersindo.feijoo@usc.es (G.F.)

* Correspondence: maite.moreira@usc.es.

Featured Application: Development and potential use of nanobiocatalysts for the removal of phenolic compounds as well as other related xenobiotics present in industrial wastewaters.

Abstract: Oxidative biocatalysis by laccase arises as a promising alternative in the development of advanced oxidation processes for the removal of xenobiotics. The aim of this work is to develop various types of nanobiocatalysts based on laccase immobilized on different superparamagnetic and non-magnetic nanoparticles to improve the stability of the biocatalysts. Several techniques of enzyme immobilization were evaluated based on ionic exchange and covalent bonding. The highest yields of laccase immobilization were achieved for the covalent laccase nanoconjugates of silica-coated magnetic nanoparticles (2.66 U mg^{-1} NPs), formed by the covalent attachment of the enzyme between the aldehyde groups of the glutaraldehyde-functionalized nanoparticle and the amino groups of the enzyme. Moreover, its application in the biotransformation of phenol as a model recalcitrant compound was tested at different pH and successfully achieved at pH 6 for 24 h. A sequential batch operation was carried out, with complete recovery of the nanobiocatalyst and minimal deactivation of the enzyme after four cycles of phenol oxidation. The major drawback associated with the use of the nanoparticles relies on the energy consumption required for their production and the use of chemicals, that account for a major contribution in the normalized index of 5.28×10^{-3}. The reduction of cyclohexane (used in the synthesis of silica-coated magnetic nanoparticles) led to a significant lower index (3.62×10^{-3}); however, the immobilization was negatively affected, which discouraged this alternative.

Keywords: laccase; nanocatalyst; immobilization; phenol; sequential batch reactor

1. Introduction

Laccase is a high potential oxidative enzyme with broad substrate specificity towards aromatic compounds, which makes it a promising candidate for the degradation of xenobiotics containing hydroxyl and amine groups [1–3]. However, the relatively low stability of the free enzyme arises as a major technical hurdle that hampers its large-scale application [4]. Beyond the potentiality of protein engineering and directed evolution to change enzyme conformation [5], enzyme immobilization can be applied to enhance the protein stability by the prevention of autolysis or proteolysis, rigidification of the enzyme structure via multipoint covalent attachment, and generation of favorable microenvironments [6–8]. This method has been demonstrated to improve the activity and stability of the biocatalyst in both aqueous and organic phases, provided that the support permits the diffusion of the substrate to the active site of the enzyme [9]. Furthermore, it may facilitate the simple recovery of the enzymes by centrifugation, sedimentation, or other physical separation methods and reuse in continuous systems. However, immobilized enzymes can also encounter several drawbacks, such as mass transfer limitations or interaction between the enzyme and the support that may reduce its catalytic potential [10].

Conceptually, there are two basic methods for enzyme immobilization, as the enzyme-support link can take place by physical or chemical interactions. Physical coupling methods include the entrapment of the enzyme within a tridimensional matrix, its encapsulation in an organic or inorganic polymer, and its adsorption to the support surface by ionic exchange [11], whereas covalent bonding assures the irreversible binding of the enzyme to the support matrix.

Among a wide range of alternatives, the large specific surface area characteristic of nanomaterials makes this type of support an ideal candidate for enzyme immobilization [12]. The efficiency of ionic exchange depends on the pH and ionic strength of the medium as well as the hydrophobic nature of the nanoparticle surface [13–15]. Regarding covalent bonding, nanoparticles may provide a homogeneous core-shell structure, which can be functionalized to react with nucleophilic groups on the enzyme [16]. Most enzymes are covalently attached to the lysine amino groups, which are typically present on the protein surface [17]. Several factors, including pH, ionic strength, protein concentration, additives, and nanoparticles structure (porous or non-porous material) may affect the biocatalyst and the effectiveness of covalent bonding between the enzyme and the support [16,18].

The immobilization of laccase on different types of nanoparticles such as silver and gold nanoparticles [19], chitosan-coated magnetic nanoparticles [20], and carbon nanotubes [21] has been demonstrated in recent years, although few processes have been used for practical applications at full-scale [22]. The main aim of this work is to perform the efficient immobilization of laccase on different types of magnetic and non-magnetic nanoparticles. Two different immobilization procedures will be followed: ionic exchange between the enzyme and the nanoparticle, and covalent bonding of the enzyme protein to the surface of the nanoparticle using glutaraldehyde or carbodiimide as cross-linkers [23,24]. Glutaraldehyde, a bifunctional and versatile agent, may react with different enzyme moieties, principally involving primary amino groups of proteins, although it may eventually react with other groups such as thiols, phenols, and imidazoles [25]. On the other hand, carbodiimide is used to form amide linkage between carboxylates and amino terminal groups from the enzyme [26]. The catalytic activity of the different nanobiocatalysts will be evaluated in terms of the biotransformation potential of phenol as the model compound. Once the successful immobilization of laccase is proved, we will aim to examine how the application of life cycle principles may be helpful in the reformulation of the production scheme of the most suitable support.

2. Materials and Methods

2.1. Chemicals and Nanoparticles for Enzyme Immobilization

(3-Aminopropyl)triethoxysilane (APTES) ($\geq$98%), 2,2′-azinobis(3-ethylbenzothiazoline-6-sulphonic acid) (ABTS) ($\geq$98%), glutaraldehyde (25%), 3-(Ethyliminomethyleneamino)-*N*,*N*-dimethylpropan-1- amine (EDC) ($\geq$98%), and fumed silica nanoparticles were purchased from Sigma-Aldrich (St. Louis, MO, USA). Non-coated magnetite nanoparticles, single-core silica-coated magnetic nanoparticles (FeO-2206W), multi-core silica-coated magnetic nanoparticles (S-57), polyacrylic acid nanoparticles (FeO-2204W and FeO-36), and polyethyleneimine-coated magnetic nanoparticles (VOZ-19) were supplied by Nanogap (Ames, Spain). Detailed characteristics of the nanoparticles evaluated are presented in Table 1.

Table 1. Characteristics of the different nanoparticles.

Type of Nanoparticles	Size (nm)	Concentration (mg NPs mL^{-1})
Fumed silica nanoparticles (fsNP)	7	59
Silica-coated magnetic nanoparticles		
FeO-2206W (single-core)	21.5 ± 2.1	5
S-57 (multi-core)	11.8 ± 2.4	10.9
Polyacrylic acid (PAA) magnetic nanoparticles		
FeO-2204W	10.1 ± 2.4	20.5
FeO-36	23.1 ± 4.9	16.2
Polyethylenimine (PEI) magnetic nanoparticles		
VOZ-19	10 ± 1.2	56
Non-coated magnetite nanoparticles	9.9 ± 1.4	17.4

2.2. Laccase Activity

Laccase activity from *Trametes versicolor* (activity ~10 U mg^{-1}, Sigma-Aldrich, St. Louis, MO, USA) was measured according to Zimmerman et al. [23]. Following this protocol, 50 μL of sample was added to 150 μL of 0.267 mM ABTS (in McIlvaine buffer; pH 3) in 96-well plates. The ABTS oxidation was monitored by measuring the absorbance at 420 nm for 7 min (with intervals of 6 s), with a molar extinction coefficient of the cation radical of 36,800 M^{-1} cm^{-1} [24]. One unit U of activity was defined as the amount of enzyme capable of producing 1 μmol of the cation radical per min.

2.3. Functionalization of Laccase onto Silica and Silica-Coated Magnetic Nanoparticles

The immobilization process for fumed silica nanoparticles (fsNP) and silica-coated magnetic nanoparticles (smNP) requires their previous functionalization, in which reactive groups are added based on the modification of their surface by the addition of (3-aminopropyl)triethoxysilane (APTES) [23]. The protocol starts with the incubation of the nanoparticles in phosphate buffer (100 mM, pH 7) and APTES (0.8 mmol APTES g^{-1} nanoparticles) under agitation (100 rpm) for 12 h at room temperature. The residual APTES concentration in the supernatants was monitored as follows: 50 μL of 5.3 mM glutaraldehyde solution was added to 150 μL supernatant. The yellow coloration due to the imine bond resulting from the chemical reaction of APTES with glutaraldehyde was measured spectrophotometrically at 390 nm. After four washing steps, no residual APTES was detected.

2.4. Immobilization of Laccase onto Silica and Polyethylenimine Nanoparticles

The amino-functionalized nanoparticles were then used to perform the immobilization of laccase according to the sorption-assisted immobilization (SAI) protocol [23], where the amino-functionalized nanoparticles and laccase (15 mg mL^{-1}) were incubated in phosphate buffer (pH 7, 100 mM) at 4 °C and 100 rpm for 2 h. Next, glutaraldehyde was added dropwise to the mixture of nanoparticles and laccase, and the solution was incubated for an additional 18 h. The unreacted glutaraldehyde and the excess and unstable bound enzymes were washed away.

The immobilization procedure for the polyethylenimine-coated magnetic nanoparticles (PEI-mNPs) was identical to the one previously described for silica nanoparticles except for the step of functionalization with APTES (not required here). The enzymatic activity of both NP-laccase conjugates and supernatants was measured in these immobilization processes as well as in the following ones to determine the activity yield, washing loss, and enzyme load. Variable concentrations of glutaraldehyde and laccase activity were used in the immobilization process: 4–8 mmol g^{-1} NPs and 0.9–1.88 U mg^{-1} NPs, respectively.

2.5. Immobilization of Laccase onto Polyacrylic Acid-Coated Magnetic Nanoparticles

The functionalization of the nanoparticles was conducted according to the method described by Nobs et al. [24]. The nanoparticles (5 mg mL^{-1}) were suspended in 2-(*N*-morpholino)ethanesulfonic acid (MES) buffer 0.1 M (pH 4.7), and EDC (12 mg mL^{-1}) and N-hydroxysuccinimide (NHS, 33 mg mL^{-1}) were added with gentle agitation (100 rpm) at room temperature (25 °C) to complete the reaction after 24 h. The unreacted NHS and EDC were removed by repeated washing and centrifugation (4000 rpm, 6 min), and were resuspended in MES buffer (0.1 M, pH 4.7). The amino-functionalized nanoparticles were then used to perform the immobilization of laccase by the aforementioned SAI method [23] with 8 mmol glutaraldehyde g^{-1} NPs and 1.88 U laccase mg^{-1} NPs.

2.6. Immobilization of Laccase by Ionic Exchange on Magnetite Nanoparticles

Laccase immobilization in magnetite nanoparticles (lacking any external coating) was carried out by ionic exchange of the enzyme with magnetite nanoparticles. Laccase was added (0.55 U mg^{-1} NPs) to previously washed nanoparticles and incubated at 4 °C, 100 rpm, and pH 5 for 4 h. After incubation, the nanobiocatalyst was washed five times in sodium phosphate buffer before storage.

2.7. Biotransformation of Phenol by Laccase Immobilized onto fsNPs and Single-Core Silica-Coated Magnetic Nanoparticles in Batch Operation

The oxidation of phenol by the enzymatic system was investigated in a reaction medium containing phenol (10 mg L^{-1}) dissolved in phosphate buffer (100 mM, pH 7) and immobilized laccase (1000 U L^{-1}) onto fsNPs or smNPs (FeO-2206W) in 10-mL flasks. In parallel, experiments with free laccase as well as controls lacking laccase with functionalized fsNPs and mNPs were also carried out. Samples were withdrawn at specific time intervals for 24 h to monitor phenol removal.

2.8. Consecutive Cycles of Batch Biotransformation of Phenol by Laccase Immobilized onto Single-Core Silica-Coated Magnetic Nanoparticles

Variable pH values (5–7) were investigated to perform the biotransformation of phenol (10 mg L^{-1}) by laccase immobilized on silica-coated magnetic nanoparticles (1000 U L^{-1}). Acetate buffer (100 mM) was applied for pH 5, while in the case of pH 6 and 7, phosphate buffer (100 mM) was used. Thereafter, the operation of the enzymatic system was conducted in a tank reactor (100 mL) under stirring at room temperature for several consecutive cycles. The reaction medium consisted of phenol (10 mg L^{-1}), phosphate buffer (100 mM, pH 6), and a single initial pulse of laccase (1000 U L^{-1}) immobilized onto FeO-2206W smNP. The effluent of the reactor was withdrawn at the end of the cycle and the nanobiocatalyst was recovered by an external magnetic field before a new cycle started.

2.9. Phenol Analysis

Phenol concentration was determined by high-performance liquid chromatography (HPLC) at a detection wavelength of 270 nm on a Jasco XLC HPLC (Jasco Analítica, Madrid, Spain). This equipment was coupled with a diode detector 3110 MD, a 4.6 × 150 nm Gemini reversed-phase column (3 µm C18 110 Å) from Phenomenex (supplied by Jasco Analítica, Madrid, Spain), and an HP ChromNav data processor. A 25-µL sample volume was injected into the column. The mobile phase contained 50% acetonitrile and 50% water. The flow rate was fixed at 0.4 mL min^{-1} under isocratic conditions.

2.10. Life Cycle Assessment Methodology

Life Cycle Assessment (LCA) is a methodology that aims to analyze products, processes, and/or services from an environmental point of view, and should be part of the decision-making process toward sustainability [27]. The guidelines established by International Organization for Standardization (ISO) standards [28] have been considered to perform the LCA study. The environmental profiles of the silica-coated mNPs production were determined according to the production route described in a previous paper [29]. In a typical synthesis of silica-coated mNPs, polyoxyethylene(5)nonylphenyl ether (Igepal CO-520) and cyclohexane are mechanically stirred before the addition of oleic-acid magnetite nanoparticles (2.5% wt in cyclohexane). Finally, ammonium hydroxide solution and tetraethyl orthosilicate (TEOS) are added consecutively to form a transparent red solution of reverse micro-emulsion. The core-shell nanoparticles are precipitated with isopropanol (IPA) to disrupt the reverse microemulsion and are then washed extensively with IPA and deionized water. Finally, the core-shell nanoparticles are re-dispersed in deionized water. In the case of the production of magnetic nanoparticles with a thin silica-coating, the procedure is similar except for the concentration of cyclohexane (0.5%, five times lower). Inventory data for the foreground systems were obtained from a semi-pilot unit and data from the background system (production of electricity, chemicals, and wastewater treatment) were taken from Ecoinvent database® version 3 [30–32] and, when possible, updated for Spain [33]. The environmental assessment was conducted using characterization factors from ReCiPe Midpoint methodology [34] and the following impact categories were considered in the analysis: climate change, ozone depletion, terrestrial acidification, freshwater eutrophication, marine eutrophication, human toxicity, photochemical oxidant formation, terrestrial ecotoxicity, freshwater ecotoxicity, marine ecotoxicity, and fossil depletion. SimaPro version 7.3.3 (PRé Consultants, Amersfoort, The Netherlands) was the software used for the computational implementation of the life cycle inventory data and the computation of the environmental profiles [35].

3. Results and Discussion

3.1. Immobilization of Laccase onto Different Types of Nanoparticles

Several strategies of laccase immobilization were evaluated on magnetic and non-magnetic nanoparticles to obtain various types of nanobiocatalysts. The enzymatic activities of both NP-laccase conjugates and supernatants were measured to determine the activity yield, washing loss, and enzyme load (Table 2). The tradeoff analysis of the different outcomes will be critical to identify the most suitable option for its further use.

Table 2. Activity yield, washing loss, and enzyme loading for the optimal doses in the immobilization processes.

Different Types of Nanoparticles	Washing Loss (%)	Activity Yield (%)	Enzyme Loading (U mg^{-1} NPs)
Covalent immobilization			
Fumed silica nanoparticles (fsNP)	5.6 ± 1.3	100 ± 6.1	1.78 ± 0.07
Silica-coated magnetic nanoparticles			
FeO-2206W (single-core)	16.4 ± 2.81	99.7 ± 0.35	2.66 ± 0.65
S-57 (multi-core)	66.63 ± 1.67	31.3 ± 0.76	0.42 ± 0.05
Polyacrylic acid (PAA) magnetic nanoparticles			
FeO-2204W	96.83 ± 1.4	2.55 ± 6.5	0.11 ± 0.34
FeO-36	99.7 ± 2.34	0.12 ± 2.3	0.01 ± 0.24
Polyethylenimine (PEI) magnetic nanoparticles			
VOZ-19	27.18 ± 0.08	80.5 ± 0.21	1.54 ± 0.03
Ionic exchange immobilization			
Non-coated magnetite nanoparticles	45.3 ± 0.9	58.5 ± 1.5	0.69 ± 0.05

Covalent bonding produces stronger bonds between the enzyme and the support, allowing its reuse more easily than with other available immobilization methods [36,37] and preventing the leaching of enzymes from the support [38,39]. In this study, different coatings as well as single- and multi-core nanoparticles were evaluated for laccase immobilization (Table 2). The covalent bonding between the supports with carboxylic groups (polyacrylic acid) did not result in satisfactory immobilization (yields lower than 5%). This may be due to excessive crosslinking of the protein molecule (due to the presence of both NH_2 and COOH groups in the enzyme and also to the instability of carbodiimide, which led to very low activity yields, as was also observed in other studies with activity yields below 14%) [40,41].

However, laccase was successfully immobilized in other nanoparticles such as PEI-mNPs, which showed an activity yield higher than 80% (Table 2). The stability of this biocatalyst was inferior to laccase immobilized onto fsNPs and FeO-2206-W smNPs (Table 2). In the specific case of PEI-mNPs, the enzyme activity after three months was 50%, which was significantly lower than those of fsNPs and FeO-2206-W (93% and 99%, respectively). Similar results were observed for a previous report with silica nanoparticles, using remarkably higher dosages of APTES and glutaraldehyde than the values considered in this research [23]. The rationale behind the high activity yields is attributed to the fact that immobilized laccases on this type of support would have high affinity for standard substrates such as ABTS. For instance, Arca-Ramos et al. [42] reported the hyperactivation of laccase from *T. versicolor* after the formation of covalent bonds with silica nanoparticles; whereas Matijosyte et al. [11] described a similar behavior for laccase from *T. villosa* (activity recovery up to 148%) after the formation of cross-linking aggregates (CLEAS®, CLEATechnologies, Delft, The Netherlands).

The process of immobilization by ionic exchange is based on the interaction of the charged groups of the enzyme with the groups of opposite charges in the support. It provides a weak bond between the enzyme and the support so that the native structure of the enzyme is unaltered. Moreover, the bonding is reversible and it is sensitive to changes in the pH and ionic strength, which can lead to the recovery of the support [9]. When this approach was considered for the immobilization of laccase, not only limited yield was evidenced, but also the change of basic pH led to enzyme desorption. When performing the immobilization of laccase at different pH values, the best results were observed when the immobilization process was performed at pH 5 with an activity yield higher than 50% (Table 2), possibly because the point of zero charge (PZC) of the magnetite is between pH 6.5–7.9 [43], while the isoelectric point of laccase is at pH 3 [44]. The main drawback is that laccase stability decreases with lower pH, which was evidenced by the reduction of the immobilized enzyme. Considering the best results of activity yield and enzyme load, fsNPs and FeO-2206W smNPs were selected for the following experiments of phenol biotransformation.

3.2. Biotransformation of Phenol by Laccase Immobilized onto fsNPs and Silica-Coated Magnetic Nanoparticles in Batch Operation

The capacity of free and immobilized enzymes onto fsNPs and FeO-2206W smNPs to transform phenol was assessed in batch operation. The results showed that the higher phenol transformation was achieved by free laccase (>75%), whereas phenol conversion was around 23% and 48% for fsNPs and FeO-2206W smNPs, respectively (Figure 1). Lower activity of immobilized laccase towards phenolic substrates has also been previously reported. For instance, Arca-Ramos et al. [42] found that bisphenol A degradation rate was much slower for immobilized laccases (from 6- to 26-fold lower than that of the free enzyme). This lower reaction rate was related to the potential aggregation of the nanoparticles which could reduce substrate accessibility. Wang et al. [45] studied phenol degradation by immobilized laccase on magnetic silica nanoparticles, and similar results were observed at pH 7. The rate of phenol conversion for laccase immobilized onto FeO-2206W smNPs (2.01 $\mu M\ h^{-1}$) is almost two times higher than that for fsNPs. Controls with phenol lacking laccase but with functionalized nanoparticles were performed, with no decrease in phenol concentration in all cases after 24 h.

Kurniawati and Nicell [46] reported that laccase can be inactivated due to the presence of free radicals in the reaction medium generated from phenol transformation (not by the substrate). However, this effect was only evident at phenol concentrations of 2000 μM (188 mg/L), almost 20-fold higher than that used in the present work. Hence, no enzymatic activity changes occurred in any of the experiments with a noticeable enzyme deactivation when performing the experiment with free laccase.

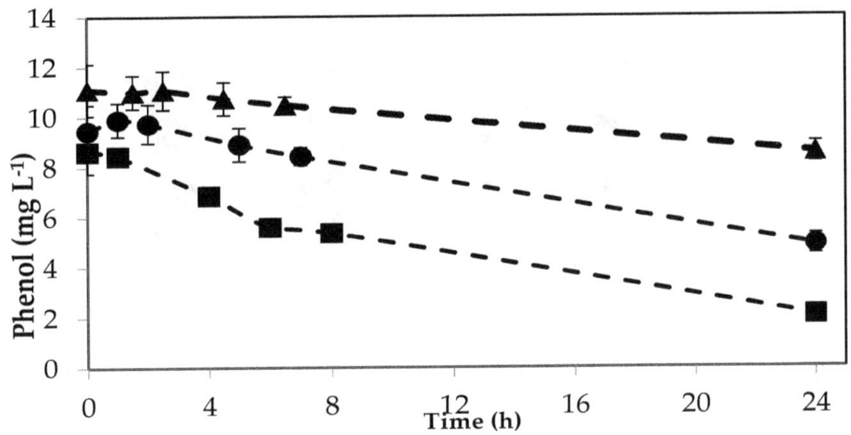

Figure 1. Phenol transformation with free enzyme (■), Tv immobilized onto fsNPs (▲), and onto FeO-2206W (•) over 24 h at pH 7.

According to the results, the nanobiocatalyst with FeO-2206W as a support seems to be the most adequate, as it achieved higher phenol biotransformation yields. Moreover, the separation of the nanobiocatalyst should be much simpler under a magnetic field, while intense centrifugation should be required when considering fumed silica nanoparticles. Accordingly, the single-core silica-coated nanobiocatalyst was used to prove its potential of reuse for phenol biotransformation in sequential batches.

3.3. Sequential Batch Biotransformation of Phenol by Laccase Immobilized onto Silica-Coated Magnetic Nanoparticles

Due to the remarkable effect of pH on phenol conversion reported in previous works [45,47], the biotransformation of phenol was assessed at different pH levels (5, 6, and 7). The conversion efficiencies and rates are shown in Table 3. An improvement of phenol conversion was observed when the pH was decreased to 6 or 5, which lead to an increase of 16%. A similar pH range was found suitable for phenol by immobilized enzyme onto silica-coated magnetic nanoparticles [45]. Furthermore, the enzymatic activity was maintained constant in all the experiments. The immobilized laccase retained 95%, 97%, and 100% of its initial activity at pH 5, 6, and 7, respectively, after incubation at room temperature for 24 h.

Table 3. Phenol biotransformation at variable pH levels by laccase immobilized onto silica-coated magnetic nanoparticles (FeO-2206W) for 24 h.

pH	Phenol Biotransformation (%)	Biotransformation Rate (mg L^{-1} h^{-1})
5	67.9	0.383
6	63.9	0.326
7	48.1	0.189

The reusability of the nanobiocatalyst was assessed in consecutive cycles of 24 h. It was observed that phenol transformation was higher than 60% and was maintained constant after four cycles (Figure 2). In other reports, phenol was almost entirely biotransformed in consecutive cycles with laccase immobilized on magnetic mesoporous silica nanoparticles with a dose of enzyme 18 times higher [45]. The immobilized laccase retained 97% of its initial activity after the consecutive batch treatments of phenol with magnetic separation.

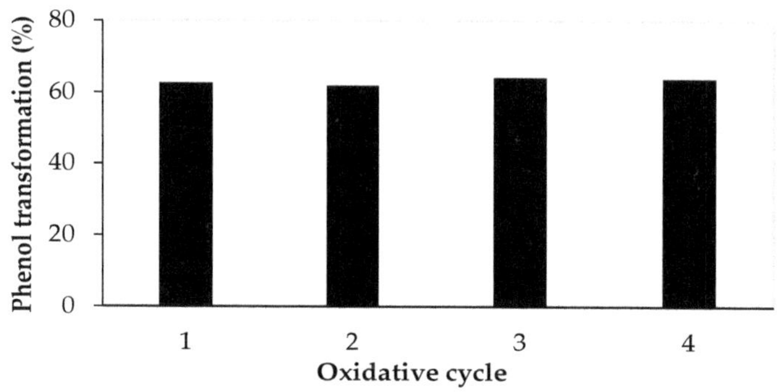

Figure 2. Phenol transformation in subsequent cycles of enzymatic treatment with laccase immobilized onto FeO-2206W (pH 6).

3.4. Environmental Indicators of the Single-Core Magnetic Nanoparticle

The development of new processes must comply with sustainability criteria. The methodology applied for the holistic assessment of the most suitable support was based on a life cycle perspective [48], which would include all aspects of activities during the life of a product, such as the extraction of raw materials and resources, production processes, use of products, recovery, recycling of some fractions, and the final disposal at the end-of-life stage.

In this study, inventory data for the foreground systems (direct inputs and outputs for each scenario) such as electricity requirements (estimated with power and operational data from the different units: reactors, dryers, heaters) as well as the use of chemicals and water were average data from semi-pilot scale experiments, obtained by on-site measurements of production processes developed for a time period of three months. The environmental assessment was conducted using characterization factors from ReCiPe Midpoint methodology [34], and the impact categories for the different mNP production routes are displayed considering one gram of FeO-2206W mNPs as a functional unit (Table 4).

Table 4. Normalized environmental impacts associated with the production of silica-coated mNPs (FeO-2206W) and silica thin shell per g of mNP.

	Scenarios	
Impact Category	**Silica-Coated mNPs**	**Silica Thin Shell**
Climate change	1.43×10^{-4}	1.03×10^{-4}
Ozone depletion	3.69×10^{-6}	2.42×10^{-6}
Terrestrial acidification	1.83×10^{-4}	1.30×10^{-4}
Freshwater eutrophication	9.11×10^{-4}	5.42×10^{-4}
Marine eutrophication	2.12×10^{-5}	1.45×10^{-5}
Human toxicity	5.93×10^{-4}	3.98×10^{-4}
Photochemical oxidant formation	1.33×10^{-4}	9.05×10^{-5}
Terrestrial ecotoxicity	6.18×10^{-6}	4.37×10^{-6}
Freshwater ecotoxicity	1.22×10^{-3}	8.36×10^{-4}
Marine ecotoxicity	1.45×10^{-3}	9.84×10^{-4}
Fossil depletion	6.20×10^{-4}	5.18×10^{-4}
Normalized index	5.28×10^{-3}	3.62×10^{-3}

The normalization results show that the impacts associated with the consumption of energy are dominant, but chemicals used in the formulations and for re-dispersion are also relevant. Regarding electricity, it is consumed for stirring, and the mechanical agitation required to obtain the transparent red solution of reverse micro-emulsion until complete reaction is remarkable (97% of the total electrical requirements). Regarding chemicals, the cyclohexane required in the formulation is the environmental critical chemical, since it is responsible for more than 85% of burdens derived from chemicals contributions regardless the impact category. Aiming to reduce the impacts, we also considered the use of a lower dose of cyclohexane to obtain a thin silica coating on the nanoparticle. Although the environmental impact was reduced, the immobilization of the enzyme was negatively affected by 40%, which was detrimental to the overall efficiency of the process. However, it should be highlighted that the production systems have been assessed at the pilot scale and optimizations should be required for large-scale application.

4. Conclusions

The study compared the immobilization of a commercially available *T. versicolor* laccase onto glutaraldehyde-activated, sulfo-NHS/EDC-activated magnetic and non-magnetic nanoparticles by covalent binding and onto magnetic nanoparticles by ion exchange, as well as its use as a nanobiocatalyst for phenol biotransformation. In summary, the most efficient biotransformation and the best activity yield was obtained by using laccase immobilized onto silica-coated magnetic nanoparticles. The magnetic nanobiocatalyst achieved a phenol biotransformation higher than 60%. One major outcome of this study is that the immobilized laccase is magnetically recoverable and can actually be reused in repeated cycles of phenol removal. The easy recovery of the nanobiocatalyst from the reaction media is a remarkable advantage from an operational perspective. Regarding the environmental impacts associated with the production of silica-coated magnetic nanoparticles, the use of energy and chemicals used in the formulations and for re-dispersion are the major contributors. Aiming to reduce the impacts, the use of a lower dose of cyclohexane implied lower environmental impact but negatively affected the immobilization yield of the enzyme, which discouraged this modification in the production process.

Acknowledgments: This work was financially supported by the Spanish Ministry of Economy and Competitiveness (CTQ2013-44762-R and CTQ2016-79461-R, program co-funded by FEDER). The authors belong to the Galician Competitive Research Group GRC 2013-032, program co-funded by FEDER. Yolanda Moldes-Diz thanks the Spanish Ministry of Economy and Competitiveness for her predoctoral fellowship.

Author Contributions: Maria Teresa Moreira, Juan M. Lema and Gumersindo Feijoo conceived and designed the experiments; Yolanda Moldes-Diz performed the experiments; Sara Feijoo performed the LCA study, Maria Teresa Moreira and Gemma Eibes analyzed the data and revised the different versions of the manuscript; Maria Teresa Moreira. and Yolanda Moldes-Diz wrote the paper.

References

1. Kunamneni, A.; Ballesteros, A.; Plou, F.J.; Alcalde, M. Fungal laccase—A versatile enzyme for biotechnological applications. In *Communicating Current Research and Educational Topics and Trends in Applied Microbiology*; Mendez-Vilas, A., Ed.; Formex: Badajoz, Spain, 2007; Volume 1, pp. 233–245. ISBN 978-84-611-9422-3.
2. Thurston, C.F. The structure and function of fungal laccases. *Microbiology* **1994**, *140*, 19–26. [CrossRef]
3. Wesenberg, D.; Kyriakides, I.; Agathos, S.N. White-rot fungi and their enzymes for the treatment of industrial dye effluents. *Biotechnol. Adv.* **2003**, *22*, 161–187. [CrossRef] [PubMed]
4. Misson, M.; Zhang, H.; Jin, B. Nanobiocatalyst advancements and bioprocessing applications. *J. R. Soc. Interface* **2015**, *12*, 20140891. [CrossRef] [PubMed]
5. Illanes, R. *Enzyme Biocatalysis : Principles and Applications*; Springer: Berlin, Germany, 2008.
6. Hernandez, R.; Fernandez-Lafuente, J.M. Control of protein immobilization: Coupling immobilization and site-directed mutagenesis to improve biocatalyst or biosensor performance. *Enzyme Microb. Technol.* **2011**, *48*, 107–122. [CrossRef] [PubMed]
7. Rodrigues, R.C.; Ortiz, C.; Berenguer-Murcia, A.; Torres, R.; Fernández-Lafuente, R. Modifying enzyme activity and selectivity by immobilization. *Chem. Soc. Rev.* **2013**, *42*, 6290–6307. [CrossRef] [PubMed]
8. Sheldon, R.A. Enzyme Immobilization: The Quest for Optimum Performance. *Adv. Synth. Catal.* **2007**, *349*, 1289–1307. [CrossRef]

9. Duran, N.; Rosa, M.A.; D´Annibale, A.; Gianfreda, L. Applications of laccases and tyrosinases(phenoloxidases) immobilized on different supports: A review. *Enzyme Microb. Technol.* **2002**, *31*, 907–931. [CrossRef]
10. Cui, J.; Jia, S.; Liang, L.; Zhao, Y.; Feng, Y. Mesoporous CLEAs-silica composite microparticles with high activity and enhanced stability. *Sci. Rep.* **2015**, *5*, 14203. [CrossRef] [PubMed]
11. Matijosyte, I.; Arends, I.W.C.E.; Vries, S.; Sheldon, R.A. Preparation and use of cross-linked enzyme aggregates (CLEAs) of laccases. *J. Mol. Catal. B Enzyme* **2010**, *62*, 142–148. [CrossRef]
12. Qu, X.; Alvarez, P.J.J.; Li, Q. Applications of nanotechnology in water and wastewater treatment. *Water Res.* **2013**, *47*, 3931–3946. [CrossRef] [PubMed]
13. Ahn, M.-Y.; Zimmerman, A.R.; Martinez, C.E.; Archibald, D.D.; Bollag, J.-M.; Dec, J. Characteristics of Trametes villosa laccase adsorbed on aluminium hydroxide. *Enzyme Microb. Technol.* **2007**, *41*, 141–148. [CrossRef]
14. Qiu, H.; Xu, C.; Huang, X.; Ding, Y.; Qu, Y.; Gao, P. Adsorption of laccase on the surface of nanoporous gold and the electron transfer between them. *J. Phys. Chem. C* **2008**, *112*, 14781–14785. [CrossRef]
15. Fernandez-Fernandez, M.; Sanroman, M.A.; Moldes, D. Recent developments and applications of immobilized laccase. *Biotechnol. Adv.* **2013**, *31*, 1808–1825. [CrossRef] [PubMed]
16. Arroyo, M. Immobilized enzymes: Theory, methods of study and applications. *ARS Pharm.* **1998**, *39*, 23–39.
17. Brady, D.; Jordaan, J. Advances in enzyme immobilisation. *Biotechnol. Lett.* **2009**, *31*, 1639–1650. [CrossRef] [PubMed]
18. García-Galán, C.; Berenguer-Murcia, A.; Fernandez-Lafuente, R.; Rodrigues, R.C. Potential of Different Enzyme Immobilization Strategies to Improve Enzyme Performance. *Adv. Synth. Catal.* **2011**, *353*, 2885–2904. [CrossRef]
19. Mazur, M.; Krysiński, P.; Michota-Kamińska, A.; Bukowska, J.; Rogalski, J.; Blanchard, G.J. Immobilization of laccase on gold, silver and indium tin oxide by zirconium–phosphonate–carboxylate (ZPC) coordination chemistry. *Bioelectrochemistry* **2007**, *71*, 15–22. [CrossRef] [PubMed]
20. Kalkan, N.; Aksoy, S.; Aksoy, E.; Hasirci, N. Preparation of chitosan-coated magnetite nanoparticles and application for immobilization of laccase. *J. Appl. Polym. Sci.* **2011**, *123*, 707–716. [CrossRef]
21. Feng, W.; Ji, P. Enzymes immobilized on carbon nanotubes. *Biotechnol. Adv.* **2011**, *29*, 889–895. [CrossRef] [PubMed]
22. DiCosimo, R.; McAuliffe, J.; Polouse, A.J.; Bohlmann, G. Industrial use of immobilized enzymes. *Chem. Soc. Rev.* **2013**, *42*, 6437–6474. [CrossRef] [PubMed]
23. Zimmermann, Y.S.; Shahgaldian, P.; Corvini, P.F.X.; Hommes, G. Sorption-assisted surface conjugation: A way to stabilize laccase enzyme. *Appl. Microbiol. Biotechnol.* **2011**, *92*, 169–178. [CrossRef] [PubMed]
24. Nobs, L.; Buchegger, F.; Gurny, R.; Allemann, E. Surface modification of poly(lactic acid) nanoparticles by covalent attachment of thiol groups by means of three methods. *Int. J. Pharm.* **2003**, *29*, 327–337. [CrossRef]
25. Barbosa, O.; Ortiz, C.; Berenguer-Murcia, A. Glutaraldehyde in bio-catalysts design: A useful crosslinker and a versatile tool in enzyme immobilization. *RSC Adv.* **2014**, *4*, 1583–1600. [CrossRef]
26. Zucca, P.; Sanjust, E. Inorganic materials as supports for covalent enzyme immobilization: Methods and mechanisms. *Molecules* **2014**, *19*, 14139–14194. [CrossRef] [PubMed]
27. Baumann, H.; Tillman, A.M. *The Hitch Hilker's Guide to LCA: An Orientation in Life Cycle Assessment Methodology and Application*; Studentlitteratur: Lund, Sweden, 2004; ISBN 9144023642.
28. *Environmental Management—Life Cycle Assessment—Principles and Framework*; ISO 14040:2006; International Organization for Standardization: Geneva, Switzerland, 2006.
29. Feijoo, S.; Gonzalez-García, S.; Moldes-Diz, Y.; Vazquez-Vazquez, C.; Feijoo, G.; Moreira, M.T. Comparative life cycle assessment of different synthesis routes of magnetic nanoparticles. *J. Clean. Prod.* **2017**, *43*, 528–538. [CrossRef]
30. Dones, R.; Bauer, C.; Bolliger, R.; Burger, B.; Faist, E.M.; Frischknecht, R.; Heck, T.; Jungbluth, N.; Röder, A.; Tuchschmid, M. *Life Cycle Inventories of Energy Systems: Results for Current Systems in Switzerland and Other UCTE Countries*; Ecoinvent Report No. 5; Swiss Centre for Life Cycle Inventories: Düberdorf, Switzerland, 2007; Available online: https://www.researchgate.net/profile/Deborah_Andrews2/publication/271710820_The_life_cycle_assessment_of_a_UK_data_centre/links/5620b9ec08aed8dd194054ea.pdf (accessed on 7 June 2016).

31. Althaus, H.J.; Chudacoff, M.; Hischier, R.; Jungbluth, N.; Osses, M.; Primas, A. *Life Cycle Inventories of Chemicals*; Ecoinvent Report No.8 v2.0 EMPA; Swiss Centre for Life Cycle Inventories: Düberdorf, Switzerland, 2007.
32. Doka, G. *Life Cycle Inventories of Waste Treatment Services*; Ecoinvent Report No. 13; Swiss Centre for Life Cycle Inventories: Dübendorf, Switzerland, 2003.
33. Red Eléctrica de España. Avance del Informe del Sistema Eléctrico Español 2014. Available online: http://www.ree.es/sites/default/files/downloadable/avance_informe_sistema_electrico_2014.pdf (accessed on 10 June 2016).
34. Goedkoop, M.J.; Heijungs, R.; Huijbregts, M.; Schryver, A.D.; Strujis, J.; Zelm, R. A Life Cycle Impact Assessment Method Which Comprises Harmonised Category Indicators at the Midpoint and the Endpoint Level. Available online: http://www.leidenuniv.nl/cml/ssp/publications/recipe_characterisation.pdf (accessed on 7 June 2016).
35. PRé Consultants. 2017. Available online: http://www.pre.nl (accessed on 1 July 2017).
36. Sheldon, R.A. Cross-linked enzyme aggregates (CLEAs): Stable and recyclable biocatalysts. *Biochem. Soc. Trans.* **2007**, *35*, 1583–1587. [CrossRef] [PubMed]
37. Ovsejevi, K.; Manta, C.; Batista-Viera, F. Reversible covalent immobilization of enzymes via disulfide bonds. *Methods Mol. Biol.* **2013**, *1051*, 89–116. [CrossRef] [PubMed]
38. Guisán, J.M. *Immobilization of Enzymes and Cells*, 2nd ed.; Humana Press Inc.: Clifton, NJ, USA, 2006; ISBN 978-1-58829-290-2.
39. Flickinger, M.; Drew, S. *Encyclopedia of Bioprocess Technology: Fermentation, Biocatalysis and Bioseparation Fermentation, Biocatalysis and Bioseparation*; Wiley: New York, NY, USA, 1999; ISBN 0-471-13822-3.
40. Majumder, A.B.; Mondal, K.; Singh, T.P.; Gupta, M.N. Designing cross-linked lipase aggregates for optimum performance as biocatalysts. *Biocatal. Biotransform.* **2008**, *26*, 235–242. [CrossRef]
41. Kumar, S.; Jana, A.K.; Maiti, M.; Dhamija, I. Carbodiimide-mediated immobilization of serratiopeptidase on amino-, carboxyl-functionalized magnetic nanoparticles and characterization for target delivery. *J. Nanopart. Res.* **2014**, *16*, 2233. [CrossRef]
42. Arca-Ramos, A.; Ammann, E.M.; Gasser, C.A.; Nastold, P.; Eibes, G.; Feijoo, G.; Lema, J.M.; Moreira, M.T.; Corvini, P.F.X. Assessing the use of nanoimmobilized laccases to remove micropollutants from wastewater. *Environ. Sci. Pollut. Res.* **2016**, *23*, 3217–3228. [CrossRef] [PubMed]
43. Corgié, S.; Kahawong, P.; Duan, X.; Bowser, D.; Edward, J.B.; Walker, L.P.; Giannelis, E.P. Self-assembled complexes of horseradish peroxidase with magnetic nanoparticles showing enhanced peroxidase activity. *Adv. Funct. Mater.* **2012**, *22*, 1940–1951. [CrossRef]
44. Claus, H.; Faber, G.; König, H. Redox-mediated decolorization of synthetic dyes by fungal laccases. *Appl. Microbiol. Biotechnol.* **2002**, *59*, 672–678. [CrossRef] [PubMed]
45. Wang, F.; Hu, Y.; Guo, C.; Huang, W.; Liu, C.-Z. Enhanced phenol degradation in cooking wastewater by immobilized laccase on magnetic mesoporous silica nanoparticles in a magnetically stabilized fluidized bed. *Bioresour. Technol.* **2012**, *110*, 120–124. [CrossRef] [PubMed]
46. Kurniawati, S.; Nicell, J.A. A Comprehensive Kinetic Model of Laccase-Catalyzed Oxidation of Aqueous Phenol. *Biotechnol. Prog.* **2009**, *25*, 763–773. [CrossRef] [PubMed]
47. Bayramoglu, G.; Arica, M.Y. Enzymatic removal of phenol and p-chlorophenol in enzyme reactor: Horseradish peroxidase immobilized on magnetic beads. *Hazard. Mater.* **2008**, *156*, 148–155. [CrossRef] [PubMed]
48. Grieger, K.D.; Laurent, A.; Miseljic, M.; Christensen, F.; Baun, A.; Olsen, S.I. Analysis of current research addressing complementary use of life-cycle assessment and risk assessment for engineered nanomaterials: Have lessons been learned from previous experience with chemicals? *J. Nanopart. Res.* **2012**, *14*, 958. [CrossRef]

Permissions

All chapters in this book were first published in MDPI; hereby published with permission under the Creative Commons Attribution License or equivalent. Every chapter published in this book has been scrutinized by our experts. Their significance has been extensively debated. The topics covered herein carry significant findings which will fuel the growth of the discipline. They may even be implemented as practical applications or may be referred to as a beginning point for another development.

The contributors of this book come from diverse backgrounds, making this book a truly international effort. This book will bring forth new frontiers with its revolutionizing research information and detailed analysis of the nascent developments around the world.

We would like to thank all the contributing authors for lending their expertise to make the book truly unique. They have played a crucial role in the development of this book. Without their invaluable contributions this book wouldn't have been possible. They have made vital efforts to compile up to date information on the varied aspects of this subject to make this book a valuable addition to the collection of many professionals and students.

This book was conceptualized with the vision of imparting up-to-date information and advanced data in this field. To ensure the same, a matchless editorial board was set up. Every individual on the board went through rigorous rounds of assessment to prove their worth. After which they invested a large part of their time researching and compiling the most relevant data for our readers.

The editorial board has been involved in producing this book since its inception. They have spent rigorous hours researching and exploring the diverse topics which have resulted in the successful publishing of this book. They have passed on their knowledge of decades through this book. To expedite this challenging task, the publisher supported the team at every step. A small team of assistant editors was also appointed to further simplify the editing procedure and attain best results for the readers.

Apart from the editorial board, the designing team has also invested a significant amount of their time in understanding the subject and creating the most relevant covers. They scrutinized every image to scout for the most suitable representation of the subject and create an appropriate cover for the book.

The publishing team has been an ardent support to the editorial, designing and production team. Their endless efforts to recruit the best for this project, has resulted in the accomplishment of this book. They are a veteran in the field of academics and their pool of knowledge is as vast as their experience in printing. Their expertise and guidance has proved useful at every step. Their uncompromising quality standards have made this book an exceptional effort. Their encouragement from time to time has been an inspiration for everyone.

The publisher and the editorial board hope that this book will prove to be a valuable piece of knowledge for researchers, students, practitioners and scholars across the globe.

List of Contributors

Ali M. Almusawy, Riyad H. Al-Anbari and Arshed Imad Al-Najar
Civil Engineering Department, University of Technology, Alsinaa Street 52, Baghdad 10066, Iraq

Qusay F. Alsalhy
Membrane Technology Research Unit, Chemical Engineering Department, University of Technology, Alsinaa Street 52, Baghdad 10066, Iraq

Mongi ben Mosbah
Materials, Environment and Energy Laboratory (UR14ES26), Faculty of Sciences of Gafsa, University of Gafsa, Gafsa 2112, Tunisia
Faculty of Sciences of Gafsa, University of Gafsa, Gafsa 2112, Tunisia

Lassaad Mechi
Materials, Environment and Energy Laboratory (UR14ES26), Faculty of Sciences of Gafsa, University of Gafsa, Gafsa 2112, Tunisia
Department of Chemistry, Faculty of Sciences, University of Hail, Saudi Arabia

Ramzi Khiari
Higher Institute of Technological Studies, Ksar Hellal 5070, Tunisia
UR13 ES 63—Research Unity of Applied Chemistry & Environment, Faculty of Sciences, University of Monastir, Monastir 5000, Tunisia
LGP2, University of Grenoble Alpes, CNRS, Grenoble INP, LGP2, F-38000 Grenoble, France

Younes Moussaoui
Faculty of Sciences of Gafsa, University of Gafsa, Gafsa 2112, Tunisia
Organic Chemistry Laboratory (LR17ES08), Faculty of Sciences of Sfax, University of Sfax, Sfax 3029, Tunisia

Rehab O. Abdel Rahman
Hot Lab. Center, Atomic Energy Authority of Egypt, Cairo, Egypt

Yung-Tse Hung
Department of Civil and Environmental Engineering, Cleveland State University, Cleveland, OH 44115, USA

Ahmed Elyahyaoui, Kawtar Ellouzi and Saidati Bouhlassa
Laboratory of Radiochemistry, Department of Chemistry, Faculty of Sciences, Mohamed V-Agdal, B.P 1014 Rabat, Morocco

Hamzeh Al Zabadi
Public Health Department, An-Najah National University, Nablus 44830, Palestine

Brahim Razzouki
Laboratory of Spectroscopy, Molecular Modeling, Materials and Environment, Department of Chemistry, Faculty of Sciences, Mohamed V, B.P 1014 Rabat, Morocco

Khalil Azzaoui and El Miloud Mejdoubi
Laboratory LMSAC, Faculty of Sciences, Mohamed 1st University, Oujda 60000, Morocco

Othman Hamed and Shehdeh Jodeh
Department of Chemistry, An-Najah National University, Nablus 44830, Palestine

Abdellatif Lamhamdi
Laboratory LMSAC, Faculty of Sciences, Mohamed 1st University, Oujda 60000, Morocco
National School of Applied Sciences Al Hoceima, Mohamed 1st University, Oujda 60000, Morocco

Simona M. Miron, Patrick Dutournié and Arnaud Ponche
Institut de Science des Materiaux de Mulhouse IS2M (UMR CNRS 7228), Universite de Haute Alsace, 68100 Mulhouse, France
Universite de Strasbourg, 3 bis rue A. Werner, 68093 Mulhouse CEDEX, France

Miguel Peña-Varón
Grupo de Saneamiento Ambiental, Instituto Cinara, Unversidad del Valle, Cali 100-00, Colombia

Andrés Toro-Vélez
Grupo de Saneamiento Ambiental, Instituto Cinara, Unversidad del Valle, Cali 100-00, Colombia
Doctorado en Ciencias Ambientales, Universidad del Cauca, Popayán 190001, Colombia

Carlos Madera-Parra
Escuela EIDENAR-Facultad de Ingeniería, Universidad del Valle, Cali 100-00, Colombia

Hector García-Hernández and Piet Lens
UNESCO-IHE Institute for Water Education, 2611 AX Delft, The Netherlands

Wen Yee Lee
Department of Chemistry, University of Texas at El Paso, El Paso, TX 79968, USA

Shane Walker
Department of Civil Engineering, University of Texas at El Paso, El Paso, TX 79968, USA

Gaogui Jing, Shuai Ren, Wei Sun and Zhiyong Gao
School of Minerals Processing and Bioengineering, Central South University, Changsha 410083, China
Key Laboratory of Hunan Province for Clean and Efficient Utilization of Strategic Calcium-Containing Mineral Resources, Central South University, Changsha 410083, China

Yuesheng Gao
Department of Chemical Engineering, Michigan Technological University, Houghton, MI 49931, USA

Marjorie Dubreuil Kristien De Sitter and Chris Dotremont
VITO-Flemish Institute for Technological Research, Boeretang 200, 2400 Mol, Belgium

Lies Eykens
VITO-Flemish Institute for Technological Research, Boeretang 200, 2400 Mol, Belgium
Department of Chemical Engineering, KU Leuven, Celestijnenlaan 200F, B-3001 Leuven, Belgium

Klaus Rose
Fraunhofer Institute for Silicate Research ISC, Neunerplatz 2, 97082 Würzburg, Germany

Luc Pinoy
Department of Chemical Engineering, Cluster Sustainable Chemical Process Technology, KU Leuven Gebroeders Desmetstraat 1, B-9000 Ghent

Bart Van der Bruggen
Department of Chemical Engineering, KU Leuven, Celestijnenlaan 200F, B-3001 Leuven, Belgium
Faculty of Engineering and the Built Environment, Tshwane University of Technology, Private Bag X680, Pretoria 0001, South Africa

Guilherme A. de Souza Reis, Michiel H. A. Michels, Ischa Lamot and Jappe H. de Best
Centre of Expertise Biobased Economy, Avans University of Applied Sciences, Lovensdijkstraat 63 4818 AJ Breda, The Netherlands

Gabriela L. Fajardo
Centre of Expertise Biobased Economy, Avans University of Applied Sciences, Lovensdijkstraat 63 4818 AJ Breda, The Netherlands
Department of Chemical Engineering, Universidade Federal de Minas Gerais, Belo Horizonte 31270-901, Brazil

Jae Wook Chung, Oghosa Charles Edewi, Jan Willem Foppen and Piet Nicolaas Luc Lens
UNESCO-IHE Institute for Water Education, 2601 DA Delft, The Netherlands

Gabriel Gerner and Rolf Krebs
Institute of Natural Resource Sciences, Zurich University of Applied Sciences, Grüental, 8820 Wädenswil, Switzerland

Anthony G. Fane
Singapore Membrane Technology Center, Nanyang Environment and Water Research Institute, Nanyang Technological University, Singapore 639798, Singapore

Ebrahim Akhondi
Singapore Membrane Technology Center, Nanyang Environment and Water Research Institute, Nanyang Technological University, Singapore 639798, Singapore
Young Researchers and Elite Club, Central Tehran Branch, Islamic Azad University, Tehran 1469669191, Iran

Farhad Zamani and Jia Wei Chew
Singapore Membrane Technology Center, Nanyang Environment and Water Research Institute, Nanyang Technological University, Singapore 639798, Singapore
School of Chemical and Biomedical Engineering, Nanyang Technological University, Singapore 637459, Singapore

Keng Han Tng and Gregory Leslie
UNESCO Centre for Membrane Science and Technology, Sydney, NSW 2052, Australia
School of Chemical Engineering, The University of New South Wales, Sydney, NSW 2052, Australia

William B. Krantz
Singapore Membrane Technology Center, Nanyang Environment and Water Research Institute, Nanyang Technological University, Singapore 639798, Singapore
Department of Chemical & Biological Engineering, University of Colorado, Boulder, CO 80309-0424, USA

Patrick Dutournié and Mejdi Jeguirim
CNRS ISM2 UMR 7361, Université de Haute Alsace, Université de Strasbourg, 68093 Mulhouse, France

Besma Khiari
National School of Engineers of Carthage, 45 rue des entrepreneurs, Charguia 2-1002, 2035 Tunis, Tunisia

Mary-Lorène Goddard
CNRS, LIMA UMR 7042, Université de Haute-Alsace, Université de Strasbourg, 68093 Mulhouse, France
LVBE, EA 3991, Université de Haute-Alsace, 68008 Colmar, France

Salah Jellali
Wastewaters and Environment Laboratory, Water research and Technologies Centre, 8020 Soliman, Tunisia

Xianjun Du
College of Electrical and Information Engineering, Lanzhou University of Technology, Lanzhou 730050, China
School of Engineering, Royal Melbourne Institute of Technology (RMIT) University, Melbourne 3000, Australia
Key Laboratory of Gansu Advanced Control for Industrial Processes, Lanzhou University of Technology, Lanzhou 730050, China
National Demonstration Center for Experimental Electrical and Control Engineering Education, Lanzhou University of Technology, Lanzhou 730050, China

Junlu Wang
College of Electrical and Information Engineering, Lanzhou University of Technology, Lanzhou 730050, China
Key Laboratory of Gansu Advanced Control for Industrial Processes, Lanzhou University of Technology, Lanzhou 730050, China
National Demonstration Center for Experimental Electrical and Control Engineering Education, Lanzhou University of Technology, Lanzhou 730050, China

Veeriah Jegatheesan
School of Engineering, Royal Melbourne Institute of Technology (RMIT) University, Melbourne 3000, Australia

Guohua Shi
Department of Energy and Power Engineering, North China Electric Power University, Baoding 071003, China

Maria Teresa Moreira, Yolanda Moldes-Diz, Sara Feijoo, Gemma Eibes, Juan M. Lema and Gumersindo Feijoo
Department of Chemical Engineering, Institute of Technology, University of Santiago de Compostela, 15782 Santiago de Compostela, Spain

Index

www.ingramcontent.com/pod-product-compliance
Lightning Source LLC
LaVergne TN
LVHW081033210926
848485LV00004B/1011